THE DISUNITY OF SCIENCE

WRITING SCIENCE

EDITORS Timothy Lenoir and Hans Ulrich Gumbrecht

CONTRIBUTORS

Mario Biagioli

Nancy Cartwright

Jordi Cat

Hasok Chang

Richard Creath

Arnold I. Davidson

John Dupré

Arthur Fine

Steve Fuller

Peter Galison

Ian Hacking

Donna J. Haraway

Evelyn Fox Keller

Karin Knorr Cetina

Timothy Lenoir

Cheryl Lynn Ross

Joseph Rouse

Simon Schaffer

David J. Stump

Alison Wylie

THE DISUNITY OF SCIENCE

Boundaries, Contexts, and Power

EDITED BY

Peter Galison and
David J. Stump

STANFORD UNIVERSITY PRESS
STANFORD, CALIFORNIA 1996

Stanford University Press
Stanford, California
© 1996 by the Board of Trustees
of the Leland Stanford Junior University
Printed in the United States of America

CIP data are at the end of the book

Stanford University Press publications are
distributed exclusively by Stanford University Press
within the United States, Canada, Mexico, and Central America;
they are distributed exclusively by Cambridge University Press
throughout the rest of the world

Preface

Is science unified or disunified? Over the last century, the question has raised the interest (and hackles) of scientists, philosophers, historians, and sociologists of science. At stake in this long-running debate is how science and, often implicitly, how society fits together. Recent years have seen a turn largely against the rhetoric of unity, in discussions ranging from the pleas of condensed matter physicists for disciplinary autonomy all the way to discussion in the humanities and social sciences of local history, feminism, multiculturalism, postmodernism, scientific relativism and realism, and social constructivism.

During 1990–91, the editors contacted a heterogeneous group of scholars representing various flavors of science studies. Disunity, amusingly enough, was one of the few topics that seemed to elicit a powerful interest from people who otherwise have not had much to say to one another. Philosophers of science have not, for example, interacted a great deal with the feminist theorists or with the sociologists of science. For several days in April 1991, a group of people gathered at Stanford University to discuss a set of seventeen precirculated papers on the subject of contextualism and disunity in science studies. On the basis of these extended discussions, many of the papers were revised, and several new ones (for example, those by Knorr Cetina, Lenoir and Ross, Hacking, and Schaffer) were commissioned. It was and is our hope that the resulting essays, including the occasionally sharp disagreements voiced in them, hold sufficient ground in common to further discussion of these issues while standing far enough apart to shed light on important arenas of contention.

We were greatly aided in the organization of the conference by Barbara Kataoka, Shannon Temple, and Jeff Rutherford at Stanford, and in the production of this volume by Jean Titilah and Pamela Butler at Harvard. For funding support, we extend our gratitude to the National Science Foundation (supplement to PLG's PYIA grant, and to DJS's postdoctoral fellowship) and to Wesleyan University. Many thanks go as well to the many critical comments and suggestions by participants at the "Disunity and Contextualism" workshop of 1991, especially to Bas van Fraassen, Bruno Latour, Lisa Lloyd, Richard Miller, Margaret Morrison, Thomas Nickles, Andrew Pickering, and Betty Smocovitis.

Contents

❖❖❖

Contributors

Mario Biagioli is Professor of History of Science at Harvard University, where he teaches the history of early modern science. He is the author of *Galileo, Courtier.*

Nancy Cartwright is Chair in Philosophy and Director of the Centre for the Philosophy of the Natural and Social Sciences at the London School of Economics. Her publications include *Nature's Capacities and Their Measurement, How the Laws of Physics Lie,* and (forthcoming) *Between Science and Politics: The Philosophy of Otto Neurath,* with Jordi Cat, Karola Fleck, and Thomas Uebel. In 1993 she was awarded a MacArthur Foundation fellowship.

Jordi Cat is a Ph.D. candidate in the Philosophy Department of the University of California at Davis. His works include a forthcoming book on the philosophy of Otto Neurath, coauthored with Nancy Cartwright, Karola Fleck, and Thomas Uebel. He is currently working in the philosophy of quantum field physics.

Hasok Chang received his Ph.D. in philosophy from Stanford University in 1993 and is now a Lecturer in Philosophy of Science at University College London. His articles in scholarly journals include "Causality and Realism in the EPR Experiment," with Nancy Cartwright, and "A Misunderstood Rebellion: The Twin-Paradox Controversy and Herbert Dingle's Vision of Science." Currently his most active interests lie along the boundary between philosophy and the history of science.

Richard Creath is Professor of Philosophy at Arizona State University. His works include papers on Carnap and in philosophy of science more generally. He is the editor of *Dear Carnap, Dear Van: The Quine-Carnap Correspondence and Related Work.*

Arnold I. Davidson is Professor of Philosophy and the Conceptual Foundations of Science at the University of Chicago and Executive Editor of *Critical Inquiry*. He is coeditor of *Questions of Evidence* and has published articles on the history of psychiatry and psychoanalysis, the history of philosophy and theology, and contemporary European philosophy. He is currently completing a book on the history and epistemology of horror and wonder.

John Dupré is Associate Professor in the Department of Philosophy at Stanford University and Cochair of the Program in the History and Philosophy of Science. He is the author of *The Disorder of Things: Metaphysical Foundations of the Disunity of Science* and editor of *The Latest on the Best: Essays on Evolution and Optimality*. His most recent work is on the philosophy of economics.

Arthur Fine is the John Evans Professor of Philosophy at Northwestern University and past president of the Philosophy of Science Association. His works include *The Shaky Game: Einstein, Realism, and the Quantum Theory*.

Steve Fuller is Chair of Sociology and Social Policy at the University of Durham. He is the founding editor of the journal *Social Epistemology* and the author of three books, the latest of which is *Philosophy, Rhetoric, and the End of Knowledge*. He is currently writing a book on the reception of Kuhn's *The Structure of Scientific Revolutions*.

Peter Galison is the Mallinckrodt Professor of the History of Science and of Physics at Harvard University. He is the author of *How Experiments End* and (forthcoming) *Image and Logic: The Material Culture of Twentieth-Century Physics*. With Bruce Hevly he coedited *Big Science: The Growth of Large-Scale Research*.

Ian Hacking is University Professor at the University of Toronto. His most recent books are *The Taming of Chance* and *Representing and Intervening*.

Donna J. Haraway is Professor in the History of Consciousness Board at the University of California at Santa Cruz. She is the author of *Crystals, Fabrics and Fields: Metaphors of Organicism in Twentieth-Century Developmental Biology*, *Primate Visions: Gender, Race, and Nature in the World of Modern Science*, and *Simians, Cyborgs, and Women: The Reinvention of Nature*.

Evelyn Fox Keller is Professor of History and Philosophy of Science in the Program in Science, Technology, and Society at the

Massachusetts Institute of Technology. Her most recent book is *Secrets of Life, Secrets of Death: Essays on Language, Gender, and Science*, and her current research is on the history of developmental biology.

Karin Knorr Cetina is Professor of Sociology and Science and Technology at the University of Bielefeld, Germany. Her books include *The Manufacture of Knowledge*, one of the first "laboratory studies," and *Advances in Social Theory and Methodology*, with Aaron Cicourel. She is currently publishing a comparative study of high-energy physics and molecular biology entitled *Epistemic Cultures* (forthcoming).

Timothy Lenoir is Professor of History and Cochair of the Program in History and Philosophy of Science at Stanford University. He is the author of *The Strategy of Life: Teleology and Mechanics in Nineteenth Century German Biology*, which examines the development of nineteenth-century German non-Darwinian theories of evolution. His other books include *Politik im Tempel der Wissenschaft: Forschung und Machtausübung im deutschen Kaiserreich*; (forthcoming) *Instituting Science*, a volume examining the formation of disciplines and the role of public institutions in the construction of scientific knowledge; and (forthcoming) *Reforming Vision: Optics, Aesthetics, and Ideology in Germany, 1845–1890*. Lenoir is currently studying the introduction of computers into biological research and the development of computer graphics and imaging devices in the biomedical sciences from the early 1960's through the 1980's. He has been a Fellow of the John Simon Guggenheim Foundation and twice a Fellow of the Institute for Advanced Studies in Berlin.

Cheryl Lynn Ross is Lecturer in the Program in Cultures, Ideas, and Values at Stanford University. She has written on theatrical and social history in seventeenth-century England, on Freud and post-Freudian religious theory, and on Virginia Woolf. Her current project is a study of naturalization and canon formation in early twentieth-century America.

Joseph Rouse is Professor of Philosophy and Science in Society at Wesleyan University. He is the author of *Knowledge and Power: Toward a Political Philosophy of Science* and (forthcoming) *Engaging Science: Science Studies after Realism, Rationality, and Social Constructivism*.

Simon Schaffer is Reader in the History and Philosophy of Science at the University of Cambridge. He is coauthor, with Steven Shapin, of *Leviathan and the Air-Pump* and coeditor of *The Uses of Experiment: Studies in the Natural Sciences*.

David J. Stump is Assistant Professor of Philosophy at the University of San Francisco. He has published papers on Henri Poincaré's conventionalism and on historicized and naturalized philosophy of science and is currently studying the formalization of mathematics at the end of the nineteenth and the beginning of the twentieth century.

Alison Wylie is Professor of Philosophy at the University of Western Ontario. Her primary interest is in the philosophy of the social sciences, specifically archaeology, and in feminist critiques of the social sciences. She is coeditor of *Critical Traditions in Archaeology* and is presently completing a book on the recent emergence of a program of gender research in archaeology, *Feminist Archaeology*.

THE DISUNITY OF SCIENCE

PETER GALISON

Introduction:
The Context of Disunity

Unity. The very term has always evoked emotions. As a political call to arms, it rouses countries to civil strife, revolution, and international war. The theme of unity is written into the history of the United States, the (former) Soviet Union, and the European nation-states as deeply as any slogan can be. So, one should immediately add, are its antitheses—independence and autonomy. Little surprise, then, if the unification of the sciences, or the autonomy of the sciences, participates in broader cultural debates.

In the interwar period, faced with the rise of fascism, the disintegration of the Hapsburg Empire, and the growing tensions between states, a movement grew up under the banner of the Unity of Science. Its roots were various; the motives of its supporters, diverse. But somehow, behind a thinly veiled facade of pure science, there lay a hope and an optimism that the fruits of modernity—the technical wonders of telephone, radio, airplane; the grids of power and railroads; and the scientific spectacles of relativity, quantum physics, and astronomy—could somehow avert calamity. Those who preached the Unity of Science saw inscribed in the new and modern *Lebensform* a rationality that, they hoped, would guard the world against the tide of fanaticism. Indeed, out of the Unity of Science movement came a series of meetings and publications, the most famous of which were the twenty or so volumes that comprised the first batch of papers of the *Encyclopedia of Unified Science*, edited and organized by the polymath philosophers of science Rudolf Carnap, Otto Neurath, and Charles Morris.

Today, across the disciplinary map, calls for the disunity of science can be heard, echoing across the fields of science and science studies. In physics, the autonomy of condensed matter physics from the guiding principles of particle physics has been debated from laboratories all the way to the halls of Congress. At stake was the future of the supercollider: if other branches of physics *follow* in some sense from "fundamental" laws, then (some argued) there were principled philosophical, if not practical, reasons for funding the more basic over the more applied endeavor. Conversely, condensed matter physicists argued that the nonreductive autonomy of their field militated against one of the single biggest science projects ever proposed. Nor is physics unique in finding unity/ disunity debates to be pressing issues. Biology departments around the world are fragmenting into new pieces, with molecular biology on one side and organismic biology on the other. To the molecular biologists the dream of a completed discipline includes the explanation of ontogenetic and phylogenetic development from primitive relations of genetic material. To the macrobiologists such reductiveness will never capture the systemic aspects of complex organisms, let alone the ecological systems in which they live and reproduce.

To scholars in science studies, such intradisciplinary splits reveal others. There are disunities within the language and practice of science between disciplines, and these divisions threaten to divide even the component parts of subfields. (We are by now used to the gap between biology and physics, but still taken aback by the difficulty a high-energy string theorist has in communicating with a high-energy particle theorist.) There are disunities to be found between the practice of science in different locales and by different groups of practitioners. Comparative studies of the conduct of physics in Japan versus that in Europe or North America, or the conduct of genetic inquiry by Barbara McClintock versus that of her (mostly male) peers, reveal a locality to scientific knowledge that makes programmatic assertions of "unity" ever harder to sustain.[1]

Most of the authors here begin with the view that there is something local about scientific knowledge and then try to explore where that intuition leads. How does the context of discovery shape knowledge? What are the philosophical consequences of a disunified science? Does, for example, an antirealism, a realism, or an arealism become defensible within a picture of local scientific knowledge? What politics lies behind and follows from a picture of the world of science more like a quilt than a pyramid? Who gains and who loses if our representation of science has its standards varying from place to place, field to field, and practitioner to practitioner? The disunity (or unity) of science is fiercely contested ground because these attributes of homogeneity and diversity are so deeply tied to the images of authority of the sciences in relation to one another—and to the broader place of science in the world.

Unity proclamations can, of course, be traced back to the Presocratics, and a thread can be pulled through the myriad pronouncements about diversity and sameness that have proliferated through the ages, and up to the present time. Such a (pre-)Plato to (post-)NATO account does a certain amount of violence to the lines of historical continuity. My own sense is that modern talk of unification in the sciences originates in the German-speaking countries of the mid-nineteenth century. For it was there, amidst the protracted political struggle for German unification, that scientific unity was raised to a scientific-philosophical ideal. The deep intertwining of political and scientific ideals is visible, for example, in the wide-ranging speeches by Rudolf Virchow, addresses with titles like his 1847 "Strivings toward Unity." As Keith Anderton has shown, Virchow maintained the unshakable conviction that political liberalism was a prerequisite for medical progress. All that was particular—tariffs and regulations—hamstrung science, particularized methodology, and impeded the advance of a unifying and standardizing technological world. To Virchow and many of his like-minded contemporaries, the universality of law became a watchword both for jurisprudence and for science.[2]

Hermann Helmholtz's famous 1869 oration, "The Aim and Progress of Physical Science," likewise tied the unification of the state to that of *Wissenschaft* but shifted attention to the physical sciences. "The ultimate aim of physical science," Helmholtz wrote, "must be to . . . reduce all phenomena to mechanics." It was a dream grounded in Virchow's of the unification of medicine and analogous to that brought to biology by Darwin. When Benjamin Thierfelder opened the Convention of Scientists and Doctors in the not-insignificant year 1871, he was thus building on a long history of unity rhetoric; indeed he could claim that the Convention itself had been a crucible in which the principle of national unity had for years survived intellectually where its political realization had been impossible.[3]

Even with (political) unification behind them, the next generation of German scientists grew up with unity as a pervasively valorized, if poorly defined, ideal of their efforts. In prewar Central Europe, boosters of the Unity of Science often held back from talking explicitly about governments. As one participant in some of the early Vienna Circle meetings once told me, there was a rather explicit agreement to "leave politics at the door."[4] Even the most politically engaged of the Vienna Circle founders, Otto Neurath, went to great lengths to separate his Austromarxism from much of his sociological and philosophical writing.

But if politics stood outside the door, as the Weimar Republic began to splinter, the terrifying presence of politics hovered like a specter over the table, only occasionally calling to the assembled. In an insightful and infrequently cited number of the *Encyclopedia of Unified Science*, John Dewey hoisted the by-then familiar standard of a unity of method, a "scientific attitude" (*wissenschaftliche Welt-auffassung*, to the Vienna Circle) that brought disparate specialties together. But he then went on beyond the coordination of scientific results to insist that "there is also a human, a cultural meaning of the unity of science." Beyond the reformation of one's own individual stance toward the scientific method, one's "efforts are hampered, oftentimes defeated, by obstructions due not merely to ignorance but to active opposition to the scientific attitude on the part of those influenced by prejudice, dogma, class interest, external authority, nationalistic and racial sentiment, and other powerful agencies."[5] For Dewey, then, the methodological unity of science became clear in its contrast with the evils of disunity—and they were obvious evils. The rising tide of intolerance could only be combated through this scientific attitude and its immediate extension to the system of education.

Everyone agreed on cooperation and unification; what these terms meant was less clear. For Carnap, unity meant something quite different from what it meant to Dewey: demonstrating unity involved an exhibition of what Carnap called the common "reduction basis" of different branches of science, such as biology and physics. By this he never meant that biological laws could be replaced by physical laws. (Carnap considered that an entirely open scientific question.) Nor did Carnap consider the reduction project to be an ontological one, in which the entities of biology, say, would turn out to be nothing but the entities of physics suitably arranged. Instead, Carnap argued that *both* biological and physical laws could be expressed in terms of everyday physical terms and procedures. There was therefore, for Carnap, a reduction of *language* quite distinct from a reduction of *laws* or *ontology*; it was the linguistic reduction that he saw as central.

Neurath took "unification" to be a desideratum not entirely unlike Carnap's. Finding a linguistic unity involved the deployment of a common language of everyday terms, the adoption of a system of universally recognizable icons, and, most important, an "encyclopedic" assemblage of scientific subjects without forcing them into a superscientific philosophical "system" such as Kantianism. For Victor Lenzen, the reduction was more explicitly nomological: "In the face of apparent disunity, developments in contemporary physics inspire the hope that quantum mechanics and the theory of relativity may be united in a single theory. And because of the basic function of generalized physics and the ever increasing development and adaptation of the techniques of specialized physics, the progress of physics augurs well for the unity of all empirical science."[6] Insofar as quantum electrodynamics represented the prototypical unification, then at least for Lenzen unification meant precisely the creation of integrated laws. To later philosophers, such as Hilary Putnam and Paul Oppenheim, the Unity of Science meant, quite explicitly, the pyramidal hierarchy that reduced one domain of science to another. The great "natural" span of the sciences built solidly one layer upon the next, from the laws of elementary particle physics up through the atomic, molecular, cellular, multicellular, psychological, and social sciences.[7]

Outside the specific context of the twentieth-century *Encyclopedia of Unified Science*, one can point to myriad other notions of unity. During the mid-nineteenth century, DuBois Reymond and many of his contemporaries saw the unification of Germany as inextricably bound with the unification of science.[8] And, in other contexts, the meaning of unification in science is different in yet other ways. Propelled by his commitment to *Naturphilosophie*, Oersted struggled for years to bind forces together before stumbling onto the power of an electric current to yank a compass needle away from the North Pole and pin it along a new alignment. Einstein's argument that electrical and magnetic fields were sides of a single entity and his later efforts to unify gravitation and electromagnetism stem from yet other philosophical and cultural roots. More recent twentieth-century physics has sought unity in a great number of ways, from Einstein's embedding of electromagnetism in a general relativistic space-time, to John Wheeler's minority bid to reduce all physics to geometry, the unified gauge theories of the 1970's, or the string theories of the 1980's and 1990's.

Given the extraordinary variety of meanings attributed to "unification," even among those lying within the same set of encyclopedic covers, one cannot help but see, with Dewey, a *cultural movement* and not merely a specific addendum to the philosophy of language or science. What these various pleas for unity had in common was a hope that an international scientific worldview could curb the divisive racial and nationalistic worldviews against which the Unity of Science saw itself in mortal struggle. This is not to suggest that unification as a theme was always or continued to be located in that particular cultural place. Quite the contrary: the antithetical drives toward unity and disunity have never stayed put. If they begin in the laboratory, they crash the boundaries going out, and, when they begin away from science, they force their way in.

Icon, signpost of truth, and siren call, "unification" has symbolized far more than a purely scientific or philosophical goal. Not surprisingly, disunity does too. In the last several decades new connotations have been affixed to unity and disunity. Instead of an affiliation between unity and internationalism, liberal democracy, and a rational worldview (opposing the disunity of national and racial fascism), the axis unity/disunity has come to polarize around ideas of cultural autonomy in opposition to forces of homogenization, hierarchy, and domination. In the sciences we find biology departments splitting into the "organismic" and the "microbiological." The organismic biologists insist that they are not merely engaged in a secondary, applied study of methods derived from the efforts of their more "fundamental" neighbors.

   Within physics, there are institutes devoted to the study of complex phenomena precisely on the premise that there are shared laws governing such phenomena not deriving from microphysics. The debate over "fundamentality" in physics carries enormous political weight because the very existence of the superconducting supercollider (SSC) depends, in good measure, on the particle physicists' claim that theirs is a basic knowledge—more so than in other disciplines. Steven Weinberg's widely read book *Dreams of a Final Theory*[9] is at one and the same time a plea for a fundamental metaphysics and a case for a multibillion-dollar collider in Texas. For Weinberg, the chains of scientific "why?" questions always come to ground in high-energy physics: ask about the blueness of the sky, and you will learn about molecules and light; ask about molecules and get atoms; ask about atoms, get nucleons and then, ultimately, to the structureless simples of the world, elementary particles, or even to the oscillations of strings. Not surprisingly now (but it would have been several decades ago), condensed matter physicists reject any notion that their discipline is less foundational, basic, or fundamental than quantum chromodynamics or superstrings. Perhaps more surprisingly, some particle physicists now demur from this ultimate argument. In the 1970's, Howard Georgi had coproduced the SU(5) Grand Unified Theory, linking the electroweak force to the strong (color) force. Twenty years later, none of the Grand Unified Theories had found broad acceptance, and string theory had lost some of the sheen it had in the 1980's. Georgi had shifted his sights to the accessible energies of particle physics, below that of GUT's or strings. Like Weinberg, Georgi was a staunch advocate of the SSC, but on utterly different grounds. For Weinberg, the SSC promised final clues to a final, unified theory. In an informal talk, Georgi argued differently:[10]

If experiment is unnecessary, if theorists can understand the structure of the universe by sitting at their desks and doing beautiful mathematics, then these limits [to our technologically fixed scientific knowledge] do not matter much. But my personal view is that nature is much more clever and imaginative than we are. We particle theorists must not get too hung up on highfalutin' theoretical ideas like unification, no matter how appealing and beautiful, or we may lose our sense of wonder at the infinite variety of nature.

Even standing on the same side of the politics of Big Science did not compel agreement on the agenda for scientific unification.

Of course, these "internal" scientific debates over fundamentality, reducibility, and so on, do not exist in a vacuum. They are profoundly embedded in a culture in which the quasi autonomy of different subcultures is valued as essential now in a way that it simply was not in the prewar years or even in the 1940's and 1950's. Contrast the language of the logical positivists with that of Patrick Suppes, who, in 1978, quite self-consciously dissented from the Unity of Science with an article, "The Plurality of Science." There Suppes identified the disunity of science with the disunity of human inquiry more generally. Opposing the "cosmological or global view of truth" and completeness, Suppes espoused a view of truth and completeness that would be fundamentally local, designed to meet and solve specific problems. "Personally," he wrote, "I applaud the divergence of language in science and find in it no grounds for skepticism or pessimism about the continued growth of science. The irreducible pluralism of languages of science is as desirable a feature as is the irreducible plurality of political views in a democracy."[11] Plurality of views, democracy—these could, in the fourth quarter of the twentieth century, find a comfortable set of associations in talk of disunity. A hundred years earlier, a Helmholtz or a Virchow could locate freedom in unity. Is there a lesson here? Perhaps it lies in a certain distance we need to take from claims of the transparency of meaning in scientific programs; the wandering valence of the disunity/unity axis can teach us a great deal about the ways we think about the aims of science in its historical realization. What we will not find is a single-valued, transhistorical function that plots assessments of unity onto a fixed political map.

Boundaries

Ian Hacking explores some of these ideas in our first essay, where he emphasizes the disunity of unity—that is, the plethora of meanings associated with the very idea of unification. At the center of his taxonomy is a distinction between singleness and harmony. Philosophers, Hacking argues, have too long fastened on singleness, the subsumption of all phenomena under a single principle, law, or language. By contrast, scientists have left a great deal of room for the harmonious adjacency of different domains of inquiry. Maxwell exemplified the scientist's unity when he pressed for an image of the book of nature as an encyclopedic aggregation of articles, fitted together but not hierarchically ordered.

Hacking goes on to argue that beyond the differing sentiments of singleness and harmony lie three different "unities" of science: a metaphysical sentiment, a practical precept, and a mode of scientific reasoning. Metaphysically, the commitment to unity amounts, typically, to the belief that there is "one scientific word, reality, truth," which includes the physicists' thesis of interconnectedness. Practically, the commitment to unity suggests a quest for the connection between phenomena, between (for example) light and magnetism, weak and electromagnetic forces, genetics and evolution. Methodologically, the commitment to unity carries the banner of one standard of reason that holds good across time, across disciplines, and across the different circumstances to which science is applied.

None of these commitments is beyond contestation. The belief that there is one scientific world blocks the move toward what Dupré will present as a multiplicity of worlds read differently by different groups and for different purposes. As Hacking himself notes, there is a feminist case for an alternative science that would be predicated on starting points different from the ones generally assumed in current conditions. Other metaphysical battles are fought over whether the world contains an ordered set of causes all deriving from a first or primary cause. Logical positivists and their followers resisted such causal talk, but insisted that the same sort of unifying hierarchy can be accepted as cascading in *logical* implication rather than causal order. Whether you believe that such unification (on the basis of cause or implication) can be ascribed to the world depends on your view of the *structural* disunity of science.

Hacking's own metaphysical sympathies appear to lie with what one might call Maxwell's Magazine of Nature: nature as disunified but interconnected. On the method front, Hacking here and elsewhere has espoused a pluralism of "styles of reasoning," based loosely on A. C. Crombie's disaggregation of six such modes of argumentation, including the historical, the statistical, the taxonomic, the mathematical, the analogical, and the experimental styles.[12] Propositions advanced that have perfectly well defined truth or falsity in one style may fail, as Hacking has often put it, even to be "candidates for truth or falsehood" in another.[13] In modern medicine certain Paracelsan claims are not such candidates. Hacking offers, as an example, the injunction that syphilis is to be treated by a salve of mercury because mercury is a sign of the planet, and the planet a sign of the marketplace, and the marketplace is where syphilis is contracted.[14] True? False? Such an utterance may well be evaluable by the Paracelsans; it would not be by twentieth-century internists or nineteenth-century hygienists. For Hacking, this internally structured "self-authentication" is characteristic of science, and while it might appear to be an invitation to Feyerabendian anarchy, Hacking insists that it is not: "My doctrine [, which] sounds like part of the current mood for skeptically undermining the sciences, turns out to be a conservative strategy explaining what is peculiar about the sciences." And what is peculiar is precisely the scientific proclivity for stabilizing belief. Each Crombian style acquires its own "realism debate," and, though Hacking doesn't say this, presumably he would be committed to arguing that a realism debate begun in the historical style of reasoning would not even make sense if it were continued in the Galilean style.

Having tackled the reasoning part of styles of reasoning within Hacking's geography of disunity, we can now turn to an exploration of the meaning of the "style." Arnold Davidson asks, Under what specific conditions can one comprehend various kinds of statements as being either true or false? This leads to an inquiry into the meaning of style itself, a subject studied nowhere with quite the intensity it has received in the history and theory of art. Following Wölfflin, Davidson declines to explore expressional content by estimating how work affects us now, in the late twentieth century. Instead, Davidson wants to know the formal possibilities at the disposal of those working within an epoch. Wölfflin: "Even the most original talent cannot proceed beyond certain limits which are fixed for it by the date of its birth. Not everything is possible at all times and certain thoughts can only be thought at certain stages of the development." To exemplify the uses of such a stylistic analysis, Wölfflin enunciates five pairs of polar categories that throw into relief the contrast between classic and baroque styles. For example, one pair contrasts distinctness of representation, in which each form is well defined and picked out completely from all else, with its stylistic opposite, in which each object shades easily into everything adjacent to it.[15]

Davidson likes Wölfflin's conceptualization of style, his deployment of polar categories, limited possibilities, breaks and discontinuities, and wants to borrow this mode of analysis for a new understanding of the emergence of a characteristic psychiatric style of reasoning, taking "the fundamental element of style, namely the categories or concepts and the way they combine with one another to constitute a style." One style of reasoning might be dubbed "anatomical," as in Moreau (du Tours)'s sixth, "genital," sense, which could be debilitated as color blindness could impair the perception of color. In just such a way, the sexual sense or instinct could be diminished, augmented, or perverted the way that other anatomical functions could be. Perversion and function were inextricably tied to one another. By Davidson's lights, we can understand what the pre–nineteenth century grouped together (or broke apart) only by recuperating their conceptual structure. When we reconstruct the organization of reasoning about sexuality based on the "natural" function of the sexual instinct, we have entered the style of psychiatric reasoning. At that point we can see the defining trait of the perversion of homosexuality as grouping it within a single set of disorders alongside masochism, sadism, and fetishism. The contrast could not be greater with the earlier anatomical style; pre-nineteenth-century reasoning set determinate locations for perversions, while late nineteenth-century psychiatric notions of sexuality prized psychiatric structures away from structures of the body. By the time of Charcot and Freud, Davidson writes, "Sexual identity is no longer exclusively linked to the anatomical structure of one's internal and external genital organs. It is now a matter of impulses, tastes, aptitudes, satisfactions, and psychic traits. There is a whole new set of concepts that make it possible to detach questions of sexual identity from facts about anatomy, a possibility that came about with the emergence of a new style of reasoning." No transcendental style of reasoning about perversions, sexuality, and impulses exists; we must give up any transhistorical unity in the psychological sciences.

Styles of reasoning address the conceptual infrastructure of scientific reasoning; that of which we cannot speak, we may well have something to say about in another stylistic regime. In such a disunified picture of science the disunity lies deeper than Carnap and Neurath feared: not only are the laws, language, and objects disunified; the very evaluative strategy for propositions is fragmented. In the 1930's, the politics of disunity appeared to be authored exclusively for the right wing—the special national and racial destinies were the exemplars of separateness. But the world has turned, as is manifest in the work of John Dupré, whose approach to such a patchwork universe resides more squarely in the domain of metaphysics, and more explicitly in the field of progressive politics. Emphasizing the lessons of recent historical studies, Dupré finds a defense of a single scientific project utterly implausible. But, as Dupré concedes, the positive case for a disunified science cannot ultimately come from any finite set of case studies; empiricism alone cannot lead inexorably to a metaphysical conclusion. Nothing traceable to air pumps or accelerators will ultimately or inevitably show that science is or must be forever disunified or unified.

More specifically, Dupré confronts two aspects of what Hacking calls "structural" unification. First, he takes on the contention that the world, in and of itself, falls into a hierarchy of entities: electrons, atoms, molecules, cells, organs, animals, cultures, and societies. Such a simple string of objects, or even its more subtle variant, the branching tree, leaves unresolved the problem of what individuation among such objects is. Certain types of optionally multicellular entities like the slime mold and behaviorally bound insect swarms raise the problem starkly—where or what is the boundary of the "individual"? Or take humans. The body can hardly be thought of as a simple composition of homogeneous entities. Ions, fluids, organs—all coexist, and there is not much sense in thinking of blood as a building block of the pituitary gland. Indeed, in the human form, in just the place where the metaphor of nature "cut at its joints" would seem to find its most apt application, it fails catastrophically.

Identifying the ways in which natural kinds should be identified is ultimately going to depend on the uses to which the classification is to be put. But here, in agreement with Hacking, Dupré is keen to distance himself from a purely "motivation"-based relativism. As he puts it: explanation must be grounded "also in terms of the objective reality that it discloses." Thus while he wants explicitly to defend a multiplicity of viewpoints about how to consider the entities (and goals) of science, he rejects the kind of symmetry between our explanations of good and bad science espoused, for example, by David Bloor. In particular, Dupré specifically wants to preserve the possibility of arguing that claims about creationism or overextended sociobiological accounts of behavior can be understood motivationally and as such are bad science, full stop. To capture this flexibility within constraints, Dupré describes "promiscuous realism," the belief (a) that there are many different kinds to which an individual belongs, none of which has priority, and (b) that these affiliations of kind are real.

The second level of Dupré's critique of unified science targets the idea of a causal hierarchy. Practically, the imposition of a causal reductionism from the macro to the micro encounters the well-known difficulties of chaotic phenomena, where even small deviations in initial conditions make later macroscopic behavior unpredictable. Similarly, the practical limits to our prediction of the behavior of even the three-body problem make deterministic forecasting of future behavior of vastly more numerous interacting entities even less precise. But beyond such obstacles lie deeper, in-principle difficulties with reductionism, especially in the supplanting of social-scientific and evolutionary biological arguments by microscopic phenomena. Dupré suspects that there simply are no determinate propensities characterizing the particular "underlying" events. Regularities, in short, might not be contained all along in the behavior of microcomponents; macroregularities might only emerge over time.[16]

My own work on disunity begins at a point of common concern with Hacking, Davidson, and Dupré: the extraordinary variety of scientific languages, practices, purposes, and forms of argumentation. But rather than focusing principally on the claim of disunity, I have attended to the liminal structure—the boundary areas ("trading zones," as I have called them) between the disunified bits of science. The argument at first sounds somewhat paradoxical: that the disunified, heterogeneous assemblage of the subcultures of science is precisely what structures its strength and coherence.[17]

More specifically, I have argued that underneath their apparent conflicts, the assumption that science splits into "frameworks" has guided widely diverse camps in the history and philosophy of science. In history, this has meant periodizing physics, for example, into blocks in which experiment and theory were fused thoroughly, both technically and sociologically, and arguing that these blocks lay in noncommunicating sectors of social and conceptual space. On the framework view, Newtonian mechanics had its theoretical mode of expression and an appropriate and mutually reinforcing set of experiments. Einsteinian mechanics did as well, and the two, it was argued, would not meet. Though interpreted differently, incommensurability claims like this one found expression in the frameworks of Carnap, in the paradigms of Thomas Kuhn, and in the conflicting interest–theoretical accounts of more recent commentators such as Barry Barnes.[18] In all three, the more-than-metaphor of nontranslatability has been the touchstone of argumentation: mere translation would never reconcile the conflict across the paradigm gap of separate ontologies, epistemologies, and nomologies. Against this framework conception of science, I argue that the different subcultures of science do, in fact, work out local trading zones in which they can coordinate their practices.

My essay here applies these ideas of local coordination to the extraordinary confluence of disciplines in the late 1940's and 1950's that formed around the notion of simulated reality, as expressed in Monte Carlos. These computer-generated realities lay between categories. They were neither experiment nor theory, or perhaps both experiment and theory. They were part of mathematical statistics, and yet often classified as part of physics. Simulations were not quite pure mathematics, not quite just a part of nuclear weapons design, yet perhaps, simultaneously, both these and more.

What happens when an H–bomb designer, a logician, an aerodynamical engineer, and a statistician sit down together? Whatever else they do, I would argue, they do not found a League of Nations with simultaneous translators (or their scientific equivalents) perched over the assemblage in metaphorical glass booths. No: they work out an intermediate language, a pidgin, that serves a local, mediating capacity. "Randomness," "experiment," and a host of other terms and actions coalesce into a more or less coherent sector of shared usefulness. While the mathematician thinks about the best definition of "random" quite differently from the physicist, in the cauldron of those early days of computer simulations a notion of "random enough for present purposes" emerged, borrowing from several cultures, yet belonging exclusively to none.

The picture of disunity I end with builds on such boundary arenas. Cultures within science differ in myriad ways. But the possibility of working out such partial, local, and specific linkages is what, I suspect, underlies the experience of continuity that these various groups feel as they work out trading zones between them. In particular, on this view there is no practice cluster that is immune to revision, break, and radical reconfiguration: theorizing, experimenting, and building instruments. But, as in the example of the simulators, practitioners of these various subcultures do not all move in synchrony. Instead of basing a picture of scientific knowledge on disjoint but internally coherent frameworks, the suggestion is that we see science as a stone wall or rope, composed of disparate and heterogeneous bits, where strength follows just from the circumstance that component parts are not precisely matched, but are intercalated.

Richard Creath argues that this view of an assemblage of subcultures brought together through trading languages is compatible with the Vienna Circle when the latter are rightly understood. Indeed, by the time Carnap defended the unity of science, he did so as a conventionalist, pragmatist, and physicalist, though not just the way Neurath was. Common to both, though, and later to Hempel was a coherentism, not a foundationalism. The language of room-sized objects was just one of several ways one could describe observation. Nor should Carnap's starting point be attributed entirely to the British empiricists, for though Carnap began his *Aufbau* with the autopsychological, his system was not (certainly not post-1930) tied to a basis of what one could describe as "my" experiences or "my" mind.

Creath emphasizes that Carnap divides the unity of science into two distinct claims toward which he has two very distinct attitudes. One assertion is that the language of science is or should be unified. For Carnap, Neurath, and their allies linguistic unification meant that the lexicon of physics (for example, the mathematical description of objects located in space-time) is or should be carried over to all other branches of science. Both Carnap and Neurath saw physicalism (the unification of scientific language) as a move in a larger battle against knowledge founded on private, introspective experience. Among the targets of Carnap's and Neurath's polemics was the *Verstehen* school of social science, predicated as it was on the private, intuitive understanding of social action that was inherently and *a priori* outside the approach of publicly available evidence. As we have already seen, Carnap views the task of linguistic unification as entirely distinct from the unification of scientific laws, an endeavor best left to the special sciences (physics, biology, chemistry, and so on). Neurath went farther, positively objecting to the hierarchical organization of the sciences even as an ideal. Thus, for both these Viennese advocates of unification, philosophy *per se* had no stake in the subsumption of many laws under a few.

What hinges on the unification of science under a public (physicalist) language? For Carnap, the virtues of such a program are purely practical; he even allows that other phenomenalist languages might have their uses, and he endorses them under his Principle of Tolerance. Neurath is not inclined to make such a pragmatic defense, and since he is not about to endorse an analytic/synthetic distinction, he turns, as Creath (and elsewhere Thomas Uebel)[19] nicely shows, to a "no private language" argument. Suppose one begins with a set of strictly private and logically distinct observations. By assumption, there would be no way to link these observations to another individual. But there is worse: because it is impossible to check these observations, we cannot even compare our observations now to our observations five minutes ago. The ensuing solipsism is so devastatingly complete, isolating the individual both temporally and spatially, that Neurath takes it as a self-refuting consequence of any form of antiphysicalism that is predicated on private language.

Finally, Creath wants to use his reading to show that the views I have defended about disunity are in fact compatible with an antifoundational, conventionalist reading of the Vienna Circle. Specifically, he shows that Carnap and Neurath do espouse a species of holism, but (according to Creath) it is a moderated holism based on fairly symmetric relations among the special sciences, and not on the extreme form of linguistic holism that subordinates observation, and by extension experiment, to theory. Thus, Creath concludes, my metaphors of an intercalated brick wall or my use of Peirce's rope analogy[20] would (retrospectively) sit well with a Carnap whose unity of science amounted to "a preference for co-herence, not current fact."

Steve Fuller addresses a metaphysical schism that has opened between two stances within science studies. On one side, he argues, stand those social constructivists, like Harry Collins and Stephen Yearley, who view their project as an application of social science. In this reading, it is crucial that those studying science can "go home" to the social sciences; there is behind every anthropological venture the fixed point of a return to social-scientific knowledge, be it sociological or social-historical. On the other side of the divide (Fuller calls it the radical side) stand Bruno Latour and Steve Woolgar, for whom the understanding of science cannot refer back to the fixed point of an external social science. "Going native" precludes going home; and if one cannot ground knowledge in the permanent base of social science, then our representations of science will necessarily be disunified in ways that are not in the worldview of the social constructivists.

In order to give a sympathetic reading of the radical position, Fuller offers two schemata that summarize the moves made by Woolgar and Latour. At root, Fuller contends, the first strategy, which he calls the Woolgar Procedure (WP), inverts "standard" metaphysics, and makes individuals instances of properties (instead of making properties dependent on preexisting individuals). To get at the radical conclusion, the WP then has us identify (1) situations where the properties are integral to what the individuals are doing and (2) situations where the properties are not integral to what the individual is doing. From (1) and (2), the WP concludes that the individual has the property only in the identified situations. Less abstractly, the WP purports that "X is a scientist" is dependent on the particular situation in which X finds himself or herself; there is no permanent attribute of being a scientist, or of a field's being a science.

Similarly, in the Latour Procedure (LP) each property of an individual is nothing but the sum of relations an individual has with other individuals. Being a scientist is a property that exists *only* insofar as the network of relations between individuals is sustained. When the network of relations stops functioning, the property stops existing: "Science exists only to the extent that the network is maintained."

Fuller argues that one of Latour's more provocative conclusions follows from this metaphysics: the inanimate must be accorded agency alongside the animate. (In Latour's words, "There are more of us than we thought.")[21] Latour's argument, as I understand it, is that if one reads the textual evidence of Pasteur's time, the microbes quite evidently are ascribed agency—they scheme, they kill, they wait, they form alliances. Instead of dismissing such talk as mere animism, Latour wants us (the readers) to refrain from imposing our own categories of the world and to treat the networks that surround microbes just the way we treat the networks with people at their nodal points. To Latour such a move is a combination of semiotic sensitivity (to the way the inanimate appears in discourse) and scientific restraint (from imposing our own metaphysics). To the social constructivist (such as Schaffer), the imputation of activity to the nonhuman is hylozoism.[22] Latour responds: representations of microbes are all we have, and these representations only have the meaning they obtain from the network of (textual) relations in which they participate.

For Latour, explanation is a representation too. Just because Latour denies any privileged social-scientific Archimedean point from which to view the world, he denies that explanation can be given in terms of a reduction from one level to a more basic level. Instead, Latour sees explanation as a translation from one (equivalent) discourse to another. And it is here that Fuller dissents. Using the notion of pidgins and trading zones, Fuller wants to challenge the notion that a translation can be effected between *explanans* and *explanandum* without touching either one. A pidgin, forming between the two, shapes both—rightly, I would say. Fuller emphasizes what he calls the "interpenetration" that a trade language effects. If Latour is really going to go all the way, he must move in some such direction as this; he must, in short, go beyond importing an older, logical positivist conception of interpretation as translation.

Contexts

Unity has to be enforced, and Simon Schaffer addresses the context of that enforcement. In particular, Schaffer wants a contextualized history of the canonical texts that structure fields and define their boundaries. In other fields, literature for example, recent scholarship by Michael McKeon reworks the meaning of texts by De Foe, Swift, Fielding, and Richardson by locating them in a tumultuous eighteenth century in which the social order was in radical transformation. As Schaffer insists, it is only through an understanding of the distinctly uncanonical paper trail of manuscripts, newspapers, correspondence, tracts, and sermons that the canonical texts find their location. The original goal of Locke's *Two Treatises* was not a purely abstract meditation on political theory; it was an intervention in the popular radical political discourse of the Glorious Revolution.

But beyond accounting for the source of the canonical, Schaffer wants contextual history to analyze the canonization process itself: How, out of the extraordinary variety of writing at a given time, is a single item extracted, supported, and legitimated? In science, particularly, the process of extraction and elevation is central to the very definition of the disciplinary map. As Schaffer puts it: "Legitimation and calibration through historical reflection are not epiphenomenal to the organization of the sciences, but fundamental for their social order. Scientific work is also the production of historical accounts that warrant achievement, define and exclude alternatives, and carefully position the author in the scientific milieu."

Schaffer is concerned with the forces that sustain the power of the canon, not just the forces that create it. In effect, he argues that science studies can fall all too easily into Barthes's "mythoclasm," a debunking endeavor designed to show that the origins of specific scientific endeavors have local and specific contexts that belie the view that scientific discovery occurs outside society and culture. Instead, he insists that the maintenance of a field and its boundaries is altogether as much in need of explanation as is its incipient form at time zero. But the work that goes into this maintenance project is often obscured, because its product, self-evidence, is precisely what effaces the effort. Like the courtier's nonchalant gestures, about which Mario Biagioli has written so eloquently, effort looks effortless.

Biagioli makes two moves against Schaffer's Context of Canonization. First, he argues that the social constructivists' relativism clashes with their other proclaimed goals. Constructivists want to challenge the "god trick" (Haraway) of judging science as if the historian were an unsituated observer gazing across the panorama of knowledge from ancient history into the infinite future in all places and all cultures. But in its place, Biagioli complains, they have left us with another god trick, this one the assertion that all knowledge is local, disconnected, and ultimately isolated, one bit from another. Instead, Biagioli contends, we need something like a model from evolutionary epistemology. Unless we want to make godlike metaphysical claims about ultimate truth, either now or in the future, we cannot talk about the "true" theory. Nor, Biagioli adds, do we want to claim that all knowledge is equivalent, all simply the result of a toss of the dice. Evolutionary epistemology *à la* Biagioli is supposed to avoid claims of relativism and claims to absolute truth—a theory "fits" in a given situation, the way a successful aardvark fits the niche of a termite-filled African plain. Neither the aardvark nor the successful theory is "true" *sub specie aeternitatis*. With this evolutionary instrumentalism, Biagioli claims we can avoid lurching to an arbitrary anarchism as we recoil from an unargued-for metaphysical realism.[23]

Biagioli then opens a second anti-Schafferian front, by challenging the reflexivity of the Context of Canonization. If the program were truly contextual and reflexive, it would self-consciously address the ways in which science studies itself has chosen, bolstered, and legitimated its canonical texts. Just as Biagioli contends that scientific texts are not arbitrarily picked out of the background of other writings, so he questions the positioning and reinforcement of texts in science studies. To Schaffer, "the canon of the human sciences provides resources for any currently possible historiography of the human sciences." Current canonical texts in science studies in some sense delimit what kind of history can be written. Just as Biagioli rejects this claim for the sciences, so he rejects it in the human sciences generally and in history of science particularly. Instead, he wants to emphasize the contingent nature of the world, the variety, the differences of approaches that can be followed both in science and in the history we write of it.

Where Schaffer wants to use social history to unravel the local character of knowledge, Biagioli wants to locate the author of historical texts, emphasizing that this very location undermines any claim that there is a canonical approach to science—or to science studies. To Schaffer, Biagioli's neo–Darwinian model avoids the labor–historical efforts that go into defining and protecting the boundaries of fields. To Biagioli, Schaffer's emphasis on canonical texts is ultimately a form of intellectual hegemony; by restricting scientific (and, by reflexivity, historical) writing to the working-out of the canon, Biagioli interprets Schaffer as denying diversity. Biagioli suggests, in effect, that if we really believe that canons "define and exclude alternatives," then the history of science will be forced into conformity with a handful of paradigmatic case studies. It seems that the antithetical poles of homogeneity and hetero-geneity arise even among discussants who agree that science and history are disunified.

Arthur Fine positions his own essay, and his antimetaphysical program, more generally, as a (sometimes critical) philosophy of contextualism and social constructivism. According to Fine there are three "planks" in the program of social constructivism: (1) beliefs are relative to prevailing social circumstances; (2) it is neces-sary to ask why a belief is held, regardless of whether that belief is held to be true or false, rational or irrational; and (3) local scientific goals are sustained in a rich field of cultural, political, and social forces. The three planks are nailed together with a consensus the-ory of truth. Instead of "facts are what make belief true," we have "facts are just consensual beliefs, and so the making of belief is the making of fact."

In general, Fine sympathizes with the disunified, context-dependent sketch of science. He finds bankruptcy in the general philosophical accounts offered over the years to describe hypoth-esis testing, explanation, and the discovery, structure, dynamics, and confirmation of theories. Since these universalist accounts fail, social constructivism ought find its day in court.

Once on the docket, however, Fine's cross-examination zeros in on the behaviorist assumptions that underlie so much of the con-structivist story. As he puts it, "In a consensus theory, the social fact of the fixation of belief is promoted from being one of the marks of truth to being the whole of it." This essentializing move, the elevation of consensus to the constitution of truth and facticity itself, is what Fine dubs "behaviorism with a vengeance." And it is there, in the preservation of an essential (social) characteristic of science, that Fine finds constructivism not to have gone far enough.

Offering a counterproposal, he advances what he calls "methodological constructivism" constituted by five points: (1) truth should be bracketed as an explanation for why science is the way it is; (2) openness, choice, and judgment are everywhere in science; (3) practices are local, and we can ignore whether or if they fit together; (4) science is human, with modalities for human action; and (5) opinion formation and dissolution can be understood only in the particularity of local conditions.

Left out of this counterproposal is any mention of a metaphysical (unified) theory of truth, and at the same time and by the same token any global pronouncement about relativism, or the eventual reduction of science to the social. There is, as Fine puts it, no assurance that scientific facts or objects are essentially objective or essentially social: let the chips fall where they may. *Methodological* constructivism, in Fine's view, does not purport to be a refutation of realism; rather, it suggests that support for realism simply cannot be found in the practice of science. Since Fine dismisses metaphysical arguments, he considers the case closed.

Like Fine, David Stump puts his chips on the disunity rather than the unity of science, maintaining a studious agnosticism toward realism. And, like Fine, Stump wants any discussion of epistemology (understood as the methods associated with securing knowledge) to reside in the domain of epistemology; he opposes attempts to solve epistemology with metaphysics. There are two categories of skeptical arguments that Stump invokes to undermine metaphysical realism. The first is that results always end up being justified by circular reasoning or unargued propositions through the assumption of a framework. Presumably this is the type of argument invoked by those who use Kuhn's paradigm shifts as a basis for relativism: no standards of argumentation transcend the paradigm. Work within a paradigm inevitably imports assumptions about the kinds of entities that can exist, the ways we learn about them, and so on. Second, distinct from what we might call "framework skepticism" is what we might call for short "abductive skepticism," the doubt that one can move from an empirical base to theoretical conclusions. Those skeptics who rest their arguments on some form of the Duhem-Quine thesis of indeterminacy presumably fall into this camp.

Stump himself has no more sympathy for the framework and

abductive skeptics than he does for the metaphysical realists. All build on the assumption that there are global (universal) claims to be made about the nature of science. But whereas Fine targets the nonlocality of the constructivists' essentializing use of behaviorism, Stump aims his arrow at the global skepticism of the social constructivists' relativism. For both Fine and Stump, disunity lies in the contingent local conditions that shape the divergent strands of scientific practice.

Karin Knorr Cetina is concerned directly with the diversity of practice. More specifically, she balks at the notion that there is such a thing as "experimental" or "empirical" practice, full stop. Instead, by drawing on an ethnographic comparison between experimental high-energy physics and molecular biology, she highlights just how separate these spheres are.

In the high-energy physics laboratory Knorr Cetina finds an overwhelming concern with the process of experimentation, a preoccupation with instruments, methods, checks and cross-checks. It is an understandable obsession, she writes, because in part the objects under investigation are too small and too short-lived to be able to confront the experimenters in anything like a direct manner. But whatever the reason, high-energy experimentalists live in an epistemic world characterized by the phrase "care of the self" that she borrows from Foucault. This reflexive concern rests on three activities, of which the first is the pursuit of "self-understanding," a stage of work devoted to mastering the instruments themselves— controlling their oddities as well as their "normal" functioning. Next there is "self-observation," the systematic monitoring of detectors, cables, and storage units; these kinds of on-line and off-line checks occupy a significant fraction of the effort that goes into an experiment. Since what emerges from the detector in the first instance is quite literally meaningless, this "massaging" of data is not an optional extra put onto preexisting results; it is in large measure a part of the results themselves. Finally, by "self-description" Knorr Cetina wants to designate the elaborate system for inscribing data and reduced information onto the multitude of media that record experimental results for future analysis.

Nothing like this electronic pulse-taking can be found in the halls of a molecular biology laboratory. Looking less as if they were in a control tower than in a kitchen (Knorr Cetina's metaphor) the biologists, unlike the physicists, are directly and viscerally connected to the phenomena themselves. Instead of engaging in a massive struggle to disentangle signal from background, the molecular biologists abandon the quest to separate wheat from chaff using calculation; they allow mutation to produce the desired organism, letting nature's selective processes kill off the unnecessary and preserve the useful. To Knorr Cetina, the physicists pursue themselves to capture that which they cannot confront, while biologists abandon the hope of detailed knowledge to confront their phenomena head-on. That both strategies are called "experimental" only obscures the fact that they represent "epistemic cultures" that a more sensitive reading would class as utterly different.

As distinct as biology and physics may be, the social sciences present a new set of problems for any overarching "unitary" model of science or scientific method. Looking at anthropology, Alison Wylie finds a cautionary tale in theories of how observation and theory should relate. More or less every twenty years, she says, the problem of interpretation emerges as a crisis in anthropology. In the 1930's, anthropologists widely argued that fact-gathering without theory was blind. Twenty years later, the crisis reemerged when Spaulding ridiculed Ford for saying that ethnographic truth might be got by polling anthropologists.

American "New Archaeologists" of the 1960's and 1970's again blasted their opponents for wallowing in inductivism. "Culture" had to play a role in the background theory of a dig if the artifacts were to gain any significance. Philosophically, the legacy of the logical positivists, the Hempelian hypotheticodeductive (HD) model, served the New Archaeologists' cause—without a hypothesis nothing would follow, and "culture" became the H of HD. Perhaps surprisingly (or perhaps not) Kuhn's challenge to the legacy of logical positivism fit smoothly into the challenge against inductivism. Instead of culture-as-hypothesis, there was now culture-as-paradigm.

As in so many fields of the philosophy of science, anti-inductivism quickly slid into a form of relativism: abductive skepticism became an argument against constrained knowledge altogether. It is at this point that Wylie's essay turns to the feminist critique of anthropology, raising questions of even wider significance. For example, she specifically confronts archaeologists' often-unstated assumption that gender divisions are timeless, and that activities with true agency are necessarily male. If horticulture is seen as more or less a passive activity, it is assigned (retrospectively) as women's work; as archaeologists revise the account to make it more interventionist and active, it is then removed to the domain of men. Shamanism is interpreted as an active and powerful role, and not surprisingly it is often, though without substantiation, taken to be self-evidently a male role.

Wylie points out that the very notion of a feminist critique on these lines carries with it some commitment to the notion of constraints to knowledge formation. As she puts it, "Political commitments do not displace evidential considerations; if anything, they enhance a commitment to empirical rigor and the critical inspection of presuppositions." Such a commitment, however, is at once in tension with a reductive social constructivism that she sees as necessarily undermining a transpolitical sense of what acceptable standards of evidence can or should be. Feminist concerns also seem to run head-on against at least some lines of postmodern deconstructive arguments that seek to dislodge any status to "the subject." For just at the moment when women and non-Western peoples have begun to claim themselves as subjects, the deconstructivists are moving to cancel the term. To sum up: as Wylie sees it in archaeology, there are two competing ideas about the relation of scientific disunity to power. One side sees a (productive) power residing in a depiction of science as maximally disunified with the various parts of science out of touch with one another, and with the very idea of the autonomous scientist dissolved. The competing political view finds power to effect change precisely in some (mitigated) concept of a science unified, at least to the point of evidentiary standards, and where previously marginal people can claim for themselves all the power that goes with being true actors in history and in science.

Power

Though it is a theme throughout this volume, power is central to the last several essays in this collection. There is no getting around political power, classically conceived, when we turn, in the essay by Cartwright, Chang, and Cat, to World War I Vienna, where unified science took its modern form. Surely no one lived the question of unity with the urgency of the sociologist–philosopher–cultural critic Otto Neurath. During the Great War, Neurath sided with the independent socialists of Bavaria, led by Kurt Eisner, who took over on 7 November 1918, set up a coalition socialist government, and founded the Bavarian Republic. Less than three months later, on 23 January 1919, Neurath came into Munich to let Eisner know his thoughts on central planning, and to propagandize for centralization with the Munich Workers' and Soldiers' Council. Just as the war economy had promoted efficiency during the fighting, so Neurath believed only full socialization (*Vollsozialisierung*) could ensure success in the young republic.

For Neurath, it was planning, not nationalization, that was important. He wanted a rationalized and coordinated distribution of goods, farming, production, and currency. After the violent suppression of the Republic by the *Freikorps* in May 1919, Neurath was arrested, tried, and sentenced, only to be rescued before he landed behind bars—his savior was an old friend and ally, Otto Bauer, the Austrian Minister of Foreign Affairs. In this political history, Cartwright, Cat, and Chang find some important roots for Neurath's outlook on unified science. Quite obviously, the virtues of centralization and the unification implied by them form a continuous set of concerns across politics and science. But more subtly, the authors point out that Neurath's political sense of unification never involved a homogenization or radical hierarchy. Quite the contrary: his vision in the Bavarian Socialist Republic was always of a *coordinated* economy, left substantially in private hands, and synchronized rather than brutally assembled into a vertically integrated whole. In science, Neurath's vision of "orchestration" similarly avoided the coercive implications of science structured from one "master law" all the way down through the nitty-gritty of applications. Neurath never believed, for example, that science now or in the future would amount to a small set of microphysical equations from which the laws of ecology, economics, and psychology might be derived.

Like Cartwright, Cat, and Chang, Timothy Lenoir and Cheryl Lynn Ross explore the link between representations of nature and the cultures in which they are embedded. But instead of looking at explicitly philosophical texts, such as those of Otto Neurath,

Lenoir and Ross take us on a walk through late nineteenth- and early twentieth-century natural history museums. In those marble halls, behind plates of glass and framed by varnished wood, is a recreation of nature altogether as complex (and disunified) as the Vienna Circle's theories of nature.

Perhaps most strikingly, Lenoir and Ross contrast the "Nature" of Richard Owen, whose exhibitions worked to combat radical socialist and Lamarckian tendencies, with the "Nature" of Henry Fairfield Osborn, whose visions stood guard against such enemies as race mixing and the decline of the West. Owen, of course, was a leading "inventor" of the whole category of dinosaur, a term he coined. Above all, Owen's ichthyosaurs, plesiosaurs, and tele- osaurs left fossil records he could point to as standing, thirty tons strong, squarely in the way of any progressive reading of history. More organized, bigger, and more sophisticated creatures pre- ceded less developed dinosaurs; worse (for the Lamarckians' faith in progress), the superior forms left not a trace of their best and most sophisticated features. By showing these creatures in all their awe-inspiring grandeur—one exhibit opened to a crowd of forty thousand—Owen popularized his vision beyond anyone's expec- tations, and by the force of his "naturalized" plea for social sta- bility earned the undying gratitude of the conservative Anglican establishment.

Osborn, too, could only have pleased his establishment peers, such as Theodore Roosevelt, when he built into New York's Mu- seum of Natural History an image of nature that captured so many contemporary concerns. In one book, Osborn contrasted the natu- ral joys experienced by a caveboy to the "Frankenstein" of the city, who eschewed the "direct vision" of nature. At the museum itself, Osborn elaborated on this direct vision of nature, which he ex- pected to convey by diorama and placards to the young visitor. One prominent theme was that of decline—of the terrible fall from the Cro-Magnons (the "Paleolithic Greeks") to the Neolithic hunt- ers, whose strength could not hide their moral and cultural in- feriority to their predecessors. The parallel with contemporary Western Europe was clear and explicit; the decline of the "Anglo- Saxon branch of the Nordic race" was the modern version of the Cro-Magnon. Only they could provide the leadership, loyalty, and self-sacrifice needed to safeguard the nation. And these blue-eyed peoples stood in mortal danger now that other stocks threatened to mix with them.

"Naturalizing" its own culture's assumptions, the museum succeeded where mere assertion could not. The glass cases, by their use of models, pictures, and scenery, conveyed an order of the world that appeared to come from the voice of nowhere while being planted firmly in the culture (be it British, American, French, or German) from which it sprang. There is no single "Nature" in these cavernous museums; there is only, as Emerson put it, "the figure of a disguised man."

Invoking Foucault, Joseph Rouse takes issue with all these historicized claims: Lenoir's and Ross's about museums, Cartwright's, Cat's, and Chang's about Neurath, or indeed about the more general project of doing history as a "single story" that binds Boyle's political actions to his laboratory work (Shapin and Schaffer's *Leviathan and the Air-Pump*),[24] or the Vienna Circle's philosophy to its politics and aesthetics (Galison's "Aufbau/Bauhaus").[25] All these projects, Rouse contends, are predicated on rationality's being located somewhere—in the liberal tradition in the soul of any rational person, in Marxism in a privileged class, and in what he calls "post-Kuhnian" science studies in the community-generated "forms of life." Grounding his opposition on an antiliberal, anti-Marxist reading of Foucault, Rouse contends that whether practitioners of science studies are preoccupied with individuals, classes, or forms of life, they have bought into the problematic of "sovereignty," and Foucault can guide us around this wrong dead end.

More specifically, Rouse wants to intertwine "power" as it appears in Foucault with "knowledge" (specifically, scientific knowledge) in such a way that what holds for Foucault's critique of power will hold, *mutatis mutandis*, for scientific knowledge. Thus Foucault opposes the "sovereign" conception of power that pinpoints authority in one place and looks for its effects somewhere else. On Rouse's reading, the Foucauldian dissent is based on six lines of reasoning: (1) power is dynamic, not a commodity, institution, structure (or other thing), and so only exists in its exercise; stability over time only exists because these exercises are reenacted. Similarly, Rouse says scientific knowledge is not possessed (cannot be possessed) and, like power, is not located in any specific place. (2) Power is disseminated throughout the body politic. It is nowhere in particular, and everywhere at once, and so (Rouse would have it) is scientific knowledge. It is not hard to see how this diffusion clashes with the work Rouse sets up in opposition: for Schaffer, myself, and others, it is one of the signal virtues of science studies to have localized the production of knowledge, and to have shown that agency is involved in the production and dissemination of that knowledge. It took work by Boyle from Gresham House to circulate the air pump to the Continent, just as it took effort by Luis Alvarez to move the bubble chamber from the halls of the Radiation Laboratory to other groups around the world. Three other theses nailed to the door by Rouse are undoubtedly less contentious: (3) that power relations are linked to one another in complex ways, and that they sometimes find disjunctions, conjunctions, or contradictions, is surely to be opposed by no one, even the proverbial liberal theorist of the sovereign; (4) that because of the diversity of modes of power and types of politics no single theory of power will ever suffice, and that an analytics of power is needed; (5) that power can be productive, not merely repressive, of knowledge is also by now a well-accepted part of our understanding of science.

But Rouse's final injunction, (6) that "power [read also "knowledge"] is always contested," would itself be contested. Reminiscent of the Latourian agonistic field, the picture of knowledge that develops appears modeled on the battlefield, and like those earlier claims, the metaphorical structure ("adversary," "target," etc.) is saturated with a particular vision of human and cultural interaction. Since much (but obviously not all) of the motivation to study recent science is precisely to understand the relation of military, industrial, and scientific sectors of our society, this absorption of vocabulary is perhaps not surprising. Not surprising—but not, as the next few authors stress, unproblematic.

Evelyn Fox Keller makes language a key component of her exploration of the scientific subject. For of the unities and disunities discussed thus far, none has addressed that most privileged unity, the object of such sustained literary-theoretical inquiry: the unity of the author. Keller begins with the story we know well—the gradual fading of the authorial "I" from scientific writing and its replacement with the anonymous, unlocated, and affectless passive voice in the mid- to late nineteenth century. By shifting the language of science, the human agency of the "manmade" quality of science is (apparently) removed. Skill, craftsmanship, and the locality of knowledge disappear in texts that have no place for them; with the social constructivists Keller underlines the importance of the machinelike objectivity that erases the work behind the building of science.

The history of the anonymous subject takes Keller into disputed ground when she argues for the continuity of this mechanization into the "postmodern." Instead of seeing computer viruses, synthetic genes, and virtual reality as harbingers of a new day in culture, she contends they are, quintessentially, the latest reports from a modern tradition extending back five hundred years. As she puts it, "The replacement of God's I/eye by a thinking and knowing machine may indeed mark a new way of speaking and thinking—perhaps neither more nor less veridical than the old—but it is one that emerges logically and with a certain inexorability out of half a millennium of history." All these novel machines, these hybrids of bones and bytes, are for Keller "enabled by and enabling" of a long chain of practices of representation that have sought to negate human agency and to flaunt mechanical surrogacy. Don't mistake the culmination of modernism with its denial, Keller in effect warns us: we must resist the assimilation of blood and machines every step of the way.

No one can write of the body/machine boundary these days, certainly not in the sphere of gender studies, with Donna Haraway's cyborgs far from view. In 1985, Haraway published her "Cyborg Manifesto: Science, Technology and Socialist-Feminism in the Late Twentieth Century," a text that is an intervention at once in feminist theory, science studies, and science fiction. "A cyborg," she argues there, "is a cybernetic organism, a hybrid of machine and organism, a creature of social reality as well as a creature of fiction."[26] Cyborgs are at once the stuff of paperback robots and also the daily implantations of heart defibrillators, artificial hormones, and command-control-communication systems of warfare. They are what is already present in medicine, the military, artificial intelligence, and virtual reality. But at the same time, Haraway wants to maintain the cyborg as a fictionalized resource reflecting what the world *could be*; what it would be like to have beings not defined by reproduction and traditional gender roles. I would put it this way: in addition to their literal instantiation, cyborgs are a placeholder for language and the imagination. The concept of cyborg allows us to talk and to think about ourselves with a complexified conception of gender and without the whole structure of Oedipal narratives of personal and collective histories. Cyborgs have no Edenic past from which they are fallen, no unalienated selves to recover, no unconscious that underlies and explains their conscious actions.

Haraway's formulation of the cyborg occupies an ambiguous arena not quite descriptive and not quite prescriptive. Insofar as the pilot of a Stealth fighter becomes a cyborg, merging with the machine to guide a smart bomb into an infrared-located tank emplacement, of course cyborgs exist. But Haraway is not flying the flag of the particular use of human-machine systems as they now stand, but rather using the culturally available imagery of such objects to suggest an expansion of what we usually consider the proper subject of action. And insofar as we do allow more than the human—reproduction, for example, as human-plus-petri-dish fertilization, etc.—then we have indeed come a long way from an essential definition of men and women in terms of "natural" reproduction. Haraway's cyborgs imaginatively extend these blurred boundaries into an asymptotic limit in which the construction of male and female may itself no longer carry much weight. This in turn gives us the space to look differently at our current relation to our bodies, gender definitions, and nature. In sum: Evelyn Fox Keller and Donna Haraway agree that the admixture of human and the technoscientific nonhuman creates a critical situation. But while Keller sees in that amalgam a crystallization of a dehumanizing, desubjectifying modern (she calls it "ultramodern"), Haraway sees a hopeful, disunified, and open-ended break with the past.

In "Modest Witness," her contribution to this volume, Haraway rocks another boat in science studies, a vessel uneasily floating Shapin and Schaffer to port and Latour to starboard. In their *Leviathan and the Air-Pump*, Shapin and Schaffer contextualized the Hobbes-Boyle dispute over scientific demonstration. Boyle, they argued, wanted a protected space in both politics and natural philosophy. Parliament offered a world where fundamental religious convictions could be left at the door and policy decided without global agreement; Gresham House allowed a miniature cosmos where the basic beliefs of vacuuism or plenism could be bracketed and matters of fact decided. Opposing this "modest" vision of consensus formation, Hobbes judged consensus to be possible only when those first principles were self-evidently and powerfully in place: self-evident authority in the political Leviathan and self-evident authority in the inexorable march from philosophical first principles through syllogistic reasoning to conclusions. The protagonist of Shapin and Schaffer's story is the "modest witness," the man who refrains from intervention and testifies without prejudice to new facts.

To Haraway the postulation of such a "modest man" demands further exploration. How does modesty, a quintessential (body-directed) feminine virtue, become a (mind-directed) male one? Or said otherwise: How does the very category of masculinity alter as the archetypal early modern natural philosopher acquires modesty as his badge of honor? On Haraway's view, through the suppression of gender questions like these, a false sense of unity emerges: the character of the natural philosopher becomes ready-made when it emerges historically and contingently. Latour, she argues, similarly naturalizes gender roles as inevitable. By depicting science (and science studies) as trials of strength, warlike struggles with victor and vanquished, Latour gives up the possibility of science's having been otherwise; he reinscribes in his writing the most problematic aspect of the scientific view of nature. But for Haraway there is no inevitability to the modest witness and certainly none to the scientist (or sociologist of science) as war hero. She argues that only a recognition of the disunified, very partial, and necessarily engaged situation of the author (scientist or critical studies scholar) can provide an avenue toward an objectivity, an objectivity better grounded than the metaphysical flight of fancy that accompanies what Nagel calls the "view from nowhere."[27]

It is now almost sixty years since the first meeting of the Unity of Science movement, and in that time the cultural position of unity itself has shifted. Not least, our wars have changed. For Germany in the second half of the nineteenth century the struggle for political unity precipitated and sustained discussion of scientific unity. For commentators on science in the 1930's, it was the threat of racial and national division that loomed largest; antifascism was the order of the day. For science studies in the 1990's, it is the potential crushing and homogenizing forces of the world that seem most threatening. Instead of World Wars, the conflicts of the post-war epoch have been myriad anti-insurgency battles, civil wars, ethnic secessions, and national autonomy movements. When Neurath began writing, the Soviet Union was completing its unity, the League of Nations stood as a beleaguered bulwark against further destruction, and the unity of science and scientists could appear as a hopeful, if frail, move against international crisis; in the 1990's, enclaves within enclaves seek the freedom from dominant powers whose names include "Greater," and "Union" or "Federation." In the 1920's, the United States faced its greatest challenge in absorbing immigrants whose cultural identities were distinct beyond question; in the 1990's, no academic issue is quite so contentious as the wars over multiculturalism, those ethnicities and subcultures that will not be "melted." Unity and disunity are markers in the history of culture whose meanings have altered almost beyond recognition. It is my hope that by tracing their trajectories, we can understand some of the power that they have had in the past, and continue to carry today.

Given the cultural resonance of disunity and unity, it is perhaps no surprise that the various authors here, despite their disagreements, explore not only the disunity of the scientific context but also, implicitly, the context of disunity.

PART I

Boundaries

IAN HACKING

The Disunities of the Sciences

What's in a name? Often, an ideology. Take the present names, in English, of the fields of specialization represented in the present collection of essays. We have the philosophy of science, the history of science, social studies of science, the history and philosophy of science—as if there were one thing, science, for there to be a philosophy or a history of social studies "of." Philosophers used to speak of the sciences, not science. In this paper I am concerned with different kinds of unity and disunity, not with different kinds of science, but it is well to begin by thinking about how an ideology, the unity of science, has affected even the names of what many of us sometimes do.

Sciences

Specialist philosophy or history of the sciences descends from two landmark bodies of work written in the 1830's. In one case the very titles make the plurality plain: William Whewell's *The History of the Inductive Sciences* (1838; three volumes) and *The Philosophy of the Inductive Sciences, Founded upon Their History* (1840; two volumes). Diversity is also firmly asserted in the other case, Auguste Comte's *Cours de philosophie positive* (1830–43; originally six volumes). His vast classificatory system of the departments of knowledge was an affirmation of difference. Comte fought long and in vain to establish the first professorial chair in the field—of *histoire des sciences*. The founding fathers of our fields of specialization differ greatly, the one adumbrating the rationalist vein still apparent, and the other prescribing the more common empiricist analysis. The role model for the rationalist Whewell was Francis Bacon, and for the empiricist Comte it was Laplace; so much for philosophical, as opposed to national, *-isms*. The two men agreed in this: both thought of themselves as philosopher-historians of the sciences, not of science.

Whewell, ever thorough, addressed the issue of science/sciences in the opening two paragraphs of his *Philosophy*. He did think that there might be something worthy of the name philosophy of science, but it was not something that he or anyone else could propound: "The Philosophy of Science, if the phrase were to be understood in the comprehensive sense which most naturally offers itself to our thoughts, would imply nothing less than a complete insight into the essence and conditions of all real knowledge, and an exposition of the best methods for the discovery of new truths."[1] As optimistic, encyclopedic, and influential as he was, he did not reject such a philosophy out of hand, but he did consider it impracticable. We should concern ourselves not with some "real knowledge" of which human beings could dream, but, as he put it, with the doctrines of solid and acknowledged certainty that do exist among us—the several sciences. He did think, what many now reject, that the very idea of a philosophy of science (in his ample understanding of the words) makes sense; but any who aspire to it "may best hope to make some progress towards the Philosophy of Science by employing [them]selves upon THE *Philosophy of the Sciences*."[2]

Comte's talk of the sciences, in the plural, was also based not on abstract principles but on what we are able to do. He spoke to this in the first lesson of the forty that constituted his *Cours*: "One cannot reduce all the sciences to a unity." Comte is not to be understood maliciously, as somehow self-refuting, saying something about all science. He wanted only to avoid being misunderstood. Since he wished to give a systematic presentation of the departments of knowledge, he feared that his course would be seen as one more of those attempts at universal explanation "that are daily hatched by those to whom scientific methods and results are entirely foreign." At that juncture he wanted to dismiss one model of unity, namely derivation of all laws from one fundamental law of nature—an "eminently chimerical" project: "I believe that the powers of the human mind are too weak, that the universe is far too complicated for such a scientific perfection to be within our powers, and I think, moreover, that one usually forms a very exaggerated idea of the benefits that would necessarily accrue, were it to be possible."[3] Despite our admirably clear forefathers, we cannot undo the terminology that has since been adopted. "Philosophy of science" is agreeably shorter than "philosophy of the sciences." Our journals are called *Philosophy of Science* and the like.[4] And why not? The stage was set long before, with, for example, the British Association for the Advancement of Science—"science" seen as a special-interest group in 1833. The American Association publishes *Science*. But because of my interest in diversity, I shall return to Whewell, Comte, and our roots, and write about the sciences.

Unities

Otto Neurath said that "the unity of science movement . . . includes scientists and persons interested in science who are conscious of the importance of a universal scientific attitude."[5] The movement was closely associated with logical positivism, but the unity of science had been an important slogan, for diverse reasons, for Helmholtz, Mach, Karl Pearson, and many others. It breaks into several parts: unity has a fairly clear meaning; unity is a good thing; the sciences are a very good thing; and the sciences are or should form a unity. The first is a little-discussed point of logic or language. The second and third are judgments of value. The fourth is hortatory, an injunction about the status or aims of the sciences: the sciences do or should form a unity.

The fourth item is usually the point of contention. I shall be concerned with its presuppositions rather than its content. The very title of this volume expresses the vogue for doubting unity of science doctrines, but the essays in it tend to emphasize the "science" side of the unity of science; to balance this I shall be concerned with the "unity" side. There are a lot of different possible types of unity. I shall seldom argue that the sciences are or fail to be unified in this or that respect. I shall mention some pros and some cons, not so much to take sides as to emphasize that there are disputes about different types of unity or disunity. Hence I do to some extent side with the disunifiers, because the hidden strengths of the unity of science movement lie in the implicit assumption that there is either a single unified way in which the sciences are or should be unified, or else that there is a simple hierarchy of increasingly strong theses about the unity of science.

Unity Is Not a Predicate

I shall not attempt a logical analysis of the concept of unity or a linguistic analysis of uses of the word "one"—although both would be in order. I shall not try to describe the complex relationships between those different words, "one," "unity," and their cognates. I do take for granted a familiar lesson from the philosophy of Kant and of Frege. Kant taught that existence is not a predicate. Frege added that being one or more in number is very much like existence—not a predicate, or at any rate not a predicate of things. Existence and number are, in their primary usage, concepts that apply to concepts. In Frege's jargon, to say that God exists is to say that something falls under the concept "God"; to say that God is one is to say that exactly one entity falls under the concept "God" (or, to eliminate the "one" in favor of the nonrelation of identity, if x and y fall under the concept, $x = y$). Existence and unity, both said by some medieval schoolmen to be perfections, are not (first-order) perfections at all, because neither is a predicate. Unity is thus like existence; as Kant said, I do not add something to the golden dollars of the merchant in saying, "and then they exist." I do not add to the properties of an apple, after saying that it is crisp and tasty, by saying that it is "one." Hence, insofar as unity connotes singleness, it cannot be a virtue or perfection. It could be a good thing that there be only one editor of this volume, but that would not be a property of the editor.

Singleness and Harmony

Logic-chopping makes us forget the emotive power of the ideas of unity and existence that make people think of them as perfections alongside omniscience and omnipotence. The unity of science was a rallying call in part because unity, in a certain framework, was a good thing. Unity has, in our tradition, been a virtue associated with many powerful ideas: the God of monotheists, the nation, the state, a people, the self. Men and women have been dying for (and against) Unions pretty much since the invention of nationalisms. Kant's transcendental unity of apperception was a necessary condition for human knowledge. The integration of a personality may be the highest aim of many psychiatrists. Hence the unity of science is one among a passionate crowd. A larger discourse on unity would make us better understand its appeal.

It is easy nowadays to be flippant about unity. It is also possible to be angry. Some of the current rage against reason is directed at an ideology of science that says there is *one* ultimate reality, *one* ultimate truth, *one* road to the truth (the scientific method), *one* sound mode of reasoning, *one* rational way of speaking. Because unity now rings in our ears as hegemonic, patriarchal, imperial, it is important not to dismiss the old virtues and values of unity. Oppressed people in the past, today, and in all the foreseeable future require those very onenesses about which we find it so easy to be lazy. People resisting despotism and its lies need ideals of one truth, one reason, one reality, and on occasion, one science. To be able to be critical of the unities is a luxury, and let us never forget it.

Unity has been an immensely powerful political tool, sometimes for what I find good, sometimes for what I find evil. To counter my logical observation that unity is not a predicate, I should notice a linguistic point about the concept of unity. With the logical point in mind, but forgetting the linguistic point, it would be hard to guess why unity could be deemed a virtue. It can be a virtue because two distinguishable although interconnected ideas are at work: not even unity is at one with itself.

The root word is *unum*, one, and for sure unity connotes singleness, oneness. Now being a singleton, that is, being the only instance of a concept, is never in itself a virtue. But there is a virtue in the offing. We can speak in context of unity's being desirable: the unity of certain concerts that we have heard, novels that we have read. A speech, a political platform, may or may not have unity. So may a character, a soul. This unity has something to do with the integration or harmony of the parts, a harmony that exists or does not exist after the item has already been individuated as one thing, one concert (starting at 8:00 and ending at 10:30, with an intermission).[6]

Let us call these two aspects of unity *singleness* and *harmonious integration*. The unity of the God of Israel or of Islam is singleness. This unity is not a property of the God and hence in itself not a perfection. Many generous souls have urged that the gods of all faiths are identical, but this is to say that we have different names and rites for the same thing, not that the same thing has the property of being identical to itself. The unity of the self, in contrast, is largely a matter of harmonious integration. That can be a real property of a person, a property that the person works hard to acquire. All traditional Western psychology regards such harmony as desirable. This may be connected with our notion of the soul. Many other peoples actively cultivate alternative personalities; individuals who do so successfully have special status, for example as shaman. Integration, I think, is something to be valued (or not) within a culture. But at least it makes sense to value it.

There are deep conceptual questions about the relation between the idea of singleness and the idea of harmony. In the unities that Aristotle introduced into the criticism of drama, the events are supposed to take place in one compressed period of time, and to be integrated. Everything happens (perhaps) in a single day. Is the integrated-harmony side of unity part of an archaic theory about what makes for singularity? For example, does it represent the idea that if there is a single entity, it will have, in some sense of "cause" (perhaps one of Aristotle's causes), a single cause, and thus betray a set of harmonious effects? Or, in contrast, is the idea of oneness to be thought of as derived from an experience of harmony? Is what makes us count something an entity, a singleton, something like the harmonious relationship of its parts? (In Leibniz's monadology, were it not for the preestablished harmony there would be infinitely many actual worlds—the monads—rather than one actual world.) If harmony is what makes for identity, then the logicians with their second-order-concept notion of unity have missed something of fundamental metaphysical and conceptual importance.

I shall not further discuss these hard questions. But the distinction between singleness and harmony matters to the very idea of the unity of science. Some of the unities of science that I shall mention have to do with singleness, and some with harmony. Not even unity is united.

Three Unities of Science

The unity of science denotes at least three distinct families of theses, each of which can be subdivided or organized in numerous ways. The first family is metaphysical, a collection of ideas about what there is. I see it as starting with a certain metaphysical sentiment, which is then followed by a number of what are, by comparison, relatively plain theses. The theses do not follow from the sentiment, and none entails the others.

The second family is a collection of practical precepts about the sciences. They correspond, in a rough and ready way, to the metaphysical theses. They have to do with method and the aims of the sciences. Each draws on a different insight about what scientists are up to. One and only one of these has deeply moved working scientists, chiefly physicists, namely the precept that one should try to find connections between important phenomena that have hitherto seemed independent. Integration and harmony are what seem to attract the scientist.

The third family forms a set of theses about scientific reasoning, and includes both logic and methodology. It seems to be almost completely independent of the preceding two. Thus the logical or methodological theses could be correct, and even the core metaphysical sentiment and all that flows from it, theoretical and practical, could be wrong. This is the strongest instance of the disunity among unity ideas, but at every possible juncture we find disunity. The lowest denominator of disunity is found at the level of methodology, where a wide spectrum of analytical philosophers of science have asserted that there is one scientific method applying across the board in the natural and human sciences (they are all on the same side in the *Positivismusstreit*)—and then produce seemingly incompatible methods, the well-known methodology disputes of Carnap, Popper, Lakatos, and the rest. But this is the least instructive kind of disunity, because it is at the level of philosophical doctrine rather than at the level of scientific activity.

Many readers will want to reorganize my kinds of unity.[7] My purpose is not to structure the distinctions but rather to display some of them in a handy way. I am cautious about every one of the unities that I shall mention. My interest is less in skepticism than in the fact that the grounds for skepticism differ from unity to unity. There are a lot of different types of unity, each of which may be called in question for its own specific set of reasons.

A Metaphysical Sentiment

The unity of science is rooted in an overarching metaphysical thought that expresses not a thesis but a sentiment. Since it is not exactly a doctrine, it lacks straightforward expression. On the one hand I could try, "There is one world, one reality, one truth." On the other: "There is one world susceptible of scientific investigation, one reality amenable to scientific description, one totality of truths equally open to all scientific inquirers who may share their techniques and experiences." I shorten the latter to, "There is one scientific world, reality, truth." Inserting the word "scientific" seems to beg some questions—Wasn't science supposed to be what got at the world, at reality, at truth? So should we not eliminate the modifier "scientific," thus going back to the first unequivocal statement?

No. On the one hand, "one world, one reality, one truth" leaves out a feeling of awe, wonder, and respect. I shall use one of our greatest unifiers, James Clerk Maxwell, three times in what follows, to very different ends. First, he is a good example of numberless reflective scientists who have thought that we have a positive duty to impress on our own minds "the extent, the order and the unity of the universe."[8] Maxwell meant harmony, not singleness. The harmony of the world is worthy of praise, or so a great many of us have thought. It has played into the argument from design for the existence of a wise Creator. So "one world, one reality, one truth" is too deflating of a serious human wonder. On the other hand, it can be made to sound too exclusive—as if we deny what is familiar to every sensitive and reflective person, and what is never questioned in much daily living: that there are a lot of realities. Here I'm not referring to stuff that today seems rather pretentious, like the "Separate Reality" of Carlos Castaneda and Don Juan.[9] Nor do I mean exaggerated scholarly talk about incommensurable cultures. I mean only the commonplace experience summed up well enough by the platitude that we all live in lots of different worlds. That is why I think that the right statement of the overarching metaphysical sentiment that bears on the unity of science is the one abbreviated as, "There is one scientific world, reality, truth."

The sentiment is neither vacuous nor hegemonous. On the one hand, the statement of it is not empty. Nelson Goodman would urge on us that once we have put in the word "world," we have gone astray. There are only world-versions, and there is no sense in which there is only one right scientific world-version. Likewise it is a serious if unresolved question of feminist criticism whether there are not only scientific mistakes in some fields that can be attributed to male blinkers (errors whose correction would still be "science") but also alternative feminist sciences, distinct in content and method. Bernard Williams's "absolute conception of reality" is at odds with both Goodman and the idea of alternative sciences.[10] So the sentiment expressed by "There is one scientific world, reality, truth" is neither universally shared nor universally spurned. Some philosophers are drawn to this sentiment quite strongly, while others think that it is unsound.

The sentiment is, by its very wording, to be sharply distinguished from the attitude that there is nothing but science, that there is no knowledge except scientific knowledge, that the only "cognitively meaningful" assertions are scientific ones. Perhaps that is the intention of Neurath's "universal scientific attitude" quoted above. Undoubtedly many powerful exponents of the unity of science can be read as if they were exclusionary in just that way. I attribute what they say more to the sociopolitical fights of Eastern Europe in the 1930's than to serious exclusion of the non-scientific. At any rate exclusiveness should not be taken to be characteristic of the metaphysical sentiment that grounds so much unity of science. Scientistic thinkers may have the prejudice that science covers all that can be known, that anything else is "mere" emotion and instinct. That is why it is important to put in that restrictive adjective "scientific." The sentence "There is one scientific world, reality, truth" does not mean that science exhausts human knowledge. To anyone who thinks otherwise, I can only pray,

> May God us keep
> From Single vision & Newton's sleep.

Not that there are exactly two visions—or some other permanent number of right visions. Blake's immediately preceding lines were:[11]

> Now I a fourfold vision see,
> And a fourfold vision is given to me;
> 'Tis fourfold in my supreme delight
> And threefold in soft Beulah's night
> And twofold Always.

Metaphysical Unity

There is one world—so goes the metaphysical sentiment—susceptible of scientific investigation, one reality amenable to scientific description, one totality of truths equally open to all scientific inquirers who may share their techniques and experiences. For short: There is one scientific world, reality, truth. For all the apparent disagreement between Nelson Goodman and Bernard Williams, it is a little hard to know what this oneness could amount to. We don't go around counting worlds.

The metaphysical sentiment translates into a number of different theses that have real content. One of these has profoundly affected the activities of many memorable scientists. For it has meant that all kinds of phenomena must be related to each other. Faraday, for example, believed that the world could not be such that light and magnetism did not somehow affect each other, and spent two decades trying to find out how. In the end he showed that magnetizing could change the polarizing properties of some substances. Einstein is the most illustrious instance of a scientist who devoted his later life to searching ways to unite phenomena. A paragon of successful unification arose from the conviction of physicists like Abdus Salam, Steven Weinberg, and Sheldon Glashow that the strong, weak, and electromagnetic forces must have some loci of interaction.

This is a thesis of *interconnectedness*. It is the clearest of the metaphysical theses. In some minds it is rooted in a religious conception of the world and how God must have made it. It is religious—in the sense of the great monotheistic religions. Faraday's life was located in the Sandamanian sect of upright Protestants, as rigorous in their lives as they conceived God to be in arranging the universe. Einstein's cultural roots in Judaism were deep. Salam's first Dirac lecture ends with a moving passage from the Quran that I shall quote later.

Philosophers tend to refine the metaphysical sentiment in another direction. The world, as Wittgenstein put it, may be made up of facts, but it is not a ragbag of facts. There is a unique fundamental structure to the truths about the world, with central truths that imply peripheral ones. Metaphysicians have tended to say that this is a structure of causes, necessary or probabilistic. In the tradition of logical positivism, which has rejected the idea that causes are fundamental, the structure is one of logical relations between laws. However qualified, this is a *structural* thesis. In my quotation above we saw that Auguste Comte, the founder of positivism, did not believe in the structural thesis. (Nor did he declare it false, it not being the business of a positivist to have views about such things.) In the present volume John Dupré forcefully argues against the causal version of the structural thesis.

There is yet another metaphysical thesis of oneness. In ordinary life we classify in ever so many different ways, but there is an underlying belief that there is one fundamental, ultimate, right system of classifying everything: nature breaks into what have been called "natural kinds." This is a *taxonomic* thesis. It has not figured greatly in unity of science debates, but has had its own recent moment in philosophy thanks to the ideas of Hilary Putnam and Saul Kripke. In the latter's version, natural-kind terms, if they refer to anything at all, are rigid designators naming real kinds, which are part of a fundamental taxonomy. The rigid-designator doctrine can itself allow of nonmeshing taxonomies, but it tends to be tied to Plato's nasty phrase (less nasty in Plato's writing than when quoted out of context) about "carving nature at the joints." There is just one set of joints, one set of kinds. Curiously the philosophers who gave us the terminology of "natural kinds," J. S. Mill and John Venn, and the one who revived it, Bertrand Russell, were far more modest and empirical in their use of the idea.[12]

One can go on to make these three theses sufficiently precise that they relate to each other; but as stated their interrelations, if any, are open-ended. They are different aspects of a certain metaphysical conception, and it is not their role to be made precise. There are no mutual entailments, except in the realms of science fictions. Suppose, for example, that there is no connection between gravitational phenomena and electromagnetic phenomena (no connection that could sensibly be compared to Faraday's connection between optical and magnetic phenomena). That would be consistent with both the taxonomic and the structural thesis. Only a very strong form of the structural thesis might be contradicted, namely the statement that there is one master law—no mere conjunction of laws—from which all other laws follow. Undoubtedly the idea of a master law, of a theory of everything, does motivate visions of interconnectedness, of structure, and even of taxonomy, but these three theses can best be treated as more modest and hence as quite distinct.

What about the metaphysical sentiment, that there is one scientific world, one reality, one truth? Could that be true, while one or even all of the three more specific theses of connectedness, structure, and taxonomy were false? Certainly. Dupré's account of the disorder of things does not entail that there is more than one world. The connection between the sentiment and the theses is not at the level of entailment but of one's feeling for what the world is like. An inclination to conceive of the world as disorderly does tend to make the idea of "one" reality increasingly trivial-or-false. Or again, what if there were two internally adequate ways of structuring "everything" that were yet incommensurable with each other? Would that entail that there are separate realities, worlds, totalities of truths? I suppose that we would have distinguishable totalities of truths, but these could readily be described as alternative nonmeshing descriptions of the world.

Finally what about the idea of branching universes, suggested by Everett and John Wheeler, which is supposed to resolve some of the philosophical difficulties about quantum mechanics? That is an idea in the grand metaphysical tradition that seems at odds with the metaphysical sentiment. There are many worlds, and for each of these, there is presumably a corresponding totality of truths. Yet as I understand this proposal, the theses of interconnectedness, structure, and taxonomy are maintained. The same laws, with the same relationships, apply in every universe; worlds differ in that different things happen, but the kinds of the things in the universe can still be described by the same uniform taxonomy. What if the laws of nature in the branching worlds themselves evolved differently?

Then we would have what cosmologists sometimes call an exotic theory. (Laws of nature are not constant.) Wheeler does in fact favor exotic universes, but that seems to be separate from his many-worlds idea. Many worlds and exotic universes taken together do indeed imply that at least the structural thesis is false. And so it goes: this plethora of obscure possibilities reminds us that there are no entailment relations among any durable metaphysical sentiments or theses of unity.

Practical Precepts

The metaphysical theses do have loose implications in practice. Collectively, they amount to this: attempt to find the metaphysical unities of interconnectedness, structure, and taxonomy. Three unifying precepts follow, but they work in different ways. For physicists like Faraday and the other great unifiers whom I have already mentioned, there is a belief that phenomena of roughly the same level of generality are connected to each other, that each one influences each other. Hence arises the injunction to try to discover the connections. For the experimenter like Faraday, this means trying to make one type of phenomenon produce a change in another type of phenomenon, as, in Faraday's case, when magnetizing glass produces a change in optical properties, namely polarization. For the theoretician such as Einstein the aim is to produce a law that connects two phenomena—an equation between energy and mass, to take only the most trite example. Experimental effect and theoretical equation feed into each other. Each is motivated by the fundamental injunction "Connect!"

Philosophers, obsessed by logic and language, tend not to emphasize connectedness but instead lean upon the metaphysical thesis of structure. They enjoin scientists to reduce the laws of one body of knowledge to the laws of another—in particular, economics to sociology, sociology to psychology, psychology to biology, biology to chemistry, geology to chemistry and geophysics, chemistry and geophysics to physics, and then unite physics and cosmology in a Grand Unified Theory. This is a *reductionist* precept.

Or rather there are two reductionist precepts. There is global reductionism, of the sort just sketched, in which the sciences are reduced to one master science, a specially general type of physics.

Then there is local reductionism, applying to what is recognized as a single science or family of sciences where one tries to find fundamental and related principles from which special laws can be derived. Local reductionism and the desire for connection go hand in hand. But I do not think that in his scientific work Faraday was a global reductionist, even though his religion preached unity—it also taught variety in God's creation. Contrary to what is commonly made out, global reductionism is not of great interest in scientific work—it is something that philosophers read into scientific work.

There is an even more purely philosophical precept, best instanced by the logical positivists, who were much preoccupied with the language of science. They thought that the global reductionist precept required that all science could be expressed in one language, for how else could the theory of one discipline be reduced to that of another? This is a *linguistic* precept. It is in a loose way related to the metaphysical thesis of taxonomy, because it implies that all our systems of classification should mesh. Note that one could subscribe to the linguistic precept but not to the taxonomic thesis. Recall, for example, Carnap's insistence on "the principle of tolerance." Although it was desirable that there should be a language of unified science, it was not so important what the language was. Carnap did not, I think, maintain that there is a uniquely right language. A student of taxonomy faithful to Carnap could have been completely indifferent as to whether a proposed language had the "right" taxonomy (mapping the natural kinds)— the need for unified language was not founded on a belief that it could be the right language.

Just as we have a rather complete disunity among the metaphysical theses, so we have a substantial disunity among the practical precepts of unity.

Logic and Method

There is an ambiguity about "method." In his lecture "Methods in Theoretical Physics," Dirac said that in the early stages of knowledge one must rely heavily upon experiment; theoretical speculation is at most suggestive. But once one "has a great deal of support to work from one can go more and more towards the mathematical procedure. With the mathematical procedure there are two main methods that one may follow, (i) to remove inconsistencies and (ii) to unite theories that were previously disjoint."[13] I shall return to this passage; evidently (i) and (ii) were seen, by Dirac, as two distinct ways of trying to honor the precept "Connect!" It is quite correct to speak of (i) and (ii) as methods, but they are a far cry from what, among philosophers, has come to be called the methodology of science. Methodology, in the philosophical sense, is about how to reason. I regard it as part of logic. There seem to be two distinct logical maxims about unity.

The chief logical maxim states that there is one and only one standard of reason by which scientific hypotheses can be investigated and judged. This is a claim about reason and evidence. Then there is a methodological maxim according to which there is one best way to find out about the world and how it works, the scientific method.

There is no longer any consensus on what the scientific method is. Is there a relationship between the logical and the methodological maxims? Not a definite one, for it depends on the specific content that is given to them. For example, the two were rather separate for Reichenbach, who sharply distinguished justification from discovery. For him, the logical maxim would demand that there was one canon of justification (deductive logic and confirmation theory, perhaps). That would be formally consistent with rejecting the methodological maxim altogether: there is no scientific method of discovery. At any rate an argument that there is one scientific method would require nonlogical premises.

As is well known, Popper rejected Reichenbach's distinction; the logic of justification is none other than *The Logic of Scientific Discovery*. There is no such thing as pure evidence outside the circumstances in which it was obtained. Only if information was collected in order to test and try to falsify a hypothesis can it corroborate the hypothesis. Thus an item of information serves as a good reason (logic) only in a particular context of discovery (methodology).

Then there is Lakatos's remarkable conclusion that there are no forward-looking good reasons for individual hypotheses or theories. He rejected the logical maxim altogether, not because he rejected logic but because he carried Popper to a natural limit, and rejected the notion of evidence for a hypothesis or theory. The logical maxim stated that there is one standard of reason by which scientific hypotheses can be judged; Lakatos held that there is none. There is only the track record of a sequence of theories, and we can only tell with hindsight what *was* reasonable. Lakatos was nevertheless a firm advocate of the thesis of methodological unity: there is a (uniquely right) methodology of scientific research programs.

Notice that although there may be great differences about the content of the methodological thesis, unifiers agree on its form: the same method (whatever it is) is to be used in all the sciences, natural, social, and human. When we turn to disunifiers—Adorno and Popper's other opponents in the *Positivismusstreit*—we disheartening find that those who insist on distinct methodologies for distinct kinds of science share the unifying methodological spirit: there is *the* method of the natural sciences, and *the* method of the human sciences. I notice no suggestion that, for example, there is an indefinite number of largely independent methodologies for use in the natural or the human sciences.

Singleness or Harmony?

I said that the idea of unity is a blend of two ideas, singleness and integrated harmony. I have stated a metaphysical sentiment, three metaphysical theses, three practical precepts, and two logical maxims. It helps to see how unrelated these are by noticing the different weights of singleness and harmony in each one. I shall score the proportion of each by percentages, which are graphic although not literally meaningful. Since every maxim, precept, thesis, or sentiment itself breaks down into variants, the numbers stand only for impressions.

Metaphysical Sentiment

"There is only one scientific world, reality, truth" is a proposition of 100 percent singleness. But of course a feel for harmony may be what prompts the whole idea: for example, Maxwell's respect for "the extent, the order and the unity of the universe." That is 100 percent harmony.

Metaphysical Theses

Interconnection. 100 percent harmony.
Structure. 50 percent–50 percent. When stated, the thesis usually sounds like a singleness doctrine—just one structure. But it is

attractive because, if true, all events would have to be understood in a harmonious way, as derived from a single structure of causes or laws.

Taxonomy. 80 percent singleness—at least when tied to Plato's abused phrase about carving nature at the joints, because there is just one set of joints. But if the taxonomic thesis were true, then best descriptions of events would in a certain way mesh, and so a type of harmony would result.

Practical Precepts and Logical Maxims

For brevity I would weight the *practical* precepts about the same as the corresponding metaphysical theses. Thus the precept "Connect!" is 100 percent harmony. Both global and local reductionism are equal parts singleness and harmonious integration. In the case of local reduction, one wants a single unified principle that harmoniously integrates diverse laws in related parts of science. The linguistic precept emphasizes singleness. As for the *logical* and methodological maxims, both seem to be 100-percent dedicated to singleness; any harmony that results from a single methodology is achieved incidentally, by silencing alternative ways of reasoning.

Philosophers and Scientists

Very roughly speaking, scientists have emphasized the harmony side of unity, and philosophers the singleness side. But we must not draw too sharp a line between philosophers and scientists.

The most enduring problematic of unity was set by one of the greatest philosopher-scientists, namely Descartes. One outstanding motivation for the unity of science has been an attempt to provide a monistic theory of mind and body, two substances sundered by Descartes. He himself at once addressed the precept "Connect!" with an elementary conjecture (the pineal gland) and much extremely difficult philosophy about the way the soul is not in the body like a pilot in a ship. Others have denied the sundering. One of the most famous instances is the unity of science manifesto issued by Helmholtz and others in 1847. Helmholz was a man who lived a certain unity of science, making major contributions to the conservation of energy, electricity, optics, and physiology. In his eyes, he worked not at several sciences but at science. His electrochemical vision of neurophysiology was an experimental and theoretical program for body-mind identity.

Ernst Mach, another notable philosopher-scientist, thought that the only real world was that of sensations, and although he believed the sciences have different objects, he thought that in the end their work could be described in a sensationalist language. Mach was intensely committed to linguistic precept, that it is an aim of science or its philosophy to produce a single language of science. In Mach's case, this was to be a completely new language, namely that of sensations. Karl Pearson was much of the same mind, with some added ethical dimensions of a scientistic sort. (Biologists are likely to be better moralists than are philosophers or politicians.)

Pearson is an exception among the enthusiastic unifiers, because he made great contributions to biometrics but none to physics. There have been very few great unifiers in biology—Aristotle, Darwin, Crick, perhaps. It is theoretical physicists who have favored unification. But even they have not been much interested in global unification, except as a sort of metaphysical picture. In his inaugural lecture, "Is the End in Sight for Theoretical Physics?" Stephen Hawking suggested (well over a decade ago) that the end is in sight—but of course that will leave most of physics still to be done. Not to mention the rest of the sciences.

Even in the case of local reduction within a science there is no one way of unifying. Recall my quotation from Dirac above. He distinguished two methods in theoretical physics, (i) to remove inconsistencies and (ii) to unite theories that were previously disjoint. One inevitably thinks that the precept connect and reduce, must mean (ii). On the contrary, we well know that Dirac's own equation arose more from (i) than from (ii). Without mentioning himself he writes that:[14]

There are many examples where the following of method (i) has led to brilliant success. Maxwell's investigation of an inconsistency in the electromagnetic equations of his time. . . . Planck's study of difficulties in the theory of black-body radiation. . . . Einstein noticed a difficulty in the theory of an atom in equilibrium. . . . But the supreme example is Einstein's discovery of his law of gravitation, which came from the need to reconcile Newtonian gravitation with special relativity.

In practice, method (ii) has not proved very fruitful.

That reminds us that there is more than one way to connect phenomena and find fundamental principles. Dirac held that for his generation (and he forecast for future ones) the direct search for unities had been fruitless. "In contrast," notes Abdus Salam, "our generation has been mainly concerned with this second method," and he found it extraordinarily fruitful.[15] His own exposition of recent unifications is magnificent, but I cannot forbear a trifling metacomment about his comment on Dirac. It shows very clearly that when one has a certain vision of unity, then it is just one unity that one sees. When Salam drew attention to Dirac's remark, he continued that Freeman Dyson "has this to say about the second of Dirac's ideas." He quotes almost all of a paragraph from Dyson, ending at the second to last sentence, "It is difficult to find among physicists any serious voices in opposition to unification." So it sounds as if Dyson is advocating method (ii) as a historical observation about physics. In fact the omitted last sentence is the point of the whole paragraph, a remark about a physicist not favoring method (ii), Emil Weichert in 1896. We then go on in the next paragraph to: "Weichert's words were ignored. His vision was too large for the time he lived in."[16] And then Dyson talks about the great diversifiers in physics. Thus on a page where Dyson is urging the complementary roles of diversifiers and unifiers, Salam sees only unifiers. And more power to him. We need, and indeed marvel at, his single vision—so long as we have workers with other visions. It is Dyson's opinion, in the same chapter,[17]

that every science needs for its healthy growth a creative balance between unifiers and diversifiers. In the physics of the last hundred years, the unifiers have had things too much their own way. Diversifiers in physics, such as Weichert in the 1890's and John Wheeler in our own time, have tended to be pushed out of the mainstream.

The Historical Postulate

There is nevertheless a straightforward historical postulate: "Unity works!" That is, the search for unifications has worked, and can be expected to go on doing so. The various sentiments, theses, precepts, and maxims of unity are in good shape, thank you very much. Science has been wonderfully successful not only in finding out new facts about the world and in creating and controlling new phenomena, but also in its unifications, bringing many facts under the wing of larger intellectual structures. Physics is exemplary. Light, optics, radiant heat, electricity, magnetism were captured by electromagnetism. Just as high-energy physics was about to drown in a plethora of its own particles, gauge theory ordered them in unified principles. Physics is not the only success story; by now we equally invoke the profound connections between organic chemistry and genetics provided by molecular biology.

Yes, unification does work—and so does diversification. But be cautious. Only in the most vicarious way can quantum chromodynamics and electroweak theory be said to be unified.[18] Even in old science matters are less well established than one might have hoped. It requires tact and tenacity to state clearly in what sense Maxwell unified electricity and magnetism.[19] But surely biology and chemistry are now unified? During the past five years there has been a vast architectural restructuring of the biology and life sciences organization on the campus of the University of California, Berkeley, breaking up groups and remolding others in the light of the changes of the past twenty-odd years. The upshot is not a unified department of biology, let alone biology and chemistry. The reorganization was grudgingly managed by one of the pioneers of molecular biology, Gunther Stent. He is head of the Department of Molecular Biology, which has six divisions that he himself did not want to bring under one superdepartment. Across campus is now a disjoint superdepartment of life sciences. Microbiology and macrobiology have been institutionalized farther apart than ever. Likewise I understand that at Harvard the sciences are now classified for administrative purposes as experimental or historical, with evolutionary biology in the second compartment and molecular biology in the first. I would describe this less as unification than as a redistribution of powers.

The reinstitutionalization may be very striking to the historian, but surely molecular biology does furnish a remarkable instance of reduction? The achievements are obviously immense, but the unifications are once again vicarious and varied. There a lot of distinctions. Alexander Rosenberg makes rather plain what has been reduced, and in what sense it has been reduced—and how much has not been reduced, and may in several senses be called irreducible.[20]

For my part I am an unabashed admirer of the great unifying

physicists. Relative to what has been done (whatever that really is), questions about unity—chromodynamics or Maxwell or recombinant DNA—are research questions, not criticisms of the thrust toward unity. But by and large in the sciences what is sought for is harmonious integration, not singleness.

It is largely the writings of philosophers that make us so constantly turn to the word "reduction" in its global sense. In my view the scientific search for harmony has been incredibly rewarding while the philosophical quest for singleness has been in vain. For the most obvious objection to the singleness doctrine, listen to the endless complaints about overspecialization of the sciences. What about the myriad papers in journals that nobody reads? What about the new scientific journals that are founded at the rate (so my librarian tells me) of four a day? This phenomenon is not incompatible with the great success stories, but it does suggest that unification is not after all the dominant trend within the sciences. I now turn to a philosopher, Patrick Suppes, for an elaboration of that theme, an elaboration that for all its iconoclasm fits snugly within the philosophical tradition of unity of science.

Unity for Logical Positivists

Positivism is in no way committed to any of the numerous unity of science theses or precepts. Comte opposed them. If he had had to opt for any of the unities I have listed, it would have been for the methodological maxim. He did think there was something worth calling positive science, a way of finding out to which the human race had progressed. Later notable positivists such as Ernst Mach and Karl Pearson had a philosophical commitment to a somewhat idealist philosophy in which the ultimate reality is sensation; Mach thought there would be an underlying language of science couched in terms of sensations. But only with the advent of logical positivism did we get what people now think of as a fully fledged unity of science movement. Just as we noted generational differences between Dirac and Salam, so it is useful to examine three generations of philosophers. I refer to Otto Neurath, writing the introduction to the first volume of the *Encyclopedia of Unified Science* (1938), the work of Paul Oppenheim, especially a paper on unified science written with the help of Hilary Putnam (1963), and finally Patrick

Suppes's enthusiasm for "the plurality of science" (1978). These form an invaluable trio. Oppenheim provides an extremely careful revision of the original program of the unity of science movement—it has become "Unity of Science as a Working Hypothesis." Suppes, the pluralist, defines his position by way of contrast with the first volume of the *Encyclopedia*. There Neurath's manifesto had the title "Unified Science as Encyclopedic Integration," but there is no evidence of an interest in what I call integrated harmony. The precept of linguistic unity strikes scientists whom I have asked as completely irrelevant to anything they do. Yet it is at the core of the logical positivist vision of unity. Internally, this owes something to Mach, but external to specialist philosophy it rides well with many cultural obsessions of Eastern Europe in the first quarter of this century.[21] Although the manifestos of the 1930's are ringing, here I shall run through their more cautious form, the product of thirty years of reflection and criticism—namely in the paper of Oppenheim and Putnam. It distinguishes "three broad concepts of Unity of Science."[22] The weakest concept is called *Unity of Language*, which is in turn broken down into a number of subconcepts. This roughly matches a version of what I call the linguistic precept. "Unity of Science in a stronger sense . . . is represented by *Unity of Laws* . . . [in] which the laws of science become reduced to the laws of some one discipline." Then there is a third and strongest sense in which the laws of the master discipline "are in some intuitive sense 'unified' or 'connected.'" In my terminology, Unity of Laws involves global reduction, and the strongest sense of Unity requires local reduction of the laws of the master discipline. It is striking that local reduction and connectedness—harmony, in short—are the concern of working scientists, but after being mentioned by Oppenheim and Putnam, they are simply dropped. The authors also importantly distinguish two senses of the term "Unity of Science" as referring,

first, to an ideal *state* of science, and, second, to a pervasive *trend* within science, seeking the attainment of that ideal. . . . In the second sense Unity of Science exists as a trend within scientific inquiry, whether or not unitary science is ever attained, and notwithstanding the simultaneous existence (and, of course, legitimacy) of other, even *incompatible* trends.

The paper calls the claim that unitary science can be attained a *working hypothesis* and argues that it is *credible* for "empirical, methodological and pragmatic reasons." The argument primarily relies upon a careful discussion of successive microreductions among "levels" of "universes of discourse."[23] Oppenheim and Putnam notice two other types of Unity of Science, which "appear[ed] doubtful" at the time they wrote. One is the Machian claim that all predicates can be reduced to sensations or to observable qualities. More surprisingly, they reject "the *Unity of Method* in science, . . . the thesis that all the empirical sciences employ the same standards of explanation, of significance, of evidence, etc."

What were the "incompatible trends" that Oppenheim and Putnam had in mind? Perhaps the tensions were sufficiently great to lead Putnam to his present view, that the very idea of The One True Theory does not make sense. But we owe to Patrick Suppes the definitive statement of these trends. He is precise and to the point, deliberately following exactly the layout of the old *Encyclopedia* and therefore that of Oppenheim's considered revision.

He starts with the Unity of Language, what I call the linguistic precept. He describes trying to read a paper in a journal to which his daughter, a student neurophysiologist, subscribed. He observes, in a droll way, how the language is just inaccessible. ("If postsynaptic adrenergic neurons in neonatal rats were chemically destroyed with 6-hydroxyudopamine . . . the normal development of presynaptic ChAc activity was prevented.")[24] The point is not that you cannot learn to read this, nor that there is any problem in mastering any one body of experimental technique. The point is that in a quite straightforward sense there is no common language of science, and that as a matter of practicability, there could not be. In 1800 it was true that an alert person, a Thomas Young, perhaps, could savor every article in the *Philosophical Transactions of the Royal Society of London*. It is no longer humanly possible to read, day by day, all the best research papers that appear in all the sciences. That is a fact about, among other things, language. And is this a bad thing? Suppes applauds

the divergence of language in science and . . . find[s] it no grounds for skepticism or pessimism about the continued growth of science. The irreducible pluralism of languages of science is as desirable a feature as is the irreducible plurality of political views in a democracy.

Dyson said the same thing with elegant brevity:[25]

The voice of science is a Babel of diverse languages and cultures. That is to me the joy and charm of science.

And what about microreduction? Even "quantum chemistry, in spite of its proximity to quantum mechanics, is and will remain an essentially autonomous discipline":

It is hopeless to try to solve the problems of quantum chemistry by applying the fundamental laws of quantum mechanics. . . . The combinatorial explosion is so drastic and so overwhelming that theoretical arguments can be given that, not only now but in the future, it will be impossible to reduce the problems of quantum chemistry to problems of ordinary quantum mechanics.

The comparison is to a chess-playing program. In principle you cannot program a computer to "foresee every possible move" in a game and mindlessly check out which moves pay off. There are about 10^{120} possible moves in a game, perhaps 10^{40} times as many as there are atoms in the universe. So you cannot have a representation, in the computer, of every possible move. Suppes is well aware that in some other sense the computer might "foresee" all the moves, by eliminating, by proof, most of the bad moves; likewise, in some nonliteral sense, the key concepts of quantum chemistry might have a physical interpretation and so, in that sense only, be "reduced" to physics.

My point in this section has not been to favor Suppes over Oppenheim and Neurath, but to emphasize how the focus of argument among philosophers of a positivist and empiricist bent is so very different from questions about unity that vex various types of scientific investigator. Oppenheim and Putnam defined two topics, exactly the two addressed by Suppes: Unity of Language and global reduction. Nowhere do I find any mention of what seems to me so central to the scientific working-out of the metaphysical sentiment of unity, namely the interconnectedness of phenomena. The philosophers who want to unify seem not just to be playing in another ballpark or even another league: they are playing a different game. Language is a central part of their game. And of course language has a historic role to play in scientific self-image. But then the role is something that the Unity of Science movement was too prim ever to mention. I mean the book of nature.

The Book of Nature

Historians of ancient civilizations used to distinguish "people of the book," namely cultures that had what they called The Book. Israelites, Christians, and Muslims are peoples of the book, and so are many others now lost, like the followers of Mani. Although peoples of the book were not necessarily strict monotheists, One Book and One God are close kin. The new scientists of the seventeenth century were also people of the book, but theirs was the Book of Nature. The metaphor is perhaps most memorable when Galileo writes of the Author of Nature writing in the Language of Mathematics. Just as God wrote one Book of words, the Bible, so God wrote one Book of Nature. One Author, one Language, one Book. Unity is built in. Or is it? One of our greatest unifiers put a remarkable twist on this image:[26]

Perhaps the book, as it has been called, of nature is regularly paged; if so, no doubt the introductory parts will explain those that follow, and the methods taught in the first chapters will be taken for granted and used as illustration in the more advanced parts of the course; but if it is not a book at all, but a magazine, nothing is more foolish than to suppose that one part can throw light on another.

James Clerk Maxwell notoriously kept his philosophical cards close to his chest. I am sure his words can be understood in several ways. They were the words of a young man talking to a Cambridge discussion club, the Apostles, and we should not hold him to them. The image of the magazine remains worthwhile on its own account. It is not the first time that a magazine has had something to do with unity, by the way. The motto of the United States, printed on its coinage, is *E Pluribus Unum*. That is not an old Latin tag. It was first used in London in 1692 as the motto of *Gentlemen's Magazine*. That was a *Reader's Digest*, taking excerpts from other magazines and printing them up in a single periodical for the busy gentleman. Out of the many (periodicals), one.

Maxwell seems to have been thinking of the Book of Nature as above all a textbook: not an axiomatic exposition beginning with the fundamental laws of nature, but as a mid-nineteenth-century physics text in the Scottish manner, in which one learns, by working through example, how to solve problems that will then be applied to more and more difficult cases, and that will be systematically extended. His talk was about analogies between different branches of science. His version of the metaphysical sentiment of unity gives new content to "There is one scientific world, reality, truth." We get something like, "All the various aspects of nature work in something like the same way, and are accessible by something like the same methods." And, says Maxwell, nature might not be like that. That would not falsify the original metaphysical sentiment. It would not even falsify the metaphysical thesis of interconnection, because the numerous natural phenomena might be connected in fundamentally different ways, so that the solution to one problem of connection bears no analogy to the next.

I think that Maxwell was rather indifferent to the metaphysical thesis of structure—as I have said, it is philosophers rather than scientists who like to think of the universe as characterized by a deductive structure of laws. The axiomatic exposition of laws would indeed constitute a Book of Nature, and Maxwell's twist is also a way of reminding ourselves that there could be a Book of Laws in which global reduction made no sense—because the Book was a magazine, with different articles for different topics. The idea of a Book of Laws is perhaps best conveyed in the writing of another great unifying philosopher-scientist, Leibniz.

His vision of knowledge and the world makes excellent sense of the metaphysical theses and practical precepts about unity. He thought of the world as brought into being by God, who contemplates the ideas of all logically possible worlds to determine which is best. It is a very small jump to making Leibniz's God not only supremely reasonable but also supremely wordy, a God who contemplates all possible world-descriptions and then creates the most perfect world answering to a possible description. The Book of Nature provides just that description. It is clear that Leibniz thinks it is essential to such descriptions that all phenomena are connected in a lawlike structure with taxonomic unity. The sciences in a sense imitate God's work. That is just because they should use what Leibniz called architectonic reasoning—they should try to figure out how the Architect would have made things, on the basis of His works. This is also a prescription, more soundly based than any in more recent philosophy, for a unified language of science, for one would be aiming at the best language, God's.[27]

Leibniz's picture makes sense of unity. Compare the situation with another supposed scientific virtue, simplicity with elegance. Theories and laws are often thought to be better if they are simple. I well understand a preference founded upon aesthetics or ease of computation, but suppose it is also urged that simple theories are more likely to be true. Why? Leibniz had a ready answer, that God preferred the simplest theory with the most diverse consequences; it was elegantly economical. I know of no other reason for thinking simplicity a guide to the truth. I suspect that many atheistical admirers of metaphysical unity likewise have, *au fond*, a thoroughly theological motivation.

But surely no modern physicist takes the Book of Nature seriously? On the contrary: Leibniz's charming story is far from antiquated. Only in terms like those of Leibniz can I understand the most widespread picture of contemporary cosmology. In the "big bang" account of the universe, it is thought that there are certain fundamental laws of nature, in which occur laws with certain parameters, fundamental constants such as the velocity of light or the gravitational constant. In one line of discussion that has acquired some notoriety of late—the cosmological anthropic principles—it is noted that most choices of values for those constants would lead to very uninteresting consequences. The universe would quickly collapse or explode. Indeed if the constants aren't just right, heavy metals would not form, galaxies would not cohere . . . and so on. So a question is posed: How come the constants are just right? I cannot understand this question outside a picture of, first, the laws of nature being written down in God's sketchbook, with free parameters, and then the blanks being filled in with particular constants of nature. More generally, I find it very difficult to grasp the force of the theses of the unities of the sciences without having in the back of my mind that wondrous image, the Book of Nature. And Maxwell's irony undoes it all in a paragraph. If you must picture the world in a wordy way, why not imagine a periodical? Salam ends his first Dirac lecture by quoting:[28]

> Though all trees on the Earth were pens and the Sea was ink,
> Seven Seas after it to replenish it,
> Yet the Words of the Lord would not be spent;
> Thy Lord is Mighty and All Wise.

I am sure the thought did not cross Salam's mind, but it did cross Maxwell's: Might not the Lord of many words have been editing a magazine?

Method and Reason

We may feel more comfortable turning from the claim that there is one unified book to the claim that there is just one reason, and one scientific method. Yet these are somewhat hollow claims. Take scientific method first. Let us turn to an ingenious piece of research by members of a team directed by Lap-Chee Tsui at my university, announced as I was finishing this essay. After seven years of unceasing toil, his group identified the genetic material that carries most cystic fibrosis. If you ask what (scientific) methods they used, you can get an answer from the daily newspaper, which, admittedly, is trying to indicate how hard the team worked.[29] Genetic linkage analysis. Pulse field electrophoresis. Chromosome jumping, a method that they pioneered. Saturation mapping. Recombinant work. Zoo blotting (you compare some isolated DNA with that of various animals to see if they have it too; if so, you think it might be genetically important). Then endless work with a copy DNA library. Finally polymerase chain reaction. It would be silly to say that they used zoo blotting, polymerase chain reaction, *and* the scientific method. Where then is this splendid specific of science, the scientific method?

Styles of Scientific Reasoning

Casting about for "the scientific method" we might try the following. Steven Weinberg, the cosmologist and theoretical physicist, writes of "the Galilean style of reasoning," namely of "making abstract models of the universe to which the physicists, at least, give a higher degree of reality than they accord the ordinary world of sensations."[30] A remarkable style, since "the universe does not seem to have been prepared with human beings in mind." This comment has been used by Noam Chomsky as part of a legitimation of his approach to linguistics, which he says uses the same style, and he writes that "we have no present alternative to pursuing the 'Galilean style' in the natural sciences at least."[31] Might this

not be the very scientific method, which some would say originated in the time of Galileo? Might we not say that Tsui and his
colleagues used the Galilean style, first postulating that there was a
certain reality there, a reality of a certain form, a genetic carrier of a
certain type for an apparently hereditary disease? What we wanted
for our thesis of methodological unity was a specification of the
method, and that is precisely the Galilean style!

A moment's reflection won't let this stand. It does not begin to
capture what Tsui and his group were doing. To turn to a different
level of generality, just when Weinberg and Chomsky wrote in
praise of the Galilean style, a distinguished historian of science,
A. C. Crombie, was writing about styles of scientific reasoning in
the European tradition, among which he enumerated six: in brief,
(a) postulation in the axiomatic mathematical sciences, (b) experimental exploration and measurement of complex detectable relations, (c) hypothetical modeling, (d) ordering of variety by comparison and taxonomy, (e) statistical analysis of populations, and
(f) historical derivation of genetic development.[32]

Evidently "style" is a risky word, because it has so many loosely
related uses, not least in describing the arts, letters, and sports.
Even in connection with the sciences there are many potential uses.
Freeman Dyson, for example, describes "two contrasting styles in
science, one welcoming diversity and the other deploring it, one
trying to diversify and the other trying to unify."[33] That is a use of
the word "style" entirely different from Weinberg's, and equally
legitimate. Crombie's encyclopedic study provides endless detailed
examples of his six "styles," thus giving in effect an ostensive
definition of his own usage of the word.

Once the idea has been defined (and one can do better than give
an ostensive definition) we can notice styles that are not in Crombie's initial list. He was primarily concerned with the development
of what he calls the Western vision of nature.[34] His study is largely
based on the history of the Western sciences from ancient Greece
until the seventeenth century—only when he gets to (f) are we
brought up to the nineteenth. So I supplement Crombie's list with
the fundamental seventeenth-century innovation, laboratory science. It is characterized by the construction of apparatus intended
to isolate and purify existing phenomena and to create new ones.

What Weinberg and Chomsky call the Galilean style is method

(c), hypothetical modeling, which is one of the things that Galileo did. It is no accident that a cosmologist and a theoretical grammarian should praise (c) and use it to legitimate their own extraordinarily powerful work, because cosmology and cognitive science are the two lively branches of natural science that (for different reasons) cannot use the laboratory style, cannot interfere in the course of nature to create new phenomena. It is one of the innumerable corollaries of Crombie's classification that we can label exactly what Chomsky opposed in most psychology—unable to use the laboratory style of reasoning, psychology and linguistics had contented themselves with experimental probing, style (b). The aptly named cognitive revolution was a switch to style (c). Chomsky is another monist, implying that the sciences have one style, (c). In fact it is only one among the most general scientific methods, along with the methods of the classificatory and of the "historical" sciences (such as philology, parts of geophysics, or evolutionary biology).

I noticed earlier the disarray among methodologists. We now know its cause. They have simply got the wrong object. They do not concern themselves with methods—in the sense of the methods of Tsui's group, or Dirac's methods of mathematical physics, or the battery of textbook methods that I believe Maxwell had in mind when writing about analogy. Nor do they concern themselves with the ways in which people reason in the various sciences. Oppenheim and Putnam dismissed, without explanation, the idea that there was one scientific method, but it has taken a historian, Crombie, to open our eyes and see that there are a number of broadly characterized and long-enduring methods of reasoning, each having its own standards of reason.

His contribution is remarkable in many ways. His inclusion of a mathematical style of reasoning is relevant to the logical positivist doctrine of the Unity of Science. (Once again I would want to supplement Crombie's analysis, and would begin by adding the Indo-Arabic "algorithmic" style to the Greek "postulational" style.) Although geometry was once the essential model of *scientia*, it was gradually excluded from science, or "natural science," because of a philosophical theory about the character of mathematical truth. The culmination was the logical positivist thesis, adapted from Wittgenstein's *Tractatus*, that mathematics has no content at all. But of course if one were to advocate the Unity of Science doctrine, it was incidentally important to exclude mathematics from science—because, according to logical positivism, mathematics and empirical science simply were not a unity, but a twosome. Crombie, restoring mathematics to the sciences, and seeing it as one (or more) style (or styles) of reasoning among others, such as experimental probing or historicogenetic derivation, gives us a sensible plurality of scientific methods, in which mathematics assumes its rightful place.

I call the plurality sensible because, unlike philosophers, Crombie does not count "one, two, many." As a historian he distinguishes a plausible initial number of six styles, which one may judge it right to supplement with a few more. We do not move to a wodge of indiscriminate plurality, but instead to a fundamental tool for both historical and philosophical analysis.[35]

Crombie is an especially valuable ally, because most older "disunifiers," with the very notable exceptions of people like Patrick Suppes and Freeman Dyson, have been skeptics about science. Many of the present youngest generation of disunifiers are quite cynical not only about established images of science but about the sciences themselves. Crombie, in contrast, is enthusiastically writing about a specifically Western vision of science that has accumulated over two and a half millennia. He takes it to be a Thoroughly Good Thing, however much we may abuse the powers it has brought us. Yet here he is denying not just the methodological maxim about the unity of science. He is also denying the logical maxim, that "There is one and only one standard of reason by which scientific hypotheses can be investigated and judged."

As I develop Crombie's ideas, he is saying that there are quite different styles of inquiry, each with its own canons of good reason. I myself go farther than he is willing to. I urge that each style brings into being new standards of reason, new criteria for what it is to be true or false. Styles of reasoning are what I call self-authenticating.[36] I am at pains to say that this is not a kind of subjectivism. My doctrine of self-authentication, which sounds like part of the current mood for skeptically undermining the sciences, turns out to be a conservative strategy explaining what is peculiar about the sciences. I shall conclude this paper by using this idea to distinguish the sciences from humanistic and ethical inquiry. Such developments are undoubtedly a philosophical exaggeration of Crombie's intentions. Nevertheless, even Crombie's own steps along this path are a radical challenge to the complacent assumption that in our own history of Western science, or in our own practice at present, there is one standard of reason, and one method or style of investigation.

Unifiers

Is science then one kind of thing at all? There is no set of features peculiar to all the sciences, and possessed only by sciences. There is no set of necessary and sufficient conditions for being a science. There are a lot of family resemblances between sciences. Importantly there are quite different kinds of "unifiers." In modern times the first among the unifiers of the natural sciences has been mathematics, to the extent that Galileo said that the Author of Nature wrote in the Language of Mathematics. During the nineteenth century the differential and integral calculi were the trademarks of much science, and at the core of Maxwell's methods, not as branches of theory but as applied tools. Analysis and its descendants governed science for the first half of the present century, be it population statistics or nuclear technology. But there are many mathematics that unify. Quantum electrodynamics became a unified theory, perhaps, only with the advent of the technique called renormalization. Electroweak theory became the paragon of unification precisely when it was cast in the form of Lagrangians, an eighteenth-century structure determining the success of an achievement of the twentieth century. Once we start looking closer at these real-life examples, the kindly common denominator "mathematics" ceases to denote one thing. Wittgenstein's phrase, "the motley of mathematics," is the one that fits. And it is just because mathematics is a motley that it does such a good job of making much science look as if it were one unified activity: if we can apply mathematics to it, it must be one thing!

There are many more unifiers, that is, tools, practices, and bodies of knowledge, that span sciences. In our day fast computation is the thing. It works at both sides of the experiment/theory divide. On the one side we can articulate theories that a decade ago would have had few intelligible or phenomenological consequences. On the other, data that no multitudes of human beings could ever have processed are transformed in short order into meaningful results. And it is not just a matter of calculating results, but of image enhancement, and of the design of apparatus, for example in the way in which an old-fashioned reflecting telescope is surpassed by many bits of the telescope, kept in exactly the right place, free of surface distortion of the sort that plagues enormous pieces of glass buffeted by winds and stressed by temperature change. Indeed to a certain extent computation replaces much experimentation by at least eliminating many possible scenarios: recombinant genetics and the acoustic architecture of the theater will increasingly be done at the console, at most stages until the final prototypes are called for. And here we get a fascinating merging of the sciences with some of the arts. Designing a building and designing a molecule both become part of computerized architecture. But the computer diversifies too, by creating new congeries of sciences. Galison's "Artificial Reality" in the present volume is the most fascinating description of this so far on record.

Scientific instruments are also among the unifiers of disparate bits of science. (Galison's example can be used to illustrate this too, with the big computer as instrument.) Throughout most of our century, regimes and practices of experimentation and instrumentation have been more powerful as a source of unity among diverse sciences than have grand unified theories. Instruments are speedily transferred from one discipline to another, not according to theoretical principles but in order to interface with phenomena. The scanning tunneling electron microscope is a relatively new device that uses an effect called quantum tunneling to produce sharp images. At first, like the transmission electron microscope, it was thought suitable only for metallurgy, but it expands into cell biology in ways not all that well thought-out, and sometimes by accident. Very few of the consumers of this device understand how it works. For that they rely on facilitators. A literally "instrumentalist" thesis suggests itself: it is not high-level theory that has stopped the innumerable branches of science from flying off in all directions, but the pervasiveness of a widely shared family of experimental practices and instrumentation.

Invention

Unity of science movements have had many practical and worldly intentions, including the survival of a free and independent spirit of inquiry during totalitarian oppression. They have nevertheless conducted themselves at the level of theory, so that my introduction of instruments and computation seems almost out of place. Yet we are completely surrounded by unifications of science incarnate, material testimony to the metaphysical thesis that all phenomena should be interconnected. Entrepreneurial inventors, not only the legendary Edison or Gates, but also myriad more humble people in machine shops or biotech labs, put together bits of purified phenomena to produce new pieces of reliable technology. Has anyone ever mentioned this in connection with the unity of science? Yes: the man whom I have called our greatest unifier, James Clerk Maxwell. He does indeed speak of "the unity of Science," but not where you'd expect it:[37]

In a University we are especially bound to recognize not only the unity of Science itself, but the communion of the workers of Science.

That was no part of a discussion of electromagnetism, but a lecture "On the Telephone," the Rede Lecture administered by the Cambridge University Board of Musical Studies. He takes

the telephone as a material symbol of the widely separated departments of human knowledge, the cultivation of which has led, by as many converging paths, to the invention of this instrument by Professor Graham Bell.

Later, instead of "the unity of Science" we find at least part of what I believe Maxwell meant by the unity of Science, a phrasing that takes us back to the sciences:

I have said the telephone is an instance of the benefit to be derived from the cross-fertilization of the sciences.

"Cross-fertilization" was not an idle phrase. Earlier in the lecture Maxwell compared scientific researchers to honeybees that crowd the flowers for the sake of the honey,

never thinking that it is the dust which they are carrying from flower to flower which is to render possible a more splendid array of flowers and a busier crowd of bees in the years to come.

The most remarkable testimony to the unifiability of the sciences is that one can combine artificial phenomena to make new devices—which often become the keys to a busier crowd of scientists and even of new sciences in the years to come.

Is Science a Natural Kind?

Very recent skeptics about the sciences tend to be disunifiers, especially those who undertake social studies of knowledge. But if we turn to the broader antiscientific tendencies in contemporary culture—those that are, in Comte's phrase, "daily hatched by those to whom scientific methods and results are entirely foreign"—then both skeptics and admirers of the sciences usually take the unity of science for granted. A valuable example is furnished by Richard Rorty's essay "Is Science a Natural Kind?"[38] "No," he answered. His immediate target was Bernard Williams, who has for some time been urging a fundamental distinction between ethics and science. Science, claims Williams, is founded upon or aims at an "absolute conception of reality" that crosses all cultures, while ethics does not.[39] To this Rorty opposed his well-known talk about the conversation of humankind, which includes moral and scientific reasoning, but with no essential difference between chemistry and literary criticism, biology and morality.

Both parties mention individual sciences, but write of science, implicitly assuming a kind of unity. That is evident in the form of Rorty's question, "Is science a natural kind?" Perhaps if we had been all along speaking of the sciences, rather than science—in the spirit of the old philosophers of the sciences with whom I began this paper—one would at least redirect Rorty's question. First I shall briefly take the natural-kind question literally, and provide a superficial but by no means irrelevant answer in terms of the sciences. Then I shall conclude with a more profound answer that draws upon the idea of styles of reasoning.

One might well deflect Rorty's question by retorting, "Natural kind of what?" Natural history, which sorts kinds into taxonomies, was the root of the idea of a natural kind, so let us use that as our model. If we think not of science but of the sciences, we should ask whether the sciences are naturally arranged upon a taxonomic tree characteristic of the kinds of natural history. Comte, with his eye on the sciences rather than science, notoriously drew up just such a tree—which, as my quotation above reminds us, was not a tree of "reduction" of one science to another. There is simply no place for morality or literary criticism upon Comte's tree.

That observation does not settle what is at issue between Williams and Rorty, but it does unexpectedly answer Rorty's question in a constructive, nonrhetorical way. It scores a tentative point in favor of Williams, without admitting any absolute conception of reality. (Comte, founder of positivism, would be delighted with this corollary of his taxonomy, and, I fear, go on to lecture us about positivist ethics, the religion of humanity.) I'm not advocating Comte's tree as a right description of twentieth-century science. We can no longer classify sciences in terms of their objects. There are constant redistributions of powers. But that leads us on to new questions about the complexities of the sciences and their interrelations, the very topic of the present collection of essays.

There is nevertheless a danger in proliferating interrelations. It makes it sound as if Rorty were absolutely right—there is nothing peculiar about the sciences. But every schoolchild knows that is just plain false! This is not because every schoolchild is possessed of Bernard Williams's absolute conception of reality. It is because one has to learn to do a certain family of recognizably "scientific" things: one must learn methods; one must, as I prefer to put it, learn a quite definite and deliberate set of styles of scientific reasoning.

If one wanted to answer the question, not "Is science a natural kind?" but, what Rorty and Williams are really arguing about, "How do the sciences differ from the humanities?" then one should attend not to what we find out but to how we find out. Of course there are lots of kinds of answer. The characterization in terms of styles of reasoning is one that proves fruitful. But here I shall have to conclude in a programmatic way, barely suggesting how the argument goes.

I claim that each of Crombie's styles of reasoning, and a few more that I have mentioned, is self-authenticating. Each style brings into being new standards for objectivity, and indeed new objects. There is an ontological or "scientific realism" dispute associated with every one of Crombie's styles of reasoning, because each style introduces its own class of new objects (abstract mathematical objects, unobservable theoretical entities, biological taxa, etc.). Each style introduces new laws, new modalities. This can be argued in some detail for each style.[40] But we can ask the general question: Why are there relatively few enduring styles of scientific reasoning? If they are self-authenticating, why can't everything be self-authenticating? Why was Feyerabend so manifestly wrong when he said, "Anything goes"?

The answer lies in a study of the techniques that enable a style to stabilize itself. I believe one can set forth a series of such techniques, each characteristic of a style, and in the end serving to characterize and indeed to constitute the style. That the techniques should work does of course depend upon some broad facts about people and our place in nature, of the sort that might be called, after Wittgenstein, philosophical anthropology. But the detailed account of self-authentication requires attention to the self-stabilizing techniques peculiar to a given style of reasoning. This talk of techniques is unfamiliar, but my main innovation is organizational. Many of the techniques that I describe are quite well known. For example, Duhem's famous thesis about how to save theories by adjusting auxiliary hypotheses is part of the story. It has to be extended to apparatus and theories about how it works. We don't just modify hypotheses; we rebuild instruments. Overall the laboratory stabilizes itself by a mutual adjustment of ideas (which include theories of different levels), materiel (which we revise as much as theories), and marks (including data and data analysis). All three are what Andrew Pickering has called plastic resources that we jointly mold into semirigid structures.[41] I should emphasize that although I mention Duhem, this account does not go in the direction of the underdetermination of theory by data (Quine's generalization of Duhem's remarks). On the contrary, we come to understand why theories are so determinate, almost inescapable. Likewise my account of the stability of the mathematical style owes much to a rereading of two seemingly incompatible authors, Lakatos and Wittgenstein. I introduce an idea of "analytification"—of how some synthetic if *a priori* propositions are made analytic. The logical positivist doctrine of the *a priori* is historicized. But we no more arrive at the radical conventionalism or constructionalism sometimes read into *Remarks on the Foundations of Mathematics* than we arrive at the underdetermination of theory by data.

A happy by-product of my analysis is not only that each style has its own self-stabilizing techniques, but also that some are more effective than others. The taxonomic and the historicogenetic styles have produced nothing like the stability of the laboratory or the mathematical style, and I claim to be able to show why. On the other hand the statistical style is so stable that it even has its own word to hint about its most persistent techniques: "robust." In the case of statistics there is an almost too-evident version of self-authentication (the use of probabilities to assess probabilities). But that is only part of the story, for I emphasize the material, institutional requirements for the stability of statistical reasoning.

In carrying out this program we shall come to speak not of the scientific method, as if it were some impenetrable lump, but instead of the different styles. Once we have a clearer understanding of what, from case to case, keeps each style stable in its own way, we shall not think that there are just endless varieties of Rortian "conversation." Only some conversations are part of a discourse that has developed techniques of self-authentication. There does not exist a set of self-stabilizing techniques for the larger parts of morality and humanistic thought. That itself is a fact about "the natural history of humankind." Thus I believe that we can get a grip on the difference between moral reasoning and scientific reasoning—without invoking any "absolute conception of reality." The sciences can be grouped together in terms of one of their disunities, their styles. In that sense only, perhaps, *E Pluribus Unum*. Otherwise science as a unity invites us to cohabit either with the skeptical Rorty or with the complacent Williams, as if there were only two addresses. Instead of "science," "the sciences" and their methods provide us with a more attractive habitus, practical but freed from Rorty's type of pragmatism, objective but liberated from Williams's type of absolutism.

ARNOLD I. DAVIDSON

Styles of Reasoning, Conceptual History, and the Emergence of Psychiatry

Let me begin by recalling the opening of an often-neglected philosophical classic, "My Philosophy," written by New York University's most famous undergraduate philosophy major, Woody Allen:[1]

The development of my philosophy came about as follows: My wife, inviting me to sample her very first souffle, accidentally dropped a spoonful of it on my foot, fracturing several small bones. Doctors were called in, X-rays taken and examined, and I was ordered to bed for a month. During this convalescence, I turned to the works of some of Western society's most formidable thinkers—a stack of books I had laid aside for just such an eventuality. . . . I was not bored, as I had feared I might be; rather I found myself fascinated by the alacrity with which these great minds unflinchingly attacked morality, arts, ethics, life, and death. I remember my reaction to a typically luminous observation of Kierkegaard's: "Such a relation which relates itself to its own self (that is to say, a self) must either have constituted itself or have been constituted by another." The concept brought tears to my eyes. . . . True, the passage was totally incomprehensible to me, but what of it as long as Kierkegaard was having fun?

This passage identifies a significant philosophical problem, the problem that will concern me in this essay. What are the conditions under which various kinds of statements come to be comprehensible? Not everything is comprehensible at all times, either for individuals or for entire historical periods. I want to examine the conditions under which a corpus of statements, those statements that help to make up the discipline of psychiatry, become comprehensible. To be more precise, I am going to examine one particular form of this problem of comprehensibility, namely under what conditions can one comprehend various kinds of statements as being either true or false? Not every statement claims the status of being either true-or-false; but those statements that claim a scientific status do claim to be part of the domain of truth-and-falsehood. So my problem is, under what conditions did the statements of psychiatry come to be possible candidates for truth-and-falsehood and so come to claim the comprehensibility of a science?

In order to approach this problem here, I shall be preoccupied with a number of methodological questions, and these methodological questions will lead me to make use of two methodologically central notions, that of a style of reasoning and of a conceptual space and its history. Both notions are, to say the least, methodologically problematic. Styles of reasoning and conceptual spaces are largely metaphorical, and one aim of this essay is to give some philosophical detail to both these ideas. My paramount aim, however, is to show how these notions, despite their metaphorical appearance, really do help us to understand the emergence of psychiatry as an autonomous medical discipline. The emergence of psychiatry involved radical epistemological transformations, and we are likely to overlook them because psychiatry is today so much a part of our framework for understanding other human beings. But the historical conditions that made psychiatry and its diseases possible were quite specific, and, as I hope to show, if we forget these conditions, we are going to find ourselves in great philosophical difficulty.

The most recent invocation of the notion of a style of reasoning or thinking in science is due to A. C. Crombie, whose book *Styles of Scientific Thinking in the European Tradition* will certainly play a critical role in the acceptance or rejection of this notion.[2] I want to spend the first part of this essay saying how one might go about understanding the notion of style in a style of reasoning, but before I do that, I should say something about how not to think about this notion. There is a sense of "style," quite common in popular discussions, that links style to individual personality, even idio-

syncrasy. It is this sense of the word that is implied by the last line of
J. L. Austin's review of Ryle's *The Concept of Mind*—"Le style, c'est
Ryle."[3] This use of "style" has perhaps its most natural home in
discussions of fashion, and it is under investigation in an essay
called "Style" by the wonderful fashion critic Kennedy Fraser when
she writes,[4]

> The world of the merely tasteful—a trim edifice of bourgeois confor-
> mities, with narrow slots to be filled and straight lines to be toed—is
> bound to barricade itself, in the end, against style, which is individualistic,
> aristocratic, and reckless. . . . "Spontaneous" is style's substitute for
> "good," and "boring" that for "bad." A touch of outrageousness, far from
> being shameful, is considered by style to be a moral obligation.

Fraser's essay, which is largely a discussion of Bianca Jagger, turns
out to be quite useful for the distinctions it makes between style,
taste, and elegance. Yet, despite its intrinsic interest, this use of
"style" is not helpful in trying to characterize styles of reasoning in
the sciences. In discussing styles of reasoning in what follows, I
shall make almost no reference to differences of individual temperament. Indeed, it is perhaps a peculiarity of my understanding of a style of reasoning that proper names function almost as
placeholders for certain central concepts, so that a style of reasoning is not primarily concerned with the ideas of individuals, but
rather with the emergence of concepts and the way that they fit
together. As much as style in fashion is linked to individuals, so is it
divorced from specific personalities when we come to consider
styles of reasoning.[5]

   In beginning to characterize styles of reasoning, I want to start,
so to speak, inside out. Rather than commencing with recent
Anglo–American history and philosophy of science, I want to start
with its somewhat more philosophically alien French analogue.
Specifically, I want to consider some suggestions of Michel Foucault, who stands in a line of distinguished French epistemologists
of science beginning with Gaston Bachelard and running through
Georges Canguilhem before it reaches Foucault. At the end of an
interview "Truth and Power" given in the late 1970's, Foucault
makes two suggestions:[6]

> "Truth" is to be understood as a system of ordered procedures for the pro-
> duction, regulation, distribution, circulation and operation of statements.

"Truth" is linked in a circular relation with systems of power which produce and sustain it, and to effects of power which it induces and which extend it. A "regime" of truth.

Since Foucault is usually his own best interpreter, I like to think of this first suggestion as his own succinct retrospective interpretation of his archaeological method, while the second suggestion is his equally succinct interpretation of his genealogical method. In attempting to understand both the notions of styles of reasoning and of conceptual spaces and their history, I shall take some clues from Foucault, concentrating on the first suggestion, which he calls the method of "archaeology."

If truth is understood as a system of ordered procedures for the production, regulation, distribution, circulation, and operation of statements, and if what is part of the domain of truth is itself variable throughout history, then it should not be surprising that Foucault has undertaken to write a history of truth. Of course, one might respond that it is one thing to write a history of truth and quite another to claim that, as Foucault's colleague at the Collège de France Paul Veyne has put it, there is no other truth than that of successive historical productions.[7] Foucault does not believe that there can be a useful epistemological theory of truth divorced from the variable historical conditions under which statements become candidates for the status of truth. Combining, in his own inimitable way, some lessons from Foucault with some from Crombie, Ian Hacking (to whom I shall return) has given us the most philosophically promising characterization of styles of reasoning to date. At the end of his paper "Language, Truth and Reason," Hacking makes some assertions and draws some inferences from them that he admits are all in need of clarification. But I shall quote all five of his claims here, since they provide the background for what I want to say about styles of reasoning:[8]

1. There are different styles of reasoning. Many of these are discernible in our own history. They emerge at definite points and have distinct trajectories of maturation. Some die out, others are still going strong.
2. Propositions of the sort that necessarily require reasoning to be substantiated have a positivity, a being true-or-false, only in consequence of the styles of reasoning in which they occur.
3. Hence many categories of possibility, of what may be true or false, are

contingent upon historical events, namely the development of certain styles of reasoning.

4. It may be inferred that there are other categories of possibility than have emerged in our tradition.

5. We cannot reason as to whether alternative systems of reasoning are better or worse than ours, because the propositions to which we reason get their sense only from the method of reasoning employed. The propositions have no existence independent of the ways of reasoning towards them.

For my purposes here, the most important claims that Hacking makes are that there are different styles of reasoning and that these styles determine what statements are possible candidates for truth-or-falsehood (with the exception of those statements that require no style of reasoning at all). As new styles of reasoning develop, they bring with them new categories of possible true–or–false statements. To take an example from Hacking, consider the following statement that you might find in a Renaissance medical textbook: "Mercury salve is good for syphilis because mercury is signed by the planet Mercury which signs the marketplace, where syphilis is contracted."[9] Hacking argues, correctly I think, that our best description of this statement is not as false or as incommensurable with current medical reasoning, but rather as not even a possible candidate for truth-or-falsehood, given our currently accepted styles of reasoning. But a style of reasoning central to the Renaissance, based on the concepts of resemblance and similitude, brings with it the candidacy of such a statement for the status of true-or-false. Categories of statements get their status as true-or-false vis-à-vis historically specifiable styles of reasoning.

If we take Hacking as having given us a preliminary characterization of styles of reasoning, where are we to look for further clarification? The most difficult problem, no doubt, is precisely that of more fully cashing out the notion of style, and the most obvious place to look for help is to characterizations of style in art history. A reading of Meyer Schapiro's well-known essay "Style" shows that there is as little agreement among art historians about the notion of style as there is among philosophers about the notions of knowledge or truth.[10] So there is no question of my applying some commonly agreed–upon conception of style in art to scientific reasoning. Rather, I will appropriate that notion of style in art-historical writings that I find most useful for my concerns about the scientific comprehensibility of psychiatry. This is, of course, not to claim that my characterization of style is the only one, nor even that it is the most useful one for other purposes. The problem is to articulate some plausible conception of style that will help us to think about styles of reasoning. Schapiro clearly sets forth many of the problems that surround the idea of style in the uses made of it in art history. I have no doubt that further problems are created by transferring this idea from art to reasoning. But if we cannot make any sense of "style" in styles of reasoning, then we ought to give up the notion, instead of pretending that it serves a fruitful method-ological purpose. I do think that some sense can be made of it, and I also think that art history provides a good, even if not perfect, guide. But before I move directly into the thickets of art history, I want to say something briefly about an extremely interesting paper that does attempt to use characterizations of style in art history to help fill out the notion of styles of scientific thinking. Winifred Wisan, in a paper called "Galileo and the Emergence of a New Scientific Style," argues that a significant aspect of Galileo's contri-bution to modern science "is the creation of a new style in the study of motion, a style that caught the imagination of those who came after him and thus inspired much of the great advance that has shaped the course of modern mathematical physics."[11] Her paper proceeds mainly by trying to demonstrate that there are two dis-tinguishable styles in the study of motion that can be found in Galileo's own writings, one in his *De motu antiquiora* of around 1590 and the other in his *De motu locali*, which began to take shape about 1604–9, but was not completed and published until 1638. Al-though concentrating on Galileo, Wisan is led to chart the "evolu-tion of successive styles in treatments of motion from Aristotle to Lagrange."[12] The categories she uses for analyzing scientific styles are structure, content, techniques, and expressive quality. Al-though I will not recount the details of her understanding of these categories, an example she uses may prove helpful here. According to Wisan, Aristotle's style in treating motion involves a structure that is discursive and classificatory, while Galileo of the *De motu locali* uses a structure that is Euclidean, that is, axiomatic and geometrical; the content of Aristotle's style consists of the defini-tion of fundamental concepts, while Galileo's consists of a unified treatment of local motions; Aristotle's technique is the use of philosophical analysis and explication of terms, and Galileo uses advanced geometry to derive mathematical consequences; finally, the expressive quality of Aristotle's style is a feeling for a world of substances and essences, while the expressive quality of Galileo's style is a new feeling for the power of mathematics.[13] Wisan herself identifies the fundamental difficulties with her analysis; she says that the categories of structure, content, and techniques are "sometimes difficult to distinguish,"[14] and it is hard to discern exactly how she wants to separate these categories. She also says that the last category, expressive quality, is "hard to articulate satisfactorily."[15] I concur with this latter reservation, but want to emphasize that the notion of expressive quality remains problematic in the art criticism and history that concerns itself with style. Since I will say nothing directly about expressive quality, I want to indicate what some of its theoretical problems are. Perhaps the most significant difficulty is one of understanding how expressive qualities correspond to formal and structural elements in art and to the way in which these elements are combined.[16] In his famous discussion in *Words and Pictures* of frontal and profile modes of representation in art, Meyer Schapiro shows how, in medieval art, presenting figures either in profile or frontally had different effects as "expressive means."[17] He documents, for instance, that a conservative thirteenth-century cleric objected to representations of the Virgin Mary in profile, since the frontal form was seen as both more sacred and more beautiful.[18] Moreover,[19]

In other arts besides the medieval Christian, profile and frontal are often coupled in the same work as carriers of opposed qualities. One of the pair is the vehicle of the higher value and the other, by contrast, marks the lesser. The opposition is reinforced in turn by differences in size, posture, costume, place, and physiognomy as attributes of the polarized individuals. The duality of the frontal and profile can signify then the distinction between good and evil, the sacred and the less sacred or profane, the heavenly and the earthly, the ruler and the ruled, the noble and the plebeian, the active and the passive, the engaged and the unengaged, the living and the dead, the real person and the image. The matching of these qualities and states with the frontal and profile varies in different cultures, but common is the notion of a polarity expressed through the contrasted positions.

In discussing Egyptian reliefs and painting, Greek vase-paintings, and the paintings in medieval Arabic illuminated manuscripts, Schapiro shows that the profile is sometimes reserved for the lesser figure in a pair and sometimes for the nobler figure. He concludes—and this is crucial for my account to follow—that "the contrast as such is more essential than a fixed value of each term in the pair."[20] I also believe that it is certain contrasts in modes of representation that are essential to an account of style, and I further believe that the problem of expressive quality or value, although certainly important, has a derivative or secondary role. Whatever else may be the case about expressive quality, it is at least partly a function of formal or structural elements, and until we are clearer about the role these latter elements play in characterizing style, we are unlikely to make any headway with the problematic notion of expressive value. As Heinrich Wölfflin advises, "Instead of asking 'How do these works affect me, the modern man?' and estimating their expressional content by that standard, the historian must realize what choice of formal possibilities the epoch had at its disposal. An essentially different interpretation will then result."[21] In what follows, I shall take Wölfflin's advice, and so shall have little more to say about expressional content.

Indeed, it is Wölfflin's *Principles of Art History: The Problem of the Development of Style in Later Art* that will provide my guide in characterizing the notion of style in styles of reasoning. Although I am not unaware of the problems in Wölfflin's account, I am less concerned with the adequacy of its details than with its methodological procedure. Despite the objections one might bring to bear on this account as a whole, his procedure is highly instructive in helping us to understand the idea of a style of reasoning.[22] Wölfflin argues that the difference between the classic and baroque styles is best characterized in terms of five pairs of polar categories or concepts. Moreover, he argues that "there can be discovered in the history of a style a substratum of concepts referring to representation as such, and one could envisage a history of the development of occidental seeing, for which the variations in individual and national characteristics would cease to have any importance."[23] That is, Wölfflin wants to write a history of the visual possibilities to which artists are bound. Thus he will argue, to take a typical example, that the impression of reserve and dignity found in

Raphael's paintings is "not entirely to be attributed to an intention born of a state of mind: it is rather a question of a representational form of his epoch which he only perfected in a certain way and used for his own ends."[24] Wölfflin's procedure in writing his history of style is, as is well known, to set forth the determining concepts of classic and baroque art in terms of five pairs of opposed concepts. His five major chapters discuss the linear and the painterly, plane and recession, closed and open form, multiplicity and unity, and clearness and unclearness. The first of each of these pairs of concepts makes up the classic style, and the second of each of the pairs makes up the baroque style. Since I cannot even begin to do justice to the richness of Wölfflin's account in this paper, I will let a quotation from the conclusion to his book serve as a summary:[25]

In its breadth, the whole process of the transformation of the imagination has been reduced to five pairs of concepts. We can call them categories of beholding without danger of confusion with Kant's categories. . . . It is possible that still other categories could be set up—I could not discover them—and those given here are not so closely related that they could not be imagined in a partly different combination. For all that, to a certain extent they involve each other and, provided we do not take the expression literally, we could call them five different views of one and the same thing. The linear-plastic is connected with the compact space-strata of the plane-style, just as the tectonically self-contained has a natural affinity with the independence of the component parts and perfected clarity. On the other hand, incomplete clarity of form and the unity of effect with depreciated component parts will of itself combine with the a-tectonic flux and find its place best in the impressionist-painterly conception. And if it looks as though the recessional style did not necessarily belong to the same family, we can reply that its recessional tensions are exclusively based on visual effects which appeal to the eye only and not to plastic feeling.

We can make the tests. Among the reproductions illustrated there is hardly one which could not be utilised from any of the other points of view.

This summary gives some indication of Wölfflin's procedure in the book. He discusses in detail each of the five pairs of polar concepts, showing how they are exemplified in a wide variety of painting, drawing, sculpture, and architecture. He also shows how the concepts are linked together to constitute what we might think of as two opposed visual spaces, those of the classic and the baroque. In this way we get a determinate conception of classic and baroque style, framed in terms of their contrasted modes of representation. A similar methodological procedure in characterizing style is applied to architecture in an early book written by Wölfflin's former student Paul Frankl. Frankl argues that the differences between the Renaissance (classic) and baroque styles of architecture can be understood in terms of four polar concepts, spatial addition and spatial division, center of force and channel of force, one image and many images, and freedom and constraint.[26] Although his four pairs of polarities differ considerably from Wölfflin's, it is his methodological agreement with Wölfflin in characterizing distinct styles that interests me.

Given his understanding of the opposition between classic and baroque style, Wölfflin can formulate his famous thesis that "Even the most original talent cannot proceed beyond certain limits which are fixed for it by the date of its birth. Not everything is possible at all times, and certain thoughts can only be thought at certain stages of the development."[27] This claim might also have been found in a book by Foucault, since he too is concerned to show the possibilities to which our distinct historical periods bind us. And we should not be surprised to discover something akin to those discontinuities for which Foucault is famous appearing in Wölfflin when he writes:[28]

False judgements enter art history if we judge from the impression which pictures of different epochs, placed side by side, make on us. We must not interpret their various types of expression merely in terms of *stimmung*. They speak a different language. Thus it is false to attempt an immediate comparison of a Bramante with a Bernini in architecture from the point of view of *stimmung*. Bramante does not only incorporate a different ideal: his *mode of thought* is from the outset *differently organized* from Bernini's.

It is this conceptualization of style, with its polar categories, realms of limited possibilities, and breaks and discontinuities, that I believe needs to be adopted in trying to understand historically the change in styles of reasoning that gave rise to the emergence of psychiatry.

Although I want to focus on the idea of a conceptual space, I should at least mention some other components of style, related to those I have already discussed, that I will only touch on indirectly, but that any complete discussion of style would have to take into account. These components are most thoroughly and brilliantly discussed in Michael Baxandall's *Painting and Experience in Fifteenth-Century Italy*.[29] Baxandall wants to reconstruct what he calls "Quattrocento cognitive style," especially as it relates to Quattrocento pictorial style. He hopes to offer "insight into what it was like, intellectually and sensibly, to be a Quattrocento person,"[30] what it was like to think and see in a Quattrocento style. He takes a cognitive style to consist of the interpreting skills, categories, model patterns, and habits of inference and analogy that one possesses.[31] He then examines fifteenth-century Italian painting and society to exhibit the details of each of these components of Quattrocento cognitive style. If I were going to try to exhibit exhaustively the style of reasoning that made the statements of psychiatry possible, I would have to say something about all the elements of style that Baxandall refers to. For example, I would have to discuss the habits of inference and analogy used by psychiatrists, and to show their distinctness from previous habits of inference and analogy. I would also have to examine the interpreting skills that are part of psychiatry, especially its diagnostic skills, and to show the differences between these diagnostic skills and those of earlier times. And I would also hope to be able to show the significance of all the changes in examples of diseases from neurological to psychiatric textbooks. But I cannot do all of that in this paper, and so I shall concentrate on what I take to be the fundamental element of style, namely the categories or concepts and the way they combine with one another to constitute a style.

In attempting to transfer Wölfflin's characterization of style to styles of reasoning, two differences are immediately evident. First, Wölfflin's five pairs of polar concepts are abstracted from painting, sculpture, and architecture. Since he is dealing with visual media, the concepts of linear/painterly, plane/recession, and so on, are obviously not used in the paintings and sculpture he discusses, but are rather Wölfflin's way of best describing the visual features of these arts. He used conceptual categories to capture the visual differences between two styles. In dealing with styles of reasoning in the sciences, however, we often find that the very concepts or categories that constitute a style are used in the sciences themselves.

It is then a question of showing which concepts are fundamental to a scientific style and how they fit together to form the statements of the science, as well as demonstrating the historical specificity of the concepts. This is most fruitfully done by following Wölfflin's procedure of using polar concepts to characterize opposed styles that are both historically specific and distinct. A second, related point is that Wölfflin was concerned with modes of visual representation, whereas in discussing styles of reasoning we are concerned with what might be called modes of conceptual representation. To be more specific, I will discuss two distinct stylistic modes of representing diseases, the anatomical and the psychiatric. In the latter, diseases are represented with a set of concepts that have no equivalent in the former. In characterizing the differences between anatomical and psychiatric styles of reasoning about disease, I will first emphasize some of the central concepts necessary to understand the emergence of psychiatry. So rather than first delimiting the polar concepts of two styles, I will concentrate on the conceptual peculiarities of psychiatry. Finally, I will conclude by offering some examples in which one can more clearly see some of the polarities that do distinguish anatomical and psychiatric reasoning.

In addition to attempting to understand the conditions of emergence of a psychiatric style of reasoning, I have two other aims in what follows. First, I hope to show what use can be made of the methodological notion of a conceptual space and its history; second, and just as important, I want to begin to try to demonstrate the advantages of this methodology, which I sometimes call conceptual history, over the kind of conceptual analysis most English-speaking philosophers have been taught to think of as occupying the center of philosophy. Unless we carefully examine the historical conditions under which our concepts emerge, we are liable to find ourselves surrounded by philosophical perplexity.

The most productive use of the idea of a conceptual space, especially as it is related to the emergence of a style of reasoning, can be found in Ian Hacking's stunning book *The Emergence of Probability*. Hacking tries to show that[32]

the preconditions for the emergence of our concept of probability determined the very nature of this intellectual object, "probability," that we still recognize and employ and which, as philosophers, we still argue about. The preconditions for the emergence of probability determined the space of possible theories of probability. This means that they determined, in part, the space of possible interpretations of quantum mechanics, of statistical inference, and of inductive logic.

Hacking then argues that these preconditions include a transforma-tion of the medieval concept of *opinio*, resulting in a new concept of "internal evidence," which makes possible our current concept of probability. Hacking is here analyzing the concept of probability, but he does this by showing that the conditions under which the concept emerged determined the content of the concept. He can also show how these conditions determined the relation between this concept and other related concepts (such as opinion and evi-dence) and finally how, given this history, the concept of proba-bility played its role in various scientific and philosophical theories. (This is the kind of undertaking that Foucault is also a master of.) Hacking's methodology implies that our philosophical and scien-tific problems are individuated by using certain concepts, and that the conditions under which these concepts emerge determine what can be done with them. So the conceptual space of a philosophical or scientific problem is significantly determined by the conditions that make it possible, and the solutions and countersolutions to problems are worked out within the constraints of this securely determined space.[33]

Hacking sometimes writes as if his methodology is merely tacked on to his account, as if the account and the methodology are external to one another. But I don't quite see how one could have produced his extraordinary history without the distinctive meth-odology. The virtues of the methodology are demonstrated by the details of the history. Without the methodological underpinnings, how would one be able to show, as Hacking does, that the ques-tions, How did probability become possible? and How did the skeptical problem of induction become possible? are two distinct questions. The concept of inductive skepticism that Hume operates with is not part of the same problematic as the concept of inductive logic known to Leibniz. The analytic problem of induction, of distinguishing good and bad inductive reasons and of classifying different degrees of evidential support, appears almost a century before the Humean skeptical problem of induction. On Hacking's reading, the formulation of Hume's problem of induction requires certain conceptual transformations in, and new conceptual connections with, the notions of cause and knowledge, conceptual transformations and connections that were not necessary to the earlier formulation of theories of inductive logic.[34] Although I have simplified a complicated story, I think that the most natural way to describe Hacking's claim is precisely to say that the analytic and skeptical problems of induction belong to related but distinct conceptual spaces; or, to use Foucault's terminology, these problems belong to distinct "archaeological territories," obeying different "rules of formation."[35]

One purpose of conceptual history lies in showing us that our concepts and their organization are marked by their historical origins, that we will not really understand the problems that many of our concepts give rise to, unless we trace out the conditions of their emergence, however remote from us those conditions may appear to be. In attempting to understand psychiatry's emergence, in the nineteenth century, as an autonomous medical discipline, and specifically its conceptual autonomy from neurology and cerebral pathology, I have been led, in other writings, to focus on the emergence of new disease categories. I have especially emphasized the decline of pathological anatomy as either an explanatory theory for so-called mental diseases and disorders, or the foundation for the classification and description of these diseases. The birth of psychiatry as a distinct medical discipline is simultaneous with the emergence of a new class of functional diseases, of which sexual perversion and hysteria were perhaps the two most prominent examples. Ultimately, these functional diseases were fully describable simply as functional deviations of some kind, diseases that did not have an anatomically localizable pathology. Elsewhere, I have tried to show some of the changes in styles of reasoning that were necessary in order for true-or-false statements about such functional diseases to become possible, and I have tried to write a history of the modern medical concept of perversion, which required showing the conditions under which perversion emerged as an object of medical knowledge.[36] Although I have claimed that there were no perverts before the later part of the nineteenth century, I do not want to repeat the arguments for that claim now. Rather, I would just like to summarize some of the results of my conceptual history of perversion, first, to show, as it were, the methodology in action, and, second, to indicate some of its advantages over the kind of conceptual analysis philosophers are accustomed to.

The best way to understand the nineteenth-century obsession with perversion is to examine the concept of the sexual instinct, for the actual conception of perversion underlying clinical thought was that of a functional disease of this instinct. That is to say, the class of diseases that affected the sexual instinct was precisely the sexual perversions. A functional understanding of the instinct allowed one to isolate a set of disorders or diseases that were disturbances of the special functions of the instinct. Moreau (du Tours), in a book that influenced the first edition of Krafft-Ebing's *Psychopathia Sexualis*, argued that the clinical facts forced one to accept, as absolutely demonstrated, the psychic existence of a sixth sense, which he called the genital sense.[37] Although the notion of a genital sense may appear ludicrous, Moreau's characterization was adopted by subsequent French clinicians, and his phrase "sens génital" was preserved, by Charcot among others, as a translation of our "sexual instinct." The genital sense is just the sexual instinct, masquerading in different words. Its characterization as a sixth sense was a useful analogy. Just as one could become blind, or have acute vision, or be able to discriminate only a part of the color spectrum, and just as one might go deaf, or have abnormally sensitive hearing, or be able to hear only certain pitches, so too this sixth sense might be diminished, augmented, or perverted. What Moreau hoped to demonstrate was that this genital sense had special functions, distinct from the functions served by other organs, and that just as with the other senses, this sixth sense could be psychically disturbed without the proper working of other mental functions, either affective or intellectual, being harmed.[38] A demonstration such as Moreau's was essential in isolating diseases of sexuality as distinct morbid entities.

The *Oxford English Dictionary* reports that the first modern medical use in English of the concept of perversion occurred in 1842 in Dunglison's *Medical Lexicon*: "*Perversion*, one of the four modifications of function in disease: the three others being augmentation, diminution, and abolition."[39] The notions of perversion and function are inextricably intertwined. Once one offers a functional characterization of the sexual instinct, perversions become a natural class of diseases; and without this characterization there is really no conceptual room for this kind of disease. Whatever words of pathological anatomy he and others offered, it is clear that Krafft-Ebing understood the sexual instinct in a functional way. In his *Textbook on Insanity* Krafft-Ebing is unequivocal in his claim that life presents two instincts, those of self-preservation and sexuality; he insists that abnormal life presents no new instincts, although the instincts of self-preservation and sexuality "may be lessened, increased or manifested with perversion."[40] The sexual instinct was often compared with the instinct of self-preservation, which manifested itself in appetite. In his section "Disturbances of the Instincts," Krafft-Ebing first discusses the anomalies of the appetites, which he divides into three different kinds. There are increases of the appetite (hyperorexia), lessening of the appetite (anorexia), and perversions of the appetite, such as a "true impulse to eat spiders, toads, worms, human blood, etc."[41] Such a classification is exactly what one should expect on a functional understanding of the instinct. Anomalies of the sexual instinct are similarly classified as lessened or entirely wanting (anaesthesia), abnormally increased (hyperaesthesia), and perverse expression (paraesthesia); in addition there is a fourth class of anomalies of the sexual instinct, which consists in its manifestation outside the period of anatomical and physiological processes in the reproductive organs (paradoxia).[42] In both his *Textbook on Insanity* and *Psychopathia Sexualis*, Krafft-Ebing further divides the perversions into sadism, masochism, fetishism, and contrary sexual instinct.[43]

In order to be able to determine precisely what phenomena are functional disturbances or diseases of the sexual instinct, one must also, of course, specify what the normal or natural function of this instinct consists in. Without knowing what the normal function of the instinct is, everything and nothing could count as a functional disturbance. There would be no principled criterion to include or exclude any behavior from the disease category of perversion. So one must first believe that there is a natural function of the sexual instinct and then believe that this function is quite determinate. One might have thought that questions as momentous as these would have received extensive discussion during the nineteenth-century heyday of perversion. But, remarkably enough, no such discussion appears. There is virtually *unargued unaminity* both on the fact that this instinct does have a natural function and on what that function is. Krafft-Ebing's view is representative here:[44]

During the time of the maturation of physiological processes in the reproductive glands, desires arise in the consciousness of the individual, which have for their purpose the perpetuation of the species (sexual instinct). . . . With opportunity for the natural satisfaction of the sexual instinct, every expression of it that does not correspond with the purpose of nature—i.e., propagation—must be regarded as perverse.

Nineteenth-century psychiatry silently adopted this conception of the function of the sexual instinct, and it was often taken as so natural as not to need explicit statement. It is not at all obvious why sadism, masochism, fetishism, and homosexuality should be treated as species of the same disease, for they appear to have no essential features in common. Yet if one takes the natural function of the sexual instinct to be propagation, it becomes possible to see why they were all classified together as perversions. They all manifest the same kind of perverse expression, the same basic kind of functional deviation. Thus this understanding of the instinct permits a *unified* treatment of perversion, allows one to place an apparently heterogeneous group of phenomena under the same natural disease-kind. Had anyone denied either that the sexual instinct has a natural function or that this function is procreation, diseases of perversion, as we understand them, would not have entered psychiatric nosology.

Although this truncates a long, convoluted story, it is enough for my purposes here. What I want to do now is contrast this kind of archaeology of the concept of perversion, buttressed as it is by its arcane methodology, with a justly famous, methodologically standard conceptual analysis. Thomas Nagel's "Sexual Perversion," published in 1969 and interestingly revised ten years later for inclusion in his collection *Mortal Questions*, is one of our most famous conceptual analyses, having spawned by itself almost an entire literature. Nagel's paper seems to me, even after all these years, still to be the best of its kind, and I would like to use it as an example of the methodological limitations of its kind of approach. Nagel's analysis proceeds, through an extraordinary coupling of Jean-Paul Sartre and Paul Grice, by way of an analysis of sexual desire, which "in the paradigm case of mutual desire . . . is a complex system of superimposed mutual perceptions."[45] Roughly put, and without any of Nagel's imaginative examples, he argues that in mutual sexual desire, X desires that Y desires that X desires Y, and Y desires that X desires that Y desires X—"It involves a desire that one's partner be aroused by the recognition of one's desire that he or she be aroused."[46] Nagel continues, "I believe that some version of this overlapping system of distinct sexual perceptions and inter-actions is the basic framework of any full–fledged sexual relation and that relations involving only part of the complex are signifi-cantly incomplete";[47] and finally "various familiar deviations con-stitute truncated or incomplete versions of the complete configura-tion, and may be regarded as perversions of the central impulse."[48] Reading this article, one is struck by the fact that, despite its interest, it seems to have very little to do with the concept of perversion. One's perplexity is increased by the first sentence of the article—"There is something to be learned about sex from the fact that we possess a concept of perversion"—and by Nagel's insis-tence that "if there are any sexual perversions, they will have to be sexual desires or practices that are in some sense unnatural, though the explanation of this natural/unnatural distinction is of course the main problem."[49] This latter claim showed that Nagel inhabited the same conceptual space as late nineteenth–century psychiatry, which was able to solve the main problem by offering a conceptually determinate and unambiguous distinction between the natural and unnatural; the theory of the sexual instinct played precisely this role. But when Nagel comes to give his account of perversion as a truncated or incomplete version of the system of superimposed mutual perceptions, the main problem of characterizing the natu-ral/unnatural distinction seems to disappear. Indeed, in the original version of this paper, this distinction plays almost no role at all, even though Nagel also says there that it is the main problem. In the later version of the paper, an extremely interesting extra para-graph and a half is added. Nagel repeats his claim that familiar deviations constitute truncated or incomplete versions of the com-plete configuration of mutual perceptions, and then immediately adds: "The concept of perversion implies that a *normal* sexual de-velopment has been turned aside by distorting influences. I have little to say about this causal condition. But if perversions are in some sense *unnatural*, they must result from interference with the development of a capacity that is there potentially."[50] The next paragraph is concerned with how difficult it is to determine what causal influences are distorting, and Nagel eventually concludes, "We appear to need an independent [noncircular] criterion for a distorting influence, and we do not have one."[51] But if we cannot say what a distorting influence of normal sexual development is, then we cannot say what is natural or unnatural. And if we cannot distinguish the natural from the unnatural, then, by Nagel's own admission, we cannot make sense of the concept of perversion. Again, the theory of the sexual instinct, which was part of nineteenth-century psychiatry, was meant to serve precisely this role.

Consider finally what Nagel says about homosexuality. In the 1969 version of his paper he says, "It is not clear whether homosexuality is a perversion if that is measured by the standard of the described configuration, but it seems unlikely";[52] after a brief further discussion he concludes, "Certainly if homosexuality is a perversion, it is so in a very different sense from that in which shoe-fetishism is a perversion, for some version of the full range of interpersonal perceptions seems perfectly possible between two persons of the same sex."[53] Here is what he says in 1979: "Homosexuality cannot similarly be classed as a perversion on phenomenological grounds. Nothing rules out the full range of interpersonal perceptions between persons of the same sex. The issue then depends on whether homosexuality is produced by distorting influences that block or displace a natural tendency to heterosexual development. . . . The question is whether heterosexuality is the natural expression of male and female sexual dispositions that have not been distorted. It is an unclear question, and I do not know how to approach it."[54] I think it is clear what is going on here. Nagel wants to analyze the concept of perversion, and he is determined by the conceptual space of which that concept is part. In order to do what he claims he wants to do he must say something about the concepts of the natural and unnatural, normal sexual development, and distorting influences. These are concepts without which there is no space for the concept of sexual perversion. But what Nagel in fact does is to present a phenomenological account of ideal sexuality that has nothing to do with the concept of perversion, that makes no real use of the concepts of the natural and unnatural (and related concepts). According to the phenomenological account, certain kinds of sexual behavior need not fall short of the ideal. But it is an entirely different question as to whether they are perversions. Hence Nagel's difficulties in dealing with homosexuality. If we examine the historical conditions under which the concept of perversion emerged, we see that Nagel's claims are determined by those conditions of emergence. On the other hand, his positive account ignores those conditions of emergence and offers us a description of sexuality that actually seems to have no room for the concept of perversion. This is no matter of mere words. Nagel's account is almost internally incoherent, purporting to analyze the concept of perversion but employing unrelated concepts; this creates deep theoretical problems for his account, as indicated by the differences in the two versions of his paper. If we undertake a conceptual history of perversion, we can explain why he has the theoretical difficulties that he does have. Both determined by the conditions of emergence of the concept and attempting to ignore them, he produces an account that cannot do what he wants it to do. Nagel is "shaped by pre-history, and only archeology can display its shape."[55]

I hope that this contrast between two methods for analyzing the concept of perversion begins to show what the significance and philosophical advantages of conceptual history are. Of course, in my own account here, I have only just begun to detail the conceptual space of psychiatry, and so to show which concepts make up the psychiatric style of reasoning. In previous work, I have discussed some of the differences between the anatomical and psychiatric styles of representing sexual diseases, and I have tried to supplement my conceptual history with a discussion of the distinctive habits of inference and analogy, and the different forms of explanation that characterize the two styles.[56] But since I want to follow Wölfflin's lead in understanding style, let me just assert that if we look at the history of neurology and psychiatry in the nineteenth century, we can begin to reconstruct some of the polar concepts that make up the two opposed styles. For example, we are presented with the polarities between organ and instinct, structure and function, and anatomical defect and perversion. The first of each of these pairs of concepts partially makes up the anatomical style of reasoning about disease, while the second of each of these pairs helps constitute the psychiatric style of reasoning. Just as Wölfflin's polarities differentiate two visual modes of representation, so these polarities distinguish two conceptual modes of representation. By figuring out exactly how the concepts combine with one another in determinate ways to form possible true-or-false statements, and by understanding the kinds of inference, analogy, evidence, and explanation that are linked to these conceptual combinations, we can reconstitute a full-fledged style of reasoning. It is only by engaging in this historical task that we will understand the nature of many of the philosophical problems we face. Crombie has described his enterprise as "a kind of comparative intellectual anthropology,"[57] and I have suggested that we understand it, in the first instance, as a comparative anthropology of concepts, sometimes labeled by its French practitioners not as anthropology but as archaeology. Call it what one will, I think it ought to play a crucial role in philosophical analysis.

Let me proceed to give some examples that I think will show how radically different anatomical and psychiatric styles of reasoning about disease actually were and still are. Like Foucault, I am concerned with how systems of knowledge shape us as subjects, how these systems literally make us subjects. Our categories and conceptualizations of the self determine not only how others view us, but also how each person conceives of him- or herself. I am interested in the history of sexuality because I think that, in modern times, categories of sexuality have partially determined how we think of ourselves, have partially determined the shape of ourselves as subjects. Moreover, it is strategically useful to focus on the history of sexuality when discussing the emergence of psychiatric reasoning, since this history allows one clearly to exhibit two distinctive styles of reasoning. If we take the example of sexual identity and its disorders, we can see two systems of knowledge, exhibiting two styles of reasoning, as they are constituted in the nineteenth century. I will consider only one particular case of the anatomical style of reasoning, a case that Foucault has made famous with his publication of the memoirs of the nineteenth-century French hermaphrodite Herculine Barbin. As Foucault points out in his introduction to the case of Herculine Barbin, in the Middle Ages both canon and civil law designated those people "hermaphrodites" in whom the two sexes were juxtaposed, in variable proportions. In some of these cases the father or godfather determined at the time of baptism which sex was to be retained. However, later, when it was time for these hermaphrodites to marry, they could decide for themselves whether they wished to retain the sex that had been assigned them, or whether they preferred instead the opposite sex. The only constraint was that they could not change their minds again, but had to keep the sex that they had chosen until the end of their lives.[58]

However, gradually in the eighteenth century, and into the nineteenth century, it came to be thought that everybody had one and only one real sex, and it became the task of the medical expert to decipher "the true sex that was hidden beneath ambiguous appearances,"[59] to find the one true sex of the so–called hermaphrodite. It is in this context that the case of Herculine Barbin must be placed. Adelaide Herculine Barbin, also known as Alexina or Abel Barbin, was raised as a girl, but was eventually recognized as really being a man. Given this determination of his true sexual identity, Barbin's civil status was changed, and being unable to adapt to his new identity, he committed suicide. The details of the case are fascinating, but my concern is with how medical science determined Herculine's real sexual identity. Here are some remarks from the doctor who first examined Barbin, and who published a report in 1860 in the *Annales d'hygiène publique et de médicine légale*. After describing Barbin's genital area, Dr. Chesnet asks:[60]

What shall we conclude from the above facts? Is Alexina a woman? She has a vulva, labia majora, and a feminine urethra. . . . She has a vagina. True, it is very short, very narrow; but after all, what is it if not a vagina? These are completely feminine attributes. Yes, but Alexina has never menstruated; the whole outer part of her body is that of a man, and my explorations did not enable me to find a womb. . . . Finally, to sum up the matter, ovoid bodies and spermatic cords are found by touch in a divided scrotum. *These are the real proofs of sex.* We can now conclude and say: Alexina is a man, hermaphroditic, no doubt, but with an obvious predominance of masculine sexual characteristics.

Notice that the real proofs of sex are to be found in the anatomical structure of Barbin's sexual organs.

Writing nine years later in the *Journal de l'anatomie et de la physiologie de l'homme*, Dr. E. Goujon definitively confirms Chesnet's conclusions by using that great technique of pathological anatomy, the autopsy. After discussing Barbin's external genital organs, Goujon offers a detailed account of his internal genital organs:[61]

Upon opening the body, one saw that the epididymis of the left testicle had passed through the ring; it was smaller than the right one; the vasa deferentia drew near each other behind and slightly below the bladder, and had normal connections with the seminal vesicles. Two ejaculatory canals, one on each side of the vagina, protruded from beneath the mucous membrane of the vagina and traveled from the vesicles to the vulvar orifice. The seminal vesicles, the right one being a little larger than the left, were distended by sperm that had a normal consistency.

All of medical science, with its style of pathological anatomy, agreed with August Tardieu when he claimed in his revealingly titled book *Question médico-légale de l'identité dans ses rapports avec les vices de conformation des organes sexuels* that, "to be sure, the appearances that are typical of the feminine sex were carried very far in his case, but both science and the law were nevertheless obliged to recognize the error and to recognize the true sex of this young man."[62]

Let me now bypass a number of decades. The year is 1913, and the great psychologist of sex Havelock Ellis has written a paper called "Sexo-Aesthetic Inversion" that appears in *Alienist and Neurologist*.[63] It begins as follows:[64]

By "sexual inversion," we mean exclusively such a change in a person's sexual impulses, the result of inborn constitution, that the impulse is turned towards individuals of the same sex, while all the other impulses and tastes may remain those of the sex to which the person by anatomical configuration belongs. There is, however, a wider kind of inversion, which not only covers much more than the direction of the sexual impulses, but may not, and indeed frequently does not, include the sexual impulse at all. This inversion is that by which a person's tastes and impulses are so altered that, if a man, he emphasizes and even exaggerates the feminine characteristics in his own person, delights in manifesting feminine aptitudes and very especially, finds peculiar satisfaction in dressing himself as a woman and adopting a woman's ways. Yet the subject of this perversion experiences the normal sexual attraction, though in some cases the general inversion of tastes may extend, it may be gradually, to the sexual impulses.

After describing some cases, Ellis writes further:[65]

The precise nature of aesthetic inversion can only be ascertained by presenting illustrative examples. There are at least two types of such cases; one, the most common kind, in which the inversion is mainly confined to the sphere of clothing, and another, less common but more complete, in which cross-dressing is regarded with comparative indifference but the subject so identifies himself with those of his physical and psychic traits which recall the opposite sex that he feels really to belong to that sex, although he has no delusion regarding his anatomical conformation.

It is significant that one name Ellis considers for this disorder, although he rejects it in favor of "sexo-aesthetic inversion," is "psychical hermaphroditism," but he believes that this latter designation is not quite accurate, because people who suffer from this anomaly "are not usually conscious of possessing the psychic dispositions of both sexes but only of one, the opposite sex."[66]

Ellis's discussion descends from the psychiatric style of reasoning that began, roughly speaking, in the second half of the nineteenth century. Sexual identity is no longer exclusively linked to the anatomical structure of one's internal and external genital organs. It is now a matter of impulses, tastes, aptitudes, satisfactions, and psychic traits. There is a whole new set of concepts that makes it possible to detach questions of sexual identity from facts about anatomy, a possibility that came about only with the emergence of a new style of reasoning. And with this new style of reasoning came entirely new kinds of sexual diseases and disorders. Psychiatric theories of sexual identity disorders were not false, but were rather not even candidates for truth-or-falsehood as little as a hundred and fifty years ago. Only with the birth of a psychiatric style of reasoning were there categories of evidence, verification, explanation, and so on, that allowed such theories to be either true-or-false. Rules for the production of true discourses about sexuality radically changed in the mid- to late nineteenth century. And lest you think that Ellis's discussion is anachronistic, I should point out that the third edition of the *Diagnostic and Statistical Manual of Mental Disorders* of the American Psychiatric Association discussed disorders of sexual identity in terms that are almost conceptually identical to those of Ellis. It calls these disorders, "characterized by the individual's feelings of discomfort and inappropriateness about his or her anatomic sex and by persistent behaviors generally associated with the other sex," Gender Identity Disorders.[67] We live with the legacy of this relatively recent psychiatric style of reasoning, so foreign to earlier medical theories of sex. So-called sex-change operations were not only technologically impossible in earlier centuries; they were conceptually unintelligible as well.

Here is one final piece of evidence about stylistic changes in representing diseases, this time some visual evidence. It was not uncommon for eighteenth- and nineteenth-century medical textbooks to include drawings depicting hermaphrodites. These poor creatures were shown exhibiting their defective anatomy, the pathological structure of their organs revealing, for all to see, the condition of their diseased sexual identity. Their ambiguous status was an ambiguous anatomical status. But not too many decades later, a new iconography of sexual diseases was to appear. It is exemplified in the frontispiece to D. M. Rozier's tract on female masturbation.[68] When one opens this book, one is confronted by a drawing of a young woman. She is pale and looks as if she is in a state of mental and physical exhaustion. Her head is stiffly tilted toward her left, and her eyes are rolled back, unfocused, the pupils barely visible. She is a habitual masturbator; her body looks normal, but you can see her psyche, her personality, disintegrating before your very eyes. She stands as an emblem of psychiatric disorders, so distinct from her anatomically represented predecessors. Stanley Cavell has accurately captured the depth of the changes that have taken place between these two styles of representation when, concerning the natural and conventional, he writes, "Perhaps the idea of a new historical period is an idea of a generation whose natural reactions—not merely whose ideas or mores—diverge from the old; it is an idea of new (human) nature."[69]

In closing let me invoke a line often attributed to Burton Dreben, actually a transformation of a similar remark made by the eminent Talmudic scholar Saul Lieberman: "Rubbish is rubbish, but the history of rubbish is scholarship."[70] Although one may not think that rubbish collecting is a dignified way to spend one's mature, adult years, an ingenious rubbish collector can virtually reconstitute the cultural life of a community. You may think that the line of Kierkegaard I quoted from Woody Allen's essay is rubbish (I myself don't), or that the definitions of psychosexual disorders offered by the American Psychiatric Association aren't rubbish. However, regardless of your opinion on these matters, examining the historical conditions of comprehensibility of these statements is the most useful method we have for understanding our own styles of reasoning.

JOHN DUPRÉ

Metaphysical Disorder and Scientific Disunity

Advocates of the disunity of science do not commonly hold this position for metaphysical reasons. One reason for this is that for those skeptical about traditional conceptions of unified science, the grand systems of traditional metaphysics are likely to seem even more dubious. More simply, recent doubts about the unity of science have developed from rather different directions. Such doubts have especially emerged from the recent tendency in science studies to seek a sharper focus on the details of science than has been customary at least in earlier philosophy of science. Thus on the one hand, to historians and sociologists looking in increasing detail at the fine grain of scientific practice, the contingency and specificity of particular projects of inquiry have made the idea of science as one grand project incredible. And on the other hand, epistemologists concerned with the claims to knowledge of particular branches of science have not easily fitted these local modes of justification into broad patterns with universal applicability.

In this essay, however, I want to argue that the picture of science as radically fractured and disunified has a role for metaphysics, and moreover that an appropriate set of metaphysical views is entirely plausible.[1] It is perhaps worth noting that the strongest antipathy to metaphysics is associated with classical positivism, which is also the source for canonical accounts of the unity of science. At any rate, the utility for a defender of the disunity of science of a metaphysics at least compatible with, and perhaps even justificatory of, that position seems to me indisputable.

One central reason why advocates of disunity should care about metaphysics is that, as I shall indicate in the course of this discussion, there is a range of metaphysical positions that both remain attractive to contemporary philosophers and argue strongly for a unified science. It is, of course, impossible for any such philosophical thesis to contradict an empirical demonstration—a demonstration derived, that is to say, from the investigation of the actual practice of science—that science is at this time in a state of radical disunity. But the deeper question is whether science is disunified simply because it has not yet been unified, or rather because disunity is its inevitable and appropriate condition. Historical or sociological investigations might indeed motivate or suggest arguments on one or the other side of this question. If the principles, methods, forms of argument, and everyday practice of different sciences, and even the same sciences at different times, prove to be radically diverse, this will certainly present difficulties for the believer in scientific unity. Nevertheless, no amount of such evidence can rule out the possibility that this diversity reveals only the immaturity of most of science. If one is interested in whether disunity is an inescapable attribute of science, one must attempt some more abstract philosophy.

Reductionism

The philosophical position most generally associated with the unity of science is reductionism. This is, in the first place, an epistemological rather than a metaphysical position. The kind of reductionism I have in mind, at any rate, is a view about the nature of scientific theories, that they must aim to explain the behavior of complex objects in terms of the behaviors of their constituents. This is associated with scientific unity because it mandates strong theoretical links between each science and the science that investigates the objects that are the structural constituents of the objects of the first science. Thus all sciences are linked in a vertical, perhaps branching, structure. In its stronger versions, theories of objects at the higher levels are shown by the process of reduction to be redundant, and thus unity is established in the very strong sense of uniqueness: in the end the only theory we need is the theory that describes the behavior of the smallest objects that there are (if, indeed, there are any smallest objects).

Although, as I have said, this kind of reductionism is primarily an epistemological thesis, it is laden with metaphysical presuppositions. Most significant among these is the hierarchical ontology into which everything—or at least everything amenable to scientific investigation—is to be confined. The world is seen as composed of objects belonging to one of a sequence of levels: elementary particles, atoms, molecules, and so on. The phrase "and so on" conceals the fact that I do not really know how to continue the series. *One* continuation, though surely not the only one, might run through biology, which is itself often supposed to have such a hierarchical structure. Thus, perhaps: cells, organs, multicellular organisms, ecological communities. But though these hierarchical classifications make a point, they are vastly oversimplified. Even the first stretch is not without its difficulties. Elementary particles have become an increasingly heterogeneous bunch since the good old days of electrons, protons, and neutrons, and some of them seem to fail to be elementary by being composed of simpler entities, quarks. The sense of "composed," however, is obscure, as quarks are generally said not to exist other than in such "compositions."

But even if the picture of hierarchical levels can be sustained at these low levels, this amounts to little more than the (admittedly important) discovery that the world is composed of a relatively homogeneous set of structural constituents.[2] At the biological level such a division into discrete levels is much more clearly artificial. The fundamental unit for those areas of biology that deal with interactions between organisms is, obviously enough, the individual organism. But individual organisms span the entire hierarchy of structural complexity, from viruses, simpler than most cells, to the most complex multicellular organisms, such as apes, elephants, and octopuses.[3] And the structural constituents of a complex organism are not remotely homogeneous with respect to their positions on such a hierarchy. While there are certainly parts of an elephant that can be neatly differentiated as organs (heart, liver, brain, etc.), there are also fluids (blood, lymph), and chemical species from the very large (hormones, hemoglobin, neurotransmitters, etc.) to the very small (various essential ions). In summary, though one can roughly *classify* objects in terms of their structural complexity, such classifications need not correspond to the basic individuals of any particular area of investigation, nor do they identify the constituents from which more complex objects are uniquely composed.

The ontological hierarchy just discussed provides a useful way of raising a rather different question. I have argued that the organization of objects into such a hierarchy is not something forced on us by the way the world is. But that does not mean that there might not be reasons for looking at the world in these terms for purposes of some scientific investigation. The question thus raised is whether scientific fields must be determined simply by recognition of the different areas of phenomena that there are, or whether there might not be a much wider range of more pragmatic reasons for distinguishing areas of scientific investigation. In the latter case, it might seem quite plausible that physical size itself could provide an abstraction suitable for defining a scientific field. And at least for the extremes of size this does indeed seem to be the case. Anything below a certain size is an object of study for particle physics, regardless of whether that field is ever likely to come up with a unified theory, or even identify a set of smallest existing objects. And anything larger than a certain size belongs in the domain of astrophysics. Things close to our own size, on the other hand, especially if they resemble us in other ways, get much more detailed attention, and an abstraction as crude as gross physical size has little relevance to an area such as biology. And this anthropocentric focus on things of about our own size would remain true, I think, even if we were to find out (somehow) that there were little people living on electrons, or that our galaxy was a tiny part of the fingernail of some gigantic being.

Natural Kinds

The preceding discussion raises the question whether, or in what sense, the kinds of things investigated by science exist independent of and antecedent to such investigation. By that I do not mean to raise a doubt as to whether the individual things exist, but only as to whether it is an objective fact that they belong to those particular kinds. Answers to this question have tended to take extreme dichotomous forms. At one extreme it is held that for every individual thing of interest to science there is an objective answer to the question what kind of thing it is. Such a view will typically require that there be some feature of a thing, its essential property, that will determine unequivocally what this kind is. In the tradition of Western science such essential properties have generally been assumed to be structural properties; thus this picture of natural kinds determined by structural essences dovetails perfectly with the ontological hierarchy presented in the preceding section. At the other extreme are various nominalistic theses that deny any objective reality to anything but the individuals themselves.

I would like to propose a view distinct from either of these. I suggest that many individual things are objectively members of many individual kinds. Thus I, for example, am a human, a primate, a male, a philosophy professor, and many other things. All, or at least many, of these are perfectly real kinds; but none of them is *the* kind to which I belong. Since I deny that any of these kinds is privileged over the others, I must, of course, deny that I have any essential property that determines what kind I *really* belong to. This I am happy to do. So far I do not take the kind of pluralism here adumbrated to be particularly novel. However, it is my impression that most theorists who advocate such pluralistic positions think that the admission of equal status to so many kinds must amount to denial of any real status to any. But I see no reason why many overlapping and intersecting kinds might not be equally and genuinely real. This would preclude the general possibility of answering one kind of question to which a theory of kinds has traditionally seemed relevant, questions as to what (unique) kind a particular individual belongs to. But I see no reason why there should be any answers to such questions, any more than there need be an answer to the question what color something is (think of rainbows or peacocks). Indeed, in the special case of the classification of people it is very important to deny that there are any answers to such questions.

This combination of pluralism and realism, what I have sometimes referred to as promiscuous realism, provides the starting point for seeing the robust metaphysical basis that I suggest underlies disunified science. For if there are numerous distinct ways of classifying objects into real kinds, any one of which schemes of classification could provide the basis for a properly grounded project of scientific inquiry, then there can be no reason to expect a convergence of these projects of inquiry onto one grand theoretical system. The question that should attract one's attention is rather what grounds justify pursuing one particular project of inquiry rather than any of the many possible alternatives. The general form of the answer to such a question seems clear: we should select that project that best serves the goals that motivate our inquiry or, at any rate, whatever other goals may be potentially served by such a project. A vast body of contemporary work in the history, philosophy, and sociology of science has shown how fruitful for the explanation of scientific activity and belief investigations of the goals and motives of scientists can be. Where I differ from many of the exponents of such research is that while such an approach does presuppose that the direction of (proper) scientific research is not simply dictated by the way things are, I do not take this as contradicting the claim that good scientific research can, nevertheless, describe the way things objectively are. It is just that a particular scientific project can describe only one of the many ways things are. I shall conclude this section by mentioning one reason why I take it to be of great importance to distinguish my position from the purely skeptical interpretation of recent research into the interests guiding and motivating science. The idea that motivation is *all* that can be investigated in looking at a contemporary or historical scientific research project necessarily treats all such projects as on an equal footing. Indeed this is a fundamental methodological precept of the sociology of knowledge movement.[4] But the conjunction of the ideas that scientific research can be explained in terms of the motivations of its instigators, but also in terms of the objective reality that it discloses, also opens up the possibility that some research that can be explained in the first way cannot be explained in the second. There can, that is to say, be both good science and bad science. Since it seems to me very clear that both these categories are well represented in the past and the present, I take this conclusion to be a very important one.

Supervenience and Causal Completeness

Reductionism as a comprehensive account of the aims of science, a program of explaining phenomena at every level in terms of the properties of their constituents, is no longer widely defended. Without going into much detail about the reasons for this, I think it is safe to generalize that the difficulties with reductionism are seen as, in some sense, merely practical. For example, it is widely perceived that macroscopic phenomena are usually, from a microscopic point of view, much too complex for there to be any hope of providing tractable analyses. To make sense of macroscopic phenomena we must categorize them in terms that are, from the microscopic point of view, radically heterogeneous. In sciences such as biology, sociology, or economics nobody seriously supposes that the relevant classificatory kinds will be microphysically homogeneous. And therefore the laws of these sciences will be untranslatable into the terminology of microphysics.

That these problems are widely perceived as "merely practical" is revealed, I think, in the proliferation of supervenience theses. Such theses are often intended to indicate precisely the failure of reductionism, as for example Davidson's well-known version in the philosophy of mind.[5] But I think it is clear that supervenience inherits the metaphysical spirit of reductionism.[6] According to supervenience theses, the microscopic determines the macroscopic, at least in the sense of providing a sufficient condition for any macroscopic property. Thus if this dependency is not to be wholly mysterious, there is presumably some set of facts that *could* be known that would permit the inference of the macroscopic from a sufficient knowledge of the microscopic. Perhaps we could not, even in principle, know these facts. But God, I suppose, would need merely to exist in order to know them.

To explain in more detail how supervenience preserves the metaphysical spirit of reductionism, I must elaborate a little on the account I have so far offered of supervenience. So far I have described supervenience in a merely instantaneous sense. That is, the macroscopic properties of a thing at time t are said to be dependent on its microscopic properties at t. But it is also widely supposed by devotees of the microphysical that the microscopic properties of a thing at time $t + 1$ are dependent on the microscopic properties at t. Then since the macroscopic properties at $t + 1$ are dependent on the microscopic properties at $t + 1$, the former must depend ultimately on the microscopic properties at t.[7] Of course, to the extent that this sounds like a statement of determinism, it is almost entirely discredited at the microphysical level. However, even within an indeterministic theory, it is possible to preserve the idea that a complete causal story can be told relating the situation at t to the situation at $t + 1$. Indeterministically, this would specify only a probability distribution over possible states at $t + 1$ as a function of the state at t. This is a complete story, in the sense that nothing other than the state at t is relevant to what happens at $t + 1$. And the way that the state at t influences that at $t + 1$ is fully and quantitatively determinate. This is what I mean by the assumption of causal completeness, an assumption of which determinism is merely an extreme limiting case.

Such dependence on antecedent microphysical states leads to a problem. For much of our scientific and everyday belief consists of more or less firmly entrenched hypotheses about the causal relations between macroscopic properties of things and events. But if the macroscopic state of a thing at a certain time is dependent on its immediately preceding microscopic state, then there is an obvious problem of reconciling such truths with the causal knowledge that we take ourselves to have at the macroscopic level. It appears that our macroscopic causal beliefs can be true, or even approximately true, only to the extent that they somehow shadow the underlying microphysical processes. It may be true, for example, that my intending to hit a certain key on my word processor causes me to hit that key. (Or more likely, some nearby key.) It seems to follow from this that my intention to hit the key causes, in addition, the movement in the direction of the keyboard of a particular electron in my fingernail. But the idea of microphysical causal completeness implies that the causes of events at the microphysical level are fully specifiable at the microlevel. So the causal efficacy of my intention in moving the electron had better be *consistent* at least with such microlevel causal facts. The consistency of the macroscopic causal statement with all the billions of such microlevel causal processes appears to require either in-principle reducibility, or divinely pre-ordained harmony. At any rate, the causal completeness of the microphysical *directly* contradicts the supposition that my intention could be *necessary* for the movement of the electron, except insofar as it is itself a necessary consequence of events at the micro-level. So even the supervenientist's denial of (in-principle) reductionism seems to leave the macroscopic realm causally inert. Thus it is natural to take supervenience to involve at least God's-eye reductionism.[8]

Perhaps in an unfashionably positivistic vein, I am suspicious of the assertion of facts that God alone could know. Since I suspect that God lacks that notorious perfection, existence, the facts alleged by the supervenientist have the peculiar property of being, if, as is generally supposed, unknowable by us, unknowable. Nevertheless having presented some widely held views that appear to assume the existence of such facts, I cannot deny them without some suggestion as to which of these views I propose to reject.

The argument for the causal imperialism of the microphysical involves two premises that might reasonably be questioned: instantaneous supervenience, and microphysical causal completeness. While instantaneous supervenience does violate my mildly positivistic intuitions about unknowable facts, it may seem to follow from a metaphysical view with which I am sympathetic, the nonexistence of immaterial things. If a thing is exhaustively composed of the particles of microphysics, it may be plausible that its properties must ultimately depend, in the sense of instantaneous supervenience, on the properties and arrangement of those particles. As a matter of fact, I think this is less plausible than it seems at first sight. The intuition on which it is grounded must surely depend on two further assumptions, that it is possible sharply to distinguish the intrinsic from the relational properties of a thing, and that the relational properties of things can be reduced to intrinsic properties of the things related. For surely the relational properties of a thing cannot supervene on the microphysical properties of that thing alone. If these additional assumptions are rejected, then we will be driven to a merely global supervenience, in which the total state of the universe is held to supervene on its microphysical state. And this would surely be a thesis of such blinding epistemological vacuity as to add nothing to the thesis of the nonexistence of the immaterial.

For present purposes, however, I am more concerned with the second premise, the causal completeness of the microphysical, and I will focus on causal completeness for the remainder of this section of the essay. Why should anyone believe that microphysics describes a realm of entities about which complete causal stories can be told? Certainly such a belief is not derived directly from scientific success, since it is admitted that these stories rapidly get too complicated to tell when, for instance, the number of characters exceeds about two. Presumably it is the impressive precision with which we can tell very short stories about very few characters that encourages us to conclude that these must be part of a larger story that itself is precise and causally complete. On the other hand, I can predict very precisely what will happen if I put a hungry fox in a small enclosure with a rabbit. Nobody would be inclined to take this as showing that there must be precise laws governing the ecology of the Amazon rain forest. I suppose that moving from the kinds of experiments performed by particle physicists to, say, the microphysics of a human brain would be a considerably larger shift in complexity.

There is one fashionable move that might seem to dispose immediately of an important part of the thesis of causal complete-ness. This is the combination of indeterminacy (say quantum inde-terminacy) with the ideas developed in chaos theory. Chaos theory has emphasized the existence of mathematical functions the evolu-tion of which is indefinitely sensitive to the initial values of param-eters (though sometimes curious or wonderfully intricate patterns emerge from such functions). If such functions should best reflect aspects of physical reality—a possibility, incidentally, discussed by Duhem—then even within a fully deterministic system, prediction could present conceptual problems even for God. If, considering quantum mechanics, initial parameter values should be indeter-ministic, God's problem would, if possible, be exacerbated.

Though I think this is a scenario that should give serious pause to supporters of even the weakest reductionist theses, I shall not attempt to pursue it here. For it must be noted that this possibility not only lacks obvious relevance to the question of instantaneous supervenience, but also does not throw doubt on causal complete-ness. Indeed, the functions studied in chaos theory are typically deterministic. It does not, therefore, reveal any way of circumvent-ing the preceding arguments about supervenience, or, therefore, of affirming the genuine causal efficacy of the macroscopic. It is, in effect, merely an extreme way of proposing *practical* obstacles to reductionism. It is, indeed, part of my thesis that there are likely to be parts of nature that are not susceptible to systematic analysis in the canonical style of science, and the existence of natural chaotic systems would concretely illustrate one form of this possibility. But for the reasons just indicated I shall not rest my argument on this possibility, but will focus instead on causal completeness.

Microphysical causal completeness, to repeat, is the idea that at least in a hypothetical divine mind there is a complete causal truth to be told about the influences acting on microphysical objects. Since this is not strictly an empirically grounded belief—nobody, for example, has tried to investigate the forces acting on an electron in my, or anybody else's, fingernail—it is perhaps most plausibly diagnosed as deriving from ideas we have about macroscopic causality. The most plausible paradigms here are the macrophysical sciences such as thermodynamics or mechanics; or perhaps the structural accounts of complex functional objects, as in physiology.

An obvious feature of such domains is that they seem amenable to analysis in terms of very general laws (Newton's laws, or Maxwell's equations, for example). One point about such laws is that their generality and abstractness by no means imply that they ever provide causally complete accounts of concrete situations. As Nancy Cartwright has emphasized, by virtue of their simple form and exclusion of many factors known to be potentially relevant, without an open-ended *ceteris absentibus* clause, such laws are typically, if not always, strictly false.[9] The move to causal completeness is rather attempted in their application to the complex task of constructing gadgets or experimental setups. And we all know that this attempt is arduous, and never fully successful. (Our cars sometimes break down, for instance.) But more important than the correct analysis of these technical-scientific projects is the question whether they provide an appropriate model for very different projects of inquiry. My thesis will be that, on the contrary, they can be highly misleading.

Parallel to the success of applied science in guiding the construction of gadgets is the attempt to explain the structural basis of the properties of complex, organized objects, notably in physiological accounts of the properties of biological organisms. Success in such investigations may also be of central importance in inspiring confidence in the unlimited potential of science, and thereby in the assumption of causal completeness. An important feature of such investigations is that the *explanandum* provides us a powerful criterion for distinguishing those events in which we are interested. For example, we know that organisms are very good at respiration, and we seek out those structural elements that make this possible. In contrast, for many classes of phenomena, those typical of evolutionary biology, economics, sociology, or meteorology, for example, no such built-in teleology is available. Part of the investigation is precisely to decide what, if anything, such systems do. Our goal may remain that of seeing how interactions at one level (people performing economic exchanges, organisms consuming one another, etc.) produce global results. In the absence of a clearly defined relevant set of effects, the assumption of causal completeness is considerably less compelling.

We need a rather different picture of how we should study interactions of this latter kind, and here again I think physics provides us with an unhelpful paradigm. The most famous paradigm of causal interaction is provided by mechanics, the collision of billiard balls. And statistical mechanics gives us a model for deriving macroscopic properties from the aggregation of similar interactions. I think certain features of this paradigm can be seen to dominate scientific investigations of the upshots of multiple interactions, and I believe that this influence is often pernicious.

There is no doubt that when dealing with enormously complex phenomena (societies, ecological communities, etc.) we will get nowhere without radical strategies for simplification. Mechanics works by giving very simple characterizations of the interacting entities (mass, velocity, position, etc.) and providing general laws for the outcomes of interactions of entities described in terms of such parameters. Just such a procedure well describes typical methodology in much of evolutionary biology and economics (not to mention the rest of the social sciences insofar as they are threatened with cannibalization by these disciplines). In major parts of economics, for example, agents are characterized by income, indifference curves, and a crude sort of instrumental rationality, and a general account of the nature of exchanges between such agents is offered. Aggregated models of these interactions offer accounts of larger-scale economic processes. In evolutionary theory we have selection coefficients, reproductive rates, and so forth. Characterizing interacting populations as homogeneous except in respect of the values of a few key parameters is certainly a powerful form of simplification, and one that sometimes works well. A metaphysical danger it tends to carry with it is that the formal relations between these abstract entities, as displayed in the formal models of economics or population biology, tend to look like universal laws, failing to be true only because of the irritating intervention of further causal influences too numerous to be conveniently included in the model. Thus such models may suggest that we have the beginnings of a causally complete account, denied us in full detail only because of practical obstacles.

My central point, though it is an admittedly controversial one that cannot be adequately defended here, is that nothing in the practice of this kind of scientific methodology either presupposes or implies that whatever regularities are found to correspond to these abstract models are the consequences of determinate propensities characterizing the particular events that constitute such regularities. It is true that the belief in such propensities would encourage optimism in the search for regularities. But first, even a guarantee that there were such probabilities would not guarantee that regularities were empirically accessible; the number of factors relevant to the strength of the propensity might prevent any intelligible regularities from emerging. Second, such optimism may be misplaced. The empirical success of abstract modeling in science has been, at best, modest. The assumption that it is an appropriate approach to any domain for which it might be feasible would be grossly premature. And third, even where empirical regularities of the right sort can be found, this in no way requires that they be grounded in underlying single-case propensities.

This last point can usefully be illustrated by looking briefly at a topic about which there has been a great deal of investigation of statistical regularities, though without much effort to construct elaborate theoretical models, the game of baseball. The performances of baseball players are subject to analysis in terms of a battery of statistical measures, the most familiar being the batting average and the pitcher's earned-run average. There are serious and well-known limitations to such statistics as measures of the skills of players. Batting averages, for example, even ignoring their obvious failure to measure various batting skills (power, knowledge of the strike zone), are sensitive to the degree of threat presented by following batters, the frequency with which preceding batters are on first base (taking the first baseman out of optimal fielding position), and various features of the home ballpark in which much of the average is compiled. Equally clearly, they convey a good deal of useful, if fallible, information. Over a number of years earned-run average will very reliably distinguish an outstanding pitcher from a marginal pitcher. The same handful of batters average over .300, or drive in 90 runs, with considerable consistency.

But whereas with sufficient time such statistics can give a good idea of the capacities of baseball players, this possibility does not depend in any way on the assumption that the particular events codified by such statistics are subject to any fully determinate and completely specifiable causal influence. Evidence in favor of causal completeness here is precluded by the fact that the number of potentially relevant factors is so large as to reduce each case to causal uniqueness. This is somewhat amusingly illustrated by the frequent production by baseball commentators of facts such as that a certain hitter is batting .750 against a certain pitcher during day games, that is, 3 for 4. There are many statistical patterns, from the larger-scale patterns concerning batters in general or batters of a certain kind to those concerning particular batters in particular situations. But as we move from the smaller-scale patterns to specific events, there is no finest-grained general pattern to be found. We do not aspire to the complete causal story; we move from general knowledge to the specificity of historical narrative. This uniqueness *might* just present an epistemological problem; but I can see no reason why it should not equally well reflect the metaphysical fact that the regularities in question emerge only over time.

There remains the last-ditch defense of causal completeness, the appeal to a God's-eye story in terms of the instantaneous physiological state of the pitcher and the hitter, the air density and movements between the two, and so on. On this I have just two comments. First, such a story plays no part in our everyday understanding of statistics and is in no way presupposed by such understanding. And second, the appeal to such a story depends on the microphysical reductionism that, I have claimed, we have excellent independent reasons for rejecting.

I suggested earlier that our belief in causal completeness could be grounded only in our understanding or experience of macrophysical phenomena. But I have just been arguing that at least for a wide range of interesting events, this conception has no useful role to play. I suggest, in fact, that we have no good reason to believe in causal completeness as characterizing either the microphysical or the macrophysical. One conclusion from this is that the consequences of supervenience I discussed do not follow. We are relieved of the threatened tyranny of the microphysical. If there is no complete and all-encompassing causal nexus determining the movements of microphysical particles, it is even possible that my intention to hit a key really does cause the movements of the electrons in my fingernail.

The most general conclusion I want to draw from the preceding discussion is the following. Rejecting all forms of reductionism, and rejecting the assumption of a complete causal nexus, leaves it entirely open how much order there may prove to be in the world. There may be many kinds of phenomena in which no interesting patterns can be found; and even when there are patterns, it remains an open question how pervasive they may be. This leads naturally to the question how we evaluate various scientific projects, and thereby leads me to the final section of this essay.

The Unity of Scientism

In this final section I shall say something about why it matters whether scientific disunity is inescapable. The central answer to this question is that the political power of science rests in considerable part on the assumption that it is a unified whole. Thus "scientific" has become an honorific applying to anything that satisfies even the thinnest sociological criteria of being a part of science. If science is instead portrayed as a miscellaneous assortment of diverse investigations with only loose relations and interconnections, then particular appeals to the authority of science must stand on their own merits. This is a major step toward increasing the social accountability of scientific claims.

The semiserious title of this section suggests two distinct though related claims. First, I want to claim that the belief in a unified project of science itself helps to support a number of projects that I wish to characterize, in a broadly derogatory way, as scientistic. Second, I think that these scientistic projects involve rather typical though misplaced assumptions about what constitutes a properly scientific approach to a domain of inquiry; thus these characteristics might semifacetiously be said to constitute a genuinely unified project of scientism. Due to limitations of space, this part of the discussion must remain programmatic. I shall outline four main points that I take to be major practical consequences of the metaphysical views just discussed.

1. A belief in the unity of science tends to distribute the epistemic credentials earned by genuinely successful scientific inquiries across the entire range of practices that satisfy merely sociological criteria of scientificity. This is particularly unfortunate in tending to legitimate resistance to powerful contemporary critiques of particular areas of scientific theory. I think especially of recent feminist critiques of substantial parts of evolutionary theory, economics, psychology, and other sciences.[10] It sometimes seems as if, to be taken seriously, such critiques must demonstrate the ideological biases in, say, quantum mechanics. Scientific disunity entails that adequate defenses of particular scientific theories must be local and specific.

2. The thesis of disorder gives no reason to suppose that there is one correct way of categorizing and describing a particular domain. There may rather be a number of possible descriptions, any of which could reveal limited degrees of order and intelligibility. The necessity of providing a criterion for choosing between such possible descriptions suggests a deep sense in which values can become embedded in scientific projects. For example, the expression "positive [i.e., nonnormative] economics" might well be considered close to an oxymoron. Income, growth, and the like are not objectively inescapable characteristics of economic systems, but represent particular choices of how to attempt to describe such systems and, implicitly, to evaluate their success. If we were to change our views as to the goals of economic systems, a description in such terms could be seen as largely irrelevant.

3. I have not discussed in this essay versions of the unity of science that are purely methodological. On the whole, apart from rather vague demands for empirical accountability, such claims have been rendered extremely problematic by recent work in the history of science. However, such theses survive in often inchoate and unarticulated forms, perhaps deriving some of their credibility from a process parallel to, or even parasitic upon, that described in (1) above. One that seems at least implicitly to be widely accepted is the idea that scientific credibility is largely contingent on the extent to which claims are expressed in a quantitative, mathematical form. This, for example, underlies the rather bizarre impression

that economics is the most "scientific" of the social sciences, and also the attempts of game theorists, optimality theorists, rational-choice theorists, and suchlike to colonize the remainder of the social sciences. I say "bizarre" because much of mathematical economics, as also various other curious mathematical practices such as formal population genetics, has done little even to meet what I referred to as the "vague demand for empirical accountability." One may well suspect that these uses of mathematics have more to do with providing barriers to entry to lucrative professions than with illuminating the natural world. The fetishistic reverence for formal methods, finally, is not merely a harmless academic foible. One might argue that the growing throng of homeless people on the streets of the United States are partly indebted for their plight to the mathematical diversions of influential economists.

4. Finally, and going one step beyond the point made in (2) above, a metaphysics of disorder implies that there is no presumption that there is *any* analysis of a particular domain in the canonical style of science. While it may always be possible to provide some illumination of the small-scale events that provide the substance of the interesting processes in a domain of scientific inquiry, there can be no *a priori* answer to the question at what point there will be convergence onto pure historical contingency. One important area to which this issue is highly relevant is the case of evolutionary theory. While there is certainly some demonstrated scope for understanding particular microevolutionary processes, generalizations about the overall patterns of macroevolution might be few and far between. Similar, and perhaps more urgent, questions arise concerning the point at which social science ends and human history begins.

As I have indicated, these concluding remarks are highly programmatic. Each of my points requires much more defense and elaboration. I hope, however, that they are sufficient to show the potential for substantial practical consequences, and for the motivation of various important critical projects in the philosophy of science, provided by what might at first sight seem a rather abstract set of metaphysical issues.

PETER GALISON

Computer Simulations and the Trading Zone

At the root of most accounts of the development of science is the covert premise that science is about ontology: What objects are there? How do they interact? And how do we discover them? It is a premise that underlies three distinct epochs of inquiry into the nature of science. Among the positivists, the later Carnap explicitly advocated a view that the objects of physics would be irretrievably bound up with what he called a "framework," a full set of linguistic relations such as is found in Newtonian or Einsteinian mechanics. That these frameworks held little in common did not trouble Carnap; prediction mattered more than progress. Kuhn molded this notion and gave it a more historical focus. Unlike the positivists, Kuhn and other commentators of the 1960's wanted to track the process by which a community abandoned one framework and adopted another. To more recent scholars, the Kuhnian categorization of group affiliation and disaffiliation was by and large correct, but its underlying dynamics were deemed deficient because they paid insufficient attention to the sociological forces that bound groups to their paradigms. All three generations embody the root assumption that the analysis of science studies (or ought to study) a science classified and divided according to the objects of its inquiry. All three assume that as these objects change, particular branches of science split into myriad disconnected parts. It is a view of scientific disunity that I will refer to as "framework relativism."

In this essay, as before, I will oppose this view, but not by invoking the old positivist pipe dreams: no universal protocol languages, no physicalism, no Comtian hierarchy of knowledge, and no radical reductionism. Instead, I will focus on what appears at first to be a chaotic assemblage of disciplines and activities: thermonuclear weapons, enhanced A-bombs, poison gas, weather prediction, pion–nucleon interactions, number theory, probability theory, industrial chemistry, and quantum mechanics. No entities bind them together; they fall into no clear framework or paradigm; they have no single history that can be narrated smoothly across time. Yet the practice of these activities was sufficiently congruent in the years just after World War II for Enrico Fermi, John von Neumann, Stanislaw Ulam, and others to move back and forth across widely divergent domains. What they shared was not common laws, and most certainly not a common ontology. They held a new cluster of skills in common, a new mode of producing scientific knowledge that was rich enough to coordinate highly diverse subject matter.

Their common activity centered around the computer. More precisely, nuclear-weapons theorists transformed the nascent "calculating machine," and in the process created alternative realities to which both theory and experiment bore uneasy ties. Grounded in statistics, game theory, sampling, and computer coding, these simulations constituted what I have been calling a "trading zone," an arena in which radically different activities could be *locally*, but not globally, coordinated.

Simulations

During and shortly after World War II, nuclear-weapons builders created a mode of inquiry to address problems too complex for theory and too remote from laboratory materials for experiment. Christened "Monte Carlo" after the gambling mecca, the method amounted to the use of random numbers (*à la* roulette) to simulate the stochastic processes too complex to calculate in full analytic glory. But physicists and engineers soon elevated the Monte Carlo above the lowly status of a mere numerical calculation scheme; it came to constitute an alternative reality—in some cases a preferred one—on which "experimentation" could be conducted. Proven on what at the time was the most complex problem ever undertaken in the history of science—the design of the first hydrogen bomb—the Monte Carlo ushered physics into a place paradoxically dislocated from the traditional reality that borrowed from both experimental and theoretical domains, bound these borrowings together, and used the resulting bricolage to create a marginalized netherland that was at once nowhere and everywhere on the usual methodological map. In the process of negotiating the relationship of the Monte Carlo to traditional categories of experiment and theory, the simulators both altered and helped define what it meant to be an experimenter or theorist in the decades following World War II.

At Los Alamos during the war, physicists soon recognized that the central problem was to understand the process by which neutrons fission, scatter, and join uranium nuclei deep in the fissile core of a nuclear weapon. Experiment could not probe the critical mass with sufficient detail; theory led rapidly to unsolvable integro-differential equations. With such problems, the artificial reality of the Monte Carlo was the only solution—the sampling method could "recreate" such processes by modeling a sequence of random scatterings on a computer. Simulations advanced the design (more particularly, the refinement) of fission weapons, but remained somewhat auxiliary to the A-bomb theorists. When, at war's end, nuclear-weapons work turned to the thermonuclear bomb, the Monte Carlo became essential. For there was no analog of Fermi's Metallurgical Laboratory nuclear reactor—no controlled instance of fusion that physicists could study in peacetime. Instead, Stanislaw Ulam, Enrico Fermi, John von Neumann, Nicholas Metropolis, and a host of their colleagues built an artificial world in which "experiments" (their term) could take place. This artificial reality existed not on the bench, but in the vacuum-tube computers—the JONIAC, the ENIAC, and, as Metropolis so aptly named it, the MANIAC.

This story proceeds on several intersecting planes. It is a tale in the history of epistemology: a new method for extracting information from physical measurements and equations. It is a narrative of the history of metaphysics: a scheme of representation that presupposed a nature composed of discrete entities interacting through irreducibly stochastic processes. It is a workplace history: traditional professional categories of experimenter and theorist challenged by an increasingly large and vocal cadre of electrical engineers, and later by computer programmers. Overall, it is an account of fundamental physics inextricably tied to the development of the Superbomb, the weapon with no limit to its potential destructive power, and a description of the calculating machine from computer-as-tool to computer-as-nature.

Bombs and Models

The computer began as an instrument, a tool like other tools around the laboratory. Von Neumann could refer to it in precisely the way that he would discuss a battery or a neutron generator. To Julian Huxley in 1946, for example, von Neumann wrote: "We want to use this machine in the same sense . . . in which a cyclotron is used in a physics laboratory."[1] As has been rehearsed endlessly in the histories of the computer, there were two fairly distinct traditions of computer building. One sought to create machines that would function in an analog fashion, recreating physical relations in media or at scales different from their natural occurrence. Among these could be counted harmonic analyzers, gunnery computers, and network analyzers, as well as analog models including model ship basins or wind tunnels.[2] Such models have long histories; one thinks, for example, of nineteenth-century British electricians, as they put together pulleys, springs, and rotors to recreate the relations embodied in electromagnetism. But it is the second tradition, devoted to digital information processing, that concerns us here. Indeed, my concern is considerably narrower still: the attempt to use the emerging digital devices of the late war and postwar periods to simulate nature in its complexity.

With the Japanese surrender, the laboratory at Los Alamos began to disperse. But before they left, scientists convened to collect and record their knowledge, should it later prove necessary to return to a major effort in nuclear weapons. One such meeting was called for mid-April 1946, to discuss the still-hypothetical fusion weapon, a device that the staid, classified report regarded with some awe—even after the devastation of Hiroshima and Nagasaki: "Thermonuclear explosions can be foreseen which are not to be compared with the effects of the fission bomb, so much as to natural events like the eruption of Krakatoa. . . . Values like 10^{25} ergs for the San Francisco earthquake may be easily obtained."[3] No longer content to mimic Krakatoa *in vitro* (as had the Victorian natural philosophers discussed by Arnold Davidson in this volume), the physicists of Los Alamos now contemplated its imitation *in vivo*.

If imagining the destructive power of the H-bomb was a matter of no great difficulty, designing it was. For though the early bomb builders had thought that heavy hydrogen could be set into self-propagating fusion with relative ease, work during the war had indicated that the problem was much more complicated. Elsewhere, the Los Alamos physicists commented that[4]

the nuclear properties of the key materials are still fundamental in that the energy is supplied by nuclear reactions; but control and understanding of the phenomenon involves purely atomic consideration to a much greater extent than was the case in the fission bomb. What the reaction depends on is the complex behavior of matter at extremely high temperatures. For prediction, then, the primary requisite is a deep insight into the general properties of matter and radiation derived from the whole theoretical structure of modern physics.

When the authors said that they would need the "whole theoretical structure of modern physics," they were not exaggerating. Not only were the nuclear physics of hydrides, the diffusion of hard and soft radiation, and the hydrodynamics of explosion difficult in themselves; they had to be analyzed simultaneously and in shock at temperatures approaching that of a stellar core. How was energy lost, what was the spatial distribution of temperature, how did deuterium-deuterium and deuterium-tritium reactions proceed? How did the resultant helium nuclei deposit their energy? These and other problems could not be solved by analytical means, nor did they lend themselves to "similarity" treatments by analog devices. Experiments appeared impossible—a hundred million degrees Kelvin put the laboratory out of the picture; there was no thermonuclear equivalent to Fermi's reactor, no slow approach to criticality obtained by assembling bricks of active material. Where theory and experiment failed, some kind of numerical modeling was necessary, and here nothing could replace the prototype computer just coming into operation in late 1945: the ENIAC (Electronic Numerical Integrator And Calculator).[5]

Here is an example from von Neumann's 1944 work on numerical methods, designed for the digital processing of hydrodynamical shocks. In question is a compressible gas or liquid in which heat conduction and viscosity are negligible. In the Lagrangian form, the equations of motion are constructed by labeling the elementary volume of substance by a quantity a, which is located by its position x at the time t. We characterize the gas by its caloric equation of state, which gives its internal energy, $U = U(V,S)$, where U is a function of specific volume, $V = \partial x/\partial a$, and specific entropy, S. Normalized mass units made the density equal unity, so the amount of substance located between x and $x + dx$ is just da. It follows that the gas density is the inverse of its specific volume, and pressure, p, and temperature, T, are given by the usual thermodynamical equations,

$$p = -\frac{\partial U}{\partial V}, \quad T = \frac{\partial U}{\partial S}. \tag{1}$$

The equations of motion are designed to provide x as a function of the piece of substance labeled by a, and time, t. Conservation of momentum ($dp/dt = 0$) has two parts: a time derivative of $m\,\partial x/\partial t$ and a term corresponding to the change of pressure as one moves along in space:

$$\frac{\partial^2 x}{\partial t^2} = -V \cdot \left(\frac{\partial p}{\partial x}\right)_{t\ =\ constant} = 0. \tag{2}$$

Substituting $V = \partial x/\partial a$ leads to

$$\frac{\partial^2 x}{\partial t^2} = -\frac{\partial}{\partial a}\, p\left(\frac{\partial x}{\partial a},\, S\right). \tag{3}$$

The conservation of total energy (internal plus external) is just

$$dE = \frac{\partial U}{\partial V}\, dV + \frac{\partial U}{\partial S}\, dS + \left(\frac{\partial^2 x}{\partial t^2}\right) dx = 0, \tag{4}$$

recalling that mass units are normalized to unity. When combined with the thermodynamic relations of Equation 1, Equation 4 yields

$$dE = -p\,dV + T\,dS + \left(\frac{\partial^2 x}{\partial t^2}\right) dx = 0. \tag{5}$$

From Equation 3, we can substitute $-\partial p/\partial a$ for $\partial^2 x/\partial t^2$. With the help of an integration by parts it follows that the first and third terms on the right-hand side of Equation 5 cancel, yielding $T\,dS = 0$, or

$$\frac{\partial S}{\partial t} = 0. \tag{6}$$

This is a crucial part of von Neumann's argument: because of the result that entropy is conserved not as a condition but as a consequence of energy conservation, the solution to Equation 3 can be made assuming p is a function of V alone. Analytic and numerical techniques work on this well-understood hyperbolic differential equation perfectly well. If, however, a shock is introduced to the system, the entropy changes as our bit of matter crosses the shock. Since changes of entropy (and therefore the coefficients in the equation of motion [3]) depend on the trajectory of the shock, the problem is rendered vastly more complicated. Instead of a hyperbolic differential equation, one has a hyperbolic differential equation with variable coefficients, which in general cannot be solved either analytically or numerically.

Von Neumann's central idea was to recapture the simplicity of Equation 3, even for the shock case (where S is not constant), by assuming that the internal energy could be divided into two noninteracting parts: $U(V,S) = U^\star(V) + U^{\star\star}(S)$. Then only $U^\star(V)$ contributes to p in the thermodynamical equation (1), and this quantity, which we can label p_0, is as free of entropy dependence as it was in the nonshock case. In plain English, von Neumann's assumption that the energy $U(V,S)$ divides up this way amounts to the postulation of an "interactionless" substance that can be precisely modeled by beads of definite mass on a chain of springs. Shocks can still propagate, but entropy does not contribute to the pressure.

Letting a run over integers from negative to positive infinity, and

$$x = x(a,t) \equiv x_a(t), \tag{7}$$

can reduce the differential equation (3) to an approximate system of difference equations, well suited for machine coding:[6]

$$\frac{d^2 x_a}{dt^2} = p_0(x_a - x_{a-1}) - p_0(x_{a+1} - x_a). \tag{8}$$

Once on the machine, the simulation could produce a graphic representation of the shock wave as it propagated.

Here is what has happened. For purely computational reasons (that is, in order to process the shock problem on an electro-mechanical calculator) von Neumann redefined the goals of his work to include the elimination of entropy dependence. Until then, the only way to do this was to cancel the shocks; his new approach was to alter the model, introducing the beads-and-springs. Thus the beads-and-springs represent a model of the real gas that has been constructed largely to conform to the new calculational apparatus. It was this model, calculable only by machine, that would in effect simulate a physical system even when analytic techniques led nowhere.

Now von Neumann had to defend the validity of this form of machine representation, and he did so with a long string of plausibility claims. First, he turned the tables on the visual "representational" argument for differential equations. In particular, he pointed out that it is the hydrodynamical equation (3) that distorts the world by using a continuum to describe what in "reality" (von Neumann's term) is a discrete set of molecules. By contrast the "beads-and-springs" "corresponds" to the "quasi-molecular description of this substance." Of course, as von Neumann acknowledged, no device is about to process 6×10^{23} of these beads, but then again even the hydrodynamical equation never explicitly demands such vast numbers.

Somewhat wistfully, von Neumann concluded: "The actual $N \sim 6 \cdot 10^{23}$ is certainly great, but much smaller numbers N may already be sufficiently great. Thus there is a chance that $N \sim 10^2$ will suffice." Other arguments for why this representation could stand in for the "true" state of affairs were more involved, and contained a defense of the simplified intermolecular forces, the process of energy degradation between forms of energy, and so on.

On these grounds, von Neumann prophesied success for more complicated equations of state and more spatial dimensions—especially for the case of spherical symmetry. Not surprisingly, inward-bound shock waves in spherical space were precisely what one would begin with in studying the blasting of plutonium toward its supercritical assembly at the heart of an atomic bomb. Metropolis and Stan Frankel presented their results to the April 1946 Superbomb meeting. Present were all the members of Teller's wartime group, along with many others, including the theoretical physicist Robert Serber, who had greatly contributed to the A-bomb; the wartime head of the Los Alamos theory group, Hans Bethe; and, to the Americans' later chagrin, the Soviet spy Klaus Fuchs. Ulam was at this meeting and, five days before, had completed a speculative Los Alamos paper with J. Tuck on how thermonuclear reactions might be initiated in deuterium by the intrusion of jets from a fission device. Now, at the meeting itself, Ulam heard Frankel and Metropolis's results, fresh from the ENIAC. Having processed over a million IBM punchcards, they reported that the program had run successfully, leaving them provisionally optimistic (if one should use that term) that the Super would detonate as designed.[7] But even with this massive cybernetic assistance, it was apparent that the Super problem far exceeded the ENIAC's capacity to model, and Ulam began to search for a more efficient way to exploit the capabilities of the new machine.

The name "Monte Carlo" was coined by Metropolis and was first printed in the formal contribution by Metropolis and Ulam in the September 1949 volume of the *American Statistical Association Journal*. Clearly following von Neumann's lead in his hydrodynamic computations, Ulam and Metropolis began by pointing to the uncharted region of mechanics between the *terra firma* of the classical mechanician, in which only very few bodies could be treated, and the newer world of the statistical mechanician, in which N was 10^{23}. Something of this mesoscopic *terra incognita* existed as well in combinatoric analysis, where the numbers were too large to calculate and not yet large enough to call in the law of large numbers. Ulam and Metropolis staked their claim on the new turf, announcing the key to their method:[8]

To calculate the probability of a successful outcome of a game of solitaire (we understand here only such games where skill plays no role) is a completely intractable task. On the other hand, the laws of large numbers and the asymptotic theorems of the theory of probabilities will not throw much light even on qualitative questions concerning such probabilities. Obviously the practical procedure is to produce a large number of examples of any given game and then to examine the relative proportion of successes.

Sampling could apply even when no stochastic element was present, as in the evaluation of a finite volume within a many- (say twenty-)dimensional space. The volume could be defined, for example, by specifying a series of inequalities along each of the axes. Dividing the unit cube into ten parts along each of the twenty axes would produce 10^{20} cubes. Systematically checking each box to see if there was a point in it would be, to put it mildly, impractical. Instead, the authors suggest, one could simply sample the cubes, say 10^4 times, and find the ratio of points falling inside the required volume to the total number of points (10^4). More physically, the same sort of difficulty arises in understanding cosmic-ray showers, in which a high-energy proton collides with a nucleus and produces a shower of other particles; these in turn give rise to more, and the chain continues in an extraordinarily complex fashion. Although each step of the process is played out according to previously given probabilities, and therefore is calculable through a specifiable algorithm, obtaining the full solution is out of reach of analytic methods.

The key process, however, as one might suspect from Metropolis and Ulam's return address, was nuclear fission. There the problem was to calculate the diffusion of neutrons through an active substance such as plutonium. In such cases the neutron can scatter from the nucleus (inelastically or elastically), be absorbed by the nucleus, or cause the nucleus to fission (with the emission of n neutrons). Each process has predetermined probabilities, and a sampling of some finite number of neutrons at given energies. Unmentioned, but undoubtedly coequal with the fission, was fusion—on which Ulam was then hard at work.

For the Monte Carlo method to work, Ulam and von Neumann would need a large collection of random numbers, preferably generated by the computer itself, to pick samples of ("perform experiments on") the infinite world of scatterings, fissions, and fusions. Borrowing thousands of "true" random digits (random numbers generated from "truly random" physical processes such as radioactive decay) from books was thoroughly impractical, so the two physicists set about fabricating "pseudorandom" numbers. In one of their earliest postwar formulations, von Neumann's scheme was to take some eight- or ten-digit number to start the process, and a sequence of further numbers would emerge by the iterative application of a function (typically a polynomial) $f(x)$. Not surprisingly, this procedure left people rather puzzled at how the

Monte Carlo players could expect a definite function, applied iteratively, to produce a truly random number. The short answer is that he and Ulam didn't expect $f(x)$ to do anything of the kind; unable to generate random numbers in the classical sense, they expanded the meaning of the term.

Von Neumann's attitude at the time emerges strikingly from a January 1948 exchange with Alston Householder, a physicist at the Oak Ridge Laboratory, who in puzzlement asked, "Is it the case that one starts with a single number, randomly selected, and generates therefrom an infinite sequence of random numbers?"[9] Work at Oak Ridge on the Monte Carlo would stop, Householder added, pending the master's reply. Far from being infinite random generators, von Neumann responded, a finite machine sequentially applying $f(x)$ would, in the fullness of time, repeat a finite cycle of numbers over and over again *ad infinitum*. Suppose, for example, that the machine had a ten-digit storage capacity. Then, even if every ten-digit number were produced before repeating, after 10^{10} numbers $f(x)$ would necessarily produce a number already generated, causing the entire cycle to repeat precisely. "Consequently," von Neumann acknowledged, "no function $f(x)$ can under these conditions produce a sequence which has the essential attributes of a random sequence, if it is continued long enough." His and Ulam's hope was more modest. They wanted their computer to fabricate somewhere between a thousand and ten thousand numbers that "up to this point look reasonably 'random.'" Such sequences need be only "random for practical purposes."[10]

By "random for practical purposes," von Neumann meant that the numbers would satisfy two conditions. First, the individual digits would be as equidistributed as one would expect for a random sample of the same size. Second, the correlation of the kth neighbors would be independent for k between 1 and 8. Procedurally, the generative process worked this way: the first x would be a randomly chosen eight-digit number. This would then be inserted into, say $f(x) = x^2$, and the middle eight digits extracted for reinsertion into $f(x)$.[11] Perhaps acknowledging the legacy of his training in "pure" mathematics, von Neumann confessed: "Anyone who considers arithmetical methods of producing random digits is, of course, in a state of sin. For, as has been pointed out several times, there is no such thing as a random number—there are only methods to produce random numbers, and a strict arithmetic procedure of course is not such a method."[12]

Von Neumann went on to specify the way the simulation would run. First, a hundred neutrons would proceed through a short time interval, and the energy and momentum they transferred to ambient matter would be calculated. With this "kick" from the neutrons, the matter would be displaced. Assuming that the matter was in the middle position between the displaced position and the original position, one would then recalculate the history of the hundred original neutrons. This iteration would then repeat until a "self-consistent system" of neutron histories and matter displacement was obtained. The computer would then use this endstate as the basis for the next interval of time, Δt. Photons could be treated in the same way, or if the simplification were not plausible because of photon-matter interactions, light could be handled through standard diffusion methods designed for isotropic, black-body radiation.[13]

Specifically, von Neumann introduced a series of velocity-dependent functions that gave the probability of neutron absorption, neutron scattering, or neutron-induced fission in each of the three materials:

Absorption in A, T, S:

$$\sum_{aA} (v), \sum_{aT} (v), \sum_{aS} (v); \tag{9}$$

Scattering in A, T, S:

$$\sum_{sA} (v), \sum_{sT} (v), \sum_{sS} (v); \tag{10}$$

Fission, which only occurs in the active material A, producing 2, 3, or 4 neutrons:

$$\overset{(2)}{\underset{fA}{\sum}} (v), \overset{(3)}{\underset{fA}{\sum}} (v), \overset{(4)}{\underset{fA}{\sum}} (v). \tag{11}$$

With these assumptions, one can specify the precise location of the stochastic element in these calculations. Define $\lambda = 10^{-fd'}$ as the probability that a neutron will travel a distance d' through zone i (in which a fraction x_i is active, y_i is tamper, and z_i is slower-down material) without colliding, and where f is given by

$$f = \left[\sum_{aA} (\nu) + \overset{(2)}{\sum_{sA}} (\nu) + \overset{(3)}{\sum_{fA}} (\nu) + \overset{(4)}{\sum_{fA}} (\nu) + \sum_{fA} (\nu) \right] x_i$$

$$+ \left[\sum_{aT} (\nu) + \sum_{sT} (\nu) \right] y_i + \left[\sum_{aS} (\nu) + \sum_{sS} (\nu) \right] z_i. \tag{12}$$

Somehow, by throwing dice, by consulting a table, or by creating a computer-generated algorithm, λ must be chosen from an equi-distributed set of points on the unit interval. Each choice specifies d' for one neutron making one excursion through the reactor.

Robert D. Richtmyer (who succeeded Hans Bethe as head of the theory division at Los Alamos) responded in ways that might be expected from his position at the weapon laboratory. For nuclear weapons, there was no need for slower-down material of a sort characteristically employed in a reactor. So it is no surprise to find him saying, "Material S could, of course, be omitted for systems [of interest to us]." Similarly, in the bomb, the tamper itself typ-ically would contain fission, especially in the new, enhanced weap-ons then under consideration. None of this is made explicit, but would have been perfectly clear to cleared readers at the time. The simple formal substitutions $\sum_{fT} (\nu) \neq 0$ and $\sum_{a,sS} (\nu) = 0$ set a reactor problem into a bomb problem; the basic Monte Carlo strategy remained the same.[14]

Ulam too began to use the method, though his efforts were even more consistently directed toward the Super than von Neu-mann's. In response to a lost letter from Ulam that probably indi-cated his tentative efforts to execute a Monte Carlo by hand for the Super (on other grounds it is clear this is what he was doing), von Neumann sent his enthusiastic endorsement in December 1947: "I am waiting with great expectations to see the details of the manual Monte-Carlo procedure. What you tell me about it is intriguing, and in any case, this is the first large scale application of the statisti-cal method to a deep-lying problem of continuum-physics, so it must be very instructive."[15]

As Los Alamos geared up for a major new simulation of the Super, this time with the Monte Carlo, von Neumann wrote Ulam in March 1949: "I did, however, work on the 'S[uper],' and about a week ago I finished the discussion of the non–M[onte] C[arlo] steps that are involved."[16] This "discussion" meant setting out the logical structure of the calculation: the equations, storage requirements, and logical steps that would be needed. At the end of the day, he wanted to be able to calculate what he called the "time economy" of the calculation—how it would run on different machines, what ought to be calculated precisely, and what through Monte Carlo. These decisions were not given unambiguously, and he reported a certain amount of "demoralization" because there remained "several, mutually interdependent and yet not mutually determining, choices as to procedure [that were] both possible and important." Hardest of all would be the treatment of photons, because they could traverse the greatest number of zones in an interval Δt. Neutral massive particles (neutrons) and charged particles cut across fewer (usually only one or two) zones per Δt, and so demanded less computer time.[17]

To make the calculation, von Neumann set up a space-time division in which Δt is $1/10$ sh (shakes are 10^{-8} seconds in bomb builder jargon), and there were 100 such intervals. Space (in radius r) was to be divided according to

$$0 = r_0 < r_1 < \ldots < r_{19} < r_{20} \sim 100; \; r_{i+1} - r_i = aq^i \; (i = 1, \ldots, 19), \quad (13)$$

where the simulated Super was described as a series of these 20 concentric zones. If one assumes that the zones grow geometrically in radius, then q must be 1.1, because $r_{20} = 100$, and the radius of the initial zone is given by $\alpha = 1.75$.

Many of the simulation steps did not involve Monte Carlo methods; they built on just the sorts of hydrodynamic calculations in which von Neumann had become expert. These deterministic operations used less than 100 multiplying steps [designated 100 (M), where (M) is a multiplying step with all accompanying operations, and M indicates a multiplying step without this complement of operations] per unit time Δt and per zone Δr. There are 20 zones and 100 time intervals. Estimating a need for about 4 iterations per step, this left $20 \times 100 \times 4 = 8{,}000$ repetitions. Multiplying the number of steps by the multiplications per step gives $8{,}000 \times 100$ (M); in other words, the non–Monte Carlo part of a Super calculation would take 0.8 mega(M). Additional multiplications are needed to check and run the Monte Carlo, but these 0.8 mega(M) give a rough guide to a comparison of "hand" computing with various computers.

If the time for a hand multiplication M is 10 seconds, then a hand (M) = 8M = 80 seconds; mega(M) = 8×10^7 "man" seconds = 2.2×10^4 "man" hours. I should pause here to point out that von Neumann put "man" into scare quotes, presumably because the people doing these millions of calculations were all women. So von Neumann's "man" = woman. Moreover, von Neumann's wife, Klara von Neumann, Herman Goldstine's wife, Adele Goldstine, Foster Evans's wife, Cedra Evans, Teller's wife, Augusta Teller, and John William Mauchly's wife, Kathleen McNulty Mauchly, along with many others, were all programming these early computers. (In fact, the word "computer" itself designated the usually female calculators, and only later did the term shift from the woman to the device.) All this is background to von Neumann's peculiar quotational terminology in which a "man" year = 50 "man" weeks = 50×40 "man" hours = 2×10^3 "man" hours. A single mega(M), therefore, would take 11 "man" years to do by hand, and so the deterministic part of the Super simulation would demand 8.8 years for a single computing person to reckon.

The estimate of 0.8 mega(M) enabled von Neumann to contrast the way a Super simulation would be calculated by the ENIAC, by the SSEC (Selective Sequence Electronic Computer) IBM computer in New York, and by "our future machine" to be built at the Princeton Institute for Advanced Study.

For the ENIAC, the time of multiplication M was a mere 5 msec, so (M) = 40 msec; mega(M) = 4×10^4 sec = 11 hours. Therefore 0.8 mega(M) = 8.8 hours, which, when compounded with typical running efficiency, could be accomplished in one 16-hour day. However, this estimate was academic, as von Neumann knew. Because the ENIAC lacked sufficient memory (a mere 20 words), the problem simply could not be executed. For the SSEC IBM machine in New York, however, there was enough memory, and, to boot, it was faster: multiplication ran about 25 percent more slowly than the ENIAC, but the speed was made up in other ways. And the future machine could run the same operation fast enough to simulate the deterministic part of an identical exploding Super in about 15 minutes.

Von Neumann next described the effect of adding in the Monte Carlo steps, which have so far been ignored. If one defines *a* as the ratio

$$a = (\text{time for MC steps})/(\text{time for non-MC steps}), \quad (14)$$

then the total time would be the non-MC time multiplied by $(1 + a)$. The Monte Carlo steps are such that a single photon would take less than 60 (M), while any other particle needs less than 40 (M). There are 300 photons and 700 particles of other species. This yields $(300 \times 60) + (700 \times 40)$ (M) = 46,000 (M). While this appears to be much greater than the 100 (M) required for the non–Monte Carlo step, there are 20 zones, each of which must be examined each time for the non–Monte Carlo steps (the hydrodynamics), whereas the Monte Carlo step (a photon interaction, for example) would take place in only one zone. So $a = 23$, and the total amount of time needed for the Super would be 24 times the non–Monte Carlo totals. This meant that were the calculation assigned to a single computer (woman), a given thermonuclear-bomb simulation would take 211.2 years. In a sense this number is a crucial factor (among others, to be sure) in the decision to devote vast resources to replace the human computer with an automatic one.

Working with these highly schematized Monte Carlo simulations, over the course of 1949 Ulam and C. J. Everett (Ulam's collaborator) did manage to put a crude version of the Super problem on the computer, but they had to invoke assumptions and simplifications at every stage to accommodate the limited capacity of the ENIAC. With the program finally installed, Ulam wrote von Neumann in March 1950: "Everett managed to formalize everything so completely that it can be worked on by a computer. . . . It still has to be based on guesses and I begin to feel like the man I know in Poland who posed as a chess champion to earn money—gave nine 'simultaneous' exhibitions in a small town playing 20 opponents—was losing all 20 games and had to escape through the window!"[18]

In the midst of one such run, Ulam penned a message to von Neumann in late January 1950. Initial results apparently were not promising for detonation, and Ulam judged that earlier fears that hydrodynamic instabilities would quench the reaction were misplaced: "Hydrodynamics, so far at least, far from being a danger, is the only hope that the thing will go!" Interwoven with his cautious technical optimism came a similar political forecast: "I think that in matters of 'politics' a victory of unbelievable proportions is preparing." He was right: within four days, President Truman delivered his decision to proceed with an intensified effort on the H-bomb. Whether Ulam had inside information to this effect, I do not know. It certainly seems so.[19] In any case, von Neumann was delighted with the outcome: "I need not tell you how I feel about the 'victory.' There are, however, plenty of problems left, and not trivial ones either."[20] Mixing the political and technical inextricably, "hope," "victory," and a "go" all stood on the side of a simulated detonation; "danger" and "problems" lurked in a computational dud.

In public, both Ulam and von Neumann kept low profiles in what had become a raging debate over whether or not the United States should build the hydrogen bomb. Albert Einstein appeared on television to warn that the weapon was bringing the world inexorably closer to total annihilation. Hans Bethe and Victor Weisskopf made public appearances arguing that the level of destruction contemplated with thermonuclear weapons was practically genocidal; in addition, Bethe made no secret of his hope that the bomb would prove scientifically impossible to build.

Ulam dutifully read these moral-political tracts, but from the midst of the immensely complex calculations that had been proceeding more or less unaffected by the outside world, this contretemps appeared laughably ill informed, even irrelevant. To von Neumann, he joked:[21]

Read with constant amusement a whole series of new articles in the press about hydrogen bomb—the statements by Zacharias, Millikan, Urey, Einstein and Edward in the Bulletin of A[tomic] S[cientists] in Chicago each contribute a share of merriment.

I propose to you hereby to write jointly a "definitive" article on this subject. It will be signed by fictitious names of two foreign born scientists, "key men" in various projects, *not* atomic scientists *but* experts on the hydrogen bomb, former scientists on radar and submarine detection work etc. The first paragraph will say how secrets and lack of free exchange thwarts progress in basic (science) and prevents development of new ideas. The next [paragraph] will document it by pointing out how there are surely no secrets in nature & how *any* scientist can figure out these secrets by himself in 5 minutes.

The next [paragraph] how hydrogen bomb is too big & very immoral, the next one that it is too big to be useful, but not really big enough to be decisive.

The next that it is not clear to anyone whether it can be built at all; after that Russia probably has it already but let us all pray that it is mathematically impossible.

After that, like Edward's, a pitiful plea for all scien[tists] to work on [it], spending 7/8 of the time on the projects and 3/4 on [other topics].

It was not that Ulam thought the bomb was necessarily even possible. In one study during 1950, Ulam and Everett showed that the initiation of fusion using deuterium and tritium would demand prohibitive quantities of the fantastically expensive hydrogen isotope tritium. Then, during the summer of 1950, the two authors used the Monte Carlo method to study the behavior of thermonuclear reactions in the mass of deuterium. In the still-classified 1950 report LA-1158 ("Considerations of the Thermonuclear Reactions"), Ulam and Fermi invoked the hydrodynamics of the motion of the material, the interaction of radiation energy with matter, and various reactions between nuclei that were dependent on temperature, density, and geometry. Though the size of the calculations was small (they were still able to use desk computers and slide rules), their conclusion was that the reaction would not propagate in a volume of deuterium. Together, the two pieces of work appeared to sound the death knell for the Super: it would neither light nor burn. Later, massive simulations on electronic computers by von Neumann and by Foster Evans and Cedra Evans confirmed their assessment.[22]

By all accounts, Teller took Ulam's fizzling-bomb news hard. Ulam's pocket diary for 10 May 1950 reads, "Fights with Edward"; by 13 June 1950, with more data in hand Ulam relished his vanquishing of Teller's greatest project: "Victorious end of fights with Edward."[23]

Here the dense, terse entries carry multiple meanings. At the larger scale, Ulam was preoccupied with "victory," where the victory was one of national politics (Truman's endorsement of the H-bomb); in Ulam's diary "victory" designates triumph in Ulam's personal struggle with Teller (Ulam showed that Teller's classical Super would not explode). Given the extraordinary publicity and resources already devoted to the project, it is not too surprising that Ulam's dim reports ended neither the technical nor the moral battles, which continued throughout the latter part of 1950. By January 1951, almost exactly a year after the presidential order, there was little prospect of technical success with the weapon, and tempers were wearing thin. On Thursday, 18 January 1951, Ulam recorded one such encounter between Bethe (who hoped and argued that the bomb was impossible) and Teller, "Amusing fights: Hans-Edward"; or a few days later, "Big fight—fairly amusing." Then, sometime between 18 and 25 January 1951, Ulam realized that a radically different configuration of the hydrogen bomb might be possible. Instead of trying to create enormous temperatures with a fission bomb, the A-bomb could be used to compress the fusionable material; high pressure would enable the reaction to proceed with much less heat.[24]

On 25 January 1951, the tone of his entries abruptly changed: "Discussion with Edward on '2 bombs.'" Apparently, Ulam had brought the idea of shock compression to Teller, and Teller had added his own idea—that the same effect could be obtained more easily by "radiation implosion," the compression of the fusion fuel by an expanding plasma created by X rays from the fission bomb. (There were therefore two bombs in question, both involving compression.)

The fights ended. Ulam recorded on 26 January, "Discussion with Bethe[,] Evanses[,] Carson [Mark] on the set-up for the cylinder propagation. Write up dis[cussion] with Johnny [von Neumann] & write to Hans [Bethe]." Over the next few weeks Ulam finished his part of the paper and sketched the introduction. Its somewhat unwieldy title appeared the next month on 15 February—"Wrote Lenses. (jointly with Teller) 'Heterocatalytic deton[ation]' Radiation lenses and hydro[genous] lenses"—and was formally issued on 9 March 1951.[25]

With this classified paper, debate over the hydrogen bomb "inside the fence" virtually ended among physicists privy to the classified breakthrough. In June 1951, faced with new simulations done on the SEAC (Standards Eastern Automatic Computer) and elsewhere, the Atomic Energy Commission's General Advisory Committee, whose startling earlier moral position against the weapon had stunned the secret community, retracted their opposition and endorsed a full-tilt effort to test the bomb. Even Bethe, who had long been the strongest public scientific voice against it, ceased to object. On 30 October 1952, the United States detonated a "Teller-Ulam" bomb on the South Pacific island of Eniwetok. The force of the 10.2 megaton explosion wiped the island from the face of the earth.

Thermonuclear and enhanced-fission weaponry had crystallized a new mode of doing scientific work. During 1948 and 1949 statisticians, mathematicians, electrical engineers, and a restricted few others were called in to join with the physicists in the creation of this new trading zone in which talk of games, pseudorandom numbers, difference equations, and logical coding were common parlance. Now the style of work grew outward, and through a series of conferences passed beyond the domain of the secret and into the wider community of researchers. As it crossed that fence, debate grew about who might count as an experimenter or as a theoretician, and what would count as an experiment or theory.

The *Tertium Quid*

"Computational physics," the physicist Keith V. Roberts wrote, "combines some of the features of both theory and experiment. Like theoretical physics it is *position-free* and *scale-free*, and it can survey phenomena in phase-space just as easily as real space. It is symbolic in the sense that a program, like an algebraic formula, can handle any number of actual calculations, but each individual calculation is more nearly analogous to a single experiment or observation and provides only numerical or graphical results."[26] But if the frankly symbolic character of labor resembled techniques of the blackboard, it was still true that the techniques of error analysis and problem shooting held more in common with techniques of the bench. Roberts continued:[27]

Diagnostic measurements are relatively easy compared to their counterparts in experiments. This enables one to obtain many-particle correlations, for example, which can be checked against theory. On the other hand, there must be a constant search for "computational errors" introduced by finite mesh sizes, finite time steps, etc., and it is preferable to think of a large scale calculation as a numerical experiment, with the program as the apparatus, and to employ all the methodology which has previously been established for real experiments (notebooks, control experiments, error estimates and so on).

In short, the daily practice of error tracking bound the Monte Carlo practitioner to the experimenter.

Other activities joined the simulator with the experimenter as well. In experimental practice it is routine to use the stability of an experimental result as a sign of its robustness: Does it vary as one repeats it, or shift around parameters that ought (on prior grounds) to be irrelevant to the outcome? From the inception of the Monte Carlo, its practitioners were equally aware that their nightmare would be the production of results without constancy. In Ulam's 1949 paper with Metropolis, the authors made this clear: a "procedure is repeated as many times as required for the duration of the real process or else, in problems where we believe a stationary distribution exists, until our 'experimental' distributions do not show significant changes from one step to the next."[28] Problems of locality, replicability, stability, and error tracking are thus some of the reasons (I will come back to other, deeper ones) that led the simulators to identify their work as experimental.

At the same time, however, other practice clusters tied the simulator to the mathematician or theoretical physicist. J. M. Hammersley and D. C. Hanscomb took this link of mathematics to the computer to demand a fundamentally new classification of mathematics, one that cut across the old typology of "pure" and "applied" mathematics:[29]

A relatively recent dichotomy contrasts the theoretical mathematician with the experimental mathematician. These designations are like those commonly used for theoretical and experimental physicists, say; they are independent of whether the objectives are pure or applied, and they do not presuppose that the theoretician sits in a bare room before a blank sheet of paper while the experimentalist fiddles with expensive apparatus in a laboratory. Although certain complicated mathematical experiments demand electronic computers, others call for no more than paper and pencil. The essential difference is that theoreticians deduce conclusions from postulates, whereas experimentalists infer conclusions from observations. It is the difference between deduction and induction.

While this remark may be naive in its dichotomy of induction and deduction, it is important as an indicator of the powerful identification of the simulator with the experimenter. This language of "theoretical experiments" or "mathematical experiments" saturates the literature. In the pages of *Physical Review* from the early to mid-1950's one finds tens of such examples; of the albedo of 1-MeV photons, two authors write: "It occurred to us that a more reliable estimate of this quantity, on which there really existed no information, could be obtained by a 'theoretical experiment' using the Monte Carlo technique."[30]

Caught between a machine life and a symbol life, the computer programmer in physics became both a pariah and an irreplaceable intermediary, establishing a precarious transactional function known to border peoples on every continent. Over the course of the 1960's, the computer utterly transformed particle physics; among the changes it brought was the creation of a category of action (data analysis) that was, in its own way, as all-embracing a career as accelerator building or field theory. The novelty of this situation was not lost on the physics community, as the French nuclear physicist Lew Kowarski made abundantly clear in the summer of 1971: "As scientists get used not only to writing their own programs, but also to sitting on-line to an operating computer, as the new kind of nuclear scientist develops—neither a theoretician, nor a data-taker, but a data-processor specialized in using computers—they become too impatient to sit and wait while their job is being attended to by computer managers and operators."[31] Quoting approvingly from a colleague at New York University, Kowarski presented mathematics as the analog of mining diamonds—finding extraordinary theorems among the dross of uninteresting observations. Computers, on the other hand, sought truth the way one mines coal; with the massive, everyday labor that methodically moves earth from pits to furnaces. "This analogy," Kowarski observed, "illustrates the difference between the spirit of mathematics and that of computer science and helps us to realize that being a computational physicist, or a computational nuclear chemist, or what not, is not at all the same thing as being a mathematical physicist and so on, so that, in fact, a new way of life in nuclear science has been opened."[32]

For Kowarski, the *Lebensform* of the computer console was one destined to be more akin to that of coal miners, oceanographers, selenologists, and archaeologists than it was to physics as it was previously understood: "There will be a lot of attempts to judge such new situations by old value criteria. What is a physicist? What is an experimenter? Is simulation an experiment? Is the man who accumulates print-outs of solved equations a mathematical physicist? And the ultimate worry: are we not going to use computers as a substitute for thinking?"[33] This anxiety over the identity of physics and the physicist was both social and cognitive. The fear that "thinking" might be destroyed must be glossed as a fear that the particular pleasure (and status) of controlling a wide range of cognitive activities previously associated with experimenting would be reduced. Underlying this worry was a very real change of structure in the physics workplace and, more generally, an altered concept of demonstration.

Take replicability. For theoretical work, the recreation of argumentation is generally considered to be unproblematic. Derivations are relatively easy to repeat, if not to believe. Experimental efforts, by contrast, present notorious difficulties as many historians and sociologists of science have so effectively illustrated: air pumps, prisms, lasers all turn out to be more profoundly local in both structure and function.[34] From the start, simulations presented a hybrid problem. On one side, the work was unattached to physical objects and so appeared to be as transportable as Einstein's derivation of the A and B coefficients for quantum emission and absorption. But in practice, this was hardly the case. Roberts, among many others, bemoaned this fact, and considerable effort began to delocalize simulations and computer-analysis programs more generally. Opening the "bottleneck" (as it was commonly called) required three concurrent efforts. First, programs had to be openly published.[35] While desirable, this proved in the 1950's (and ever since, I should add) a rather wistful desire, since most interesting programs were far too large to allow distribution on the printed page. Second, Roberts pressed for what he called "portability": the use of universal languages and the physical distribution of data tapes. While the former encountered difficulties associated with variations between machine types and local programming customs, the latter ran into a host of property-rights difficulties. For example, in the 1980's questions arose as to whether the distribution of data tapes could be considered analogous to the distribution of cell samples in biology. Finally, Roberts insisted that "modularity" ought to be considered a goal for programmers, analogous to the routine elements of practice in theoretical physics such as Laplace's equation, group theory, vector algebra, or the tensor calculus.[36]

Each of these programmatic responses—advocacy of publication, portability, and modularity—was partial; none could truly universalize a set of practices that bore the deep stamp of its localized creation. Lamenting this state of affairs Roberts commented: "There are many good programs that can only be used in one or two major laboratories (notably the Los Alamos hydrodynamics codes), and others which have gone out of use because their originators moved on to other work."[37] In more recent times this phenomenon has become known as "program rot"—people and machines move on, leaving older programs dysfunctional, often irretrievably so.

At the same time that physicists struggled to uproot the world of simulated realities from a particular place, others applauded the deracination already achieved. Kowarski, for example, began to speak about a simultaneous "liberation in space" and a "liberation in time" afforded by the new modality of research: "Perhaps, when links as comprehensive as those used in television become available at long distance, there will be even less reason for the user to spend a lot of his time on the site where his physical events are being produced. This may even abolish the kind of snobbery which decrees today that only those may be considered as physicists who are bodily present at the kill, that is at the place and time when the particle is actually coming out of the accelerator and hitting the detector."[38] Kowarski had in mind events originally encoded from particle collisions at a centralized laboratory. But just as events produced at Brookhaven or Berkeley could be issued as a stack of cards or spool of tape, so too could events created by simulations. In both cases, analysis, still the central stage of an experiment, was removed from any single place.

Liberation in time similarly mixed experiment and the life of the experimenter. Kowarski extolled the fact that the computer expanded the timescale of events—often 10–31 seconds—to the scale of minutes, hours, and days in which we live. Second, the time became repeatable as physicists reprocessed the same set of events in ever-different ways to reveal different patterns of order. Finally, the computer allowed the physicists to break the ties between the accelerator "beamtime" and their own time, time to live the lives of university-based scholars with teaching and departmental and familial duties. It therefore links, in yet one more way, the different worlds of beamtimes and lifetimes (to borrow the evocative title of

Sharon Traweek's anthropological study of the Stanford Linear Accelerator Center).[39]

Artificial Reality

As Monte Carlo simulations developed, it became clear that their practitioners shared a great deal with experimenters—I gave the examples of a shared concern with error tracking, locality, replicability, and stability. But the self-representation of Monte Carlo users as experimenters is so pervasive that I now want to zero in on this notion in two different ways in an effort to uncover the practices underlying this talk of "experiment" done on keyboards rather than lab benches.

The first point about the relation of the Monte Carlo to experimentation is that the simulator spends time processing as well as generating "data"—the scare quotes denoting that the term has now been expanded to include those generated from pseudorandom numbers in Monte Carlo simulations. These practices, as the Monte Carlo folks immediately recognized, held more in common with experimental than with theoretical activity. As much as anyone, Herman Kahn of the Rand Corporation (later famous for his *On Thermonuclear War*) continuously emphasized that the simulator had no business simply reporting a probability, say the probability that a neutron will penetrate the concrete shielding wall of a nuclear reactor. Instead, the only meaningful statement would be a probability p along with a certainty m: "This situation is clearly not unknown to the experimental physicist, as the results of measurement are in this form." If the problem is surprising, Kahn insisted, it is because the need to reduce variance is not a situation theorists typically encounter. Variance reduction arises in the context of the Monte Carlo because "one is not carrying out a mathematical computation in the usual (analytic) sense, but is carrying out a mathematical experiment with the aid of tables of random digits."[40]

With a limited sample of "particles" that the computer can track, it is frequently the case that the interesting phenomena occur so rarely that uncertainty runs riot. If, for example, a thousand particles are sent out from a simulated reactor and only ten penetrate the barrier, then the accuracy of statements about these ten particles will be extremely slight. Kahn and others (following on early work by Ulam and von Neumann themselves) particularly pressed three strategies for reducing this uncertainty known as splitting, statistical estimation, and importance sampling. For example, the idea of the importance sample is to augment examination of a particular region of phenomena of interest, get a reduced variance, and then compensate for this bias in the final estimate of the probability. More realistically, neutrons might be given an extra large probability to scatter toward that part of the outside of a reactor shield where the people are in an effort to reduce the error associated with the estimate of neutrons penetrating the concrete barrier.

I conclude from these variance–reduction techniques, and from the earlier discussion of error tracking, that there are two sides of the simulators' concern with error. The first (error tracking) bears on the ability of the Monte Carlo to get the "correct" expectation values; it is experienced by the simulators as akin to the experimentalists' search for accuracy by quashing systematic errors in the apparatus itself. The second (variance-reduction techniques) is the direct analogue of the experimentalists' search to increase the precision of their examinations. Taken together, these day-to-day commonalities between the practices of the simulator and those of the bench experimentalist tended to press the two groups together in their self-identification. Some simulators went farther, arguing that, because they could control the precise conditions of their runs, they in fact had an edge on experimentalists; in other words, it was the simulator, not the experimentalist, who should be seen as the central figure in balancing theory. "The Monte Carlo methods," Kahn concluded, "are more useful . . . than experiments, since there exists the certainty that the comparison of Monte Carlo and analytic results are based on the same physical data and assumptions."[41] Simulators, not bench workers, could make the green-eyed, six-toed, curly-haired pigs that science now demanded.

All the forms of assimilation of Monte Carlos to experimentation that I have presented so far (stability, error tracking, variance reduction, replicability, and so on) have been fundamentally epistemic. That is, they are all means and practices by which the researchers can argue toward the validity and robustness of their conclusions. Now I want to turn in a different direction, toward what amounted to a *metaphysical case* for the validity of Monte Carlos as a form of natural philosophical inquiry. The argument, as it was presented by a variety of people (including occasionally Ulam himself), was based on a purportedly fundamental affinity between the Monte Carlo and the statistical underpinnings of the world itself. In other words, because both Monte Carlo and nature were stochastic, the method could offer a glimpse of reality previously hidden to the analytically minded mathematician. As Ulam himself once put it, his and von Neumann's hunt for the Monte Carlo had been a quest for a *homomorphic image* of a physical problem—where the particles would be represented by fictitious "particles" in computation.[42]

Gilbert King, a chemist at Arthur D. Little who had been in operations analysis at the Office of Scientific Research and Development during the war, is a good spokesman for this *simulacrum* interpretation of the Monte Carlo. Already in late 1949 at the IBM Seminar on Computation, he insisted that "from the viewpoint of a physicist or a chemist, there really is no differential equation connected with the problem. That is just an abstraction."[43] Two years later he amplified on these comments, arguing that the computer should "not be considered as a glorified slide rule" but as an "organism" that could treat a problem in an entirely new way. To King, it was clear that the directness of the Monte Carlo gave it a role vastly more important than just another approximation method:[44]

Classical mathematics is only a tool for engineers and physicists and is not inherent in the realities with which they attempt to deal. It has been customary to idealize and simplify the mechanisms of the physical world in the form of differential and other types of equations of classical mathematics, because solutions or methods of attack have been discovered during the last few hundred years with means generally available—namely, pencil, paper, and logarithm tables.

Engineering was far too complex for such traditional paper-and-pencil solutions; engineers substituted difference equations for the differential equations, and sought approximate solutions by numerical methods. The classical mathematical physicist looked down on such "crude" methods as a "poor man's solution" adopted in the absence of an aesthetically satisfying abstract and analytic one. King's worldview entirely inverted the mathematical physicists' epistemic hierarchy.

Refusing the mathematical physicists' invitation to the sacred realm of partial differential equations, King argued that such expressions refracted the world through a distorting prism, and insisted that the engineer's tools mapped directly onto something deeper:[45]

There is no fundamental reason to pass through the abstraction of the differential equation. Any model of an engineering or physical process involves certain assumptions and idealizations which are more or less openly implied in setting up the mathematical equation. By making other simplifications, sometimes less stringent, the situation to be studied can be put directly to the computing machines, and a more realistic model is obtained than is permissible in the medium of differential or integral equations.

King's claim is radical. For, contrary to a long tradition supervaluating the differential and integral equations as reflecting a Platonic metaphysics hidden behind appearances, it is King's view that it is the engineer, not the mathematical physicist, who has something to say about reality. This "more realistic model" is so because nature is at root statistical, and the representative schemes such as integrodifferential equations that eschewed the statistical were bound to fail.

Consider the diffusion equation, the bread and butter of industrial chemists like King:

$$\frac{\partial \mu}{\partial t} = D \, \frac{\partial^2 \mu}{\partial x^2}. \tag{15}$$

For the specific, simple case of a capillary tube and dye released at its center point, the solution is known in closed form. According to the solution, the dye molecules will move a specifiable mean distance Δx in a time Δt. This can be simulated by a simple Monte Carlo conducted with a coin: we increase x by Δx if the coin comes up heads, and decrease x by Δx if the coin reads tails. A random generator in a computer can proceed in a similar manner by using even numbers as the basis for a positive increment and an odd number for a decrement. In this way, repeated over many runs, one obtains a distribution. King celebrated this sequential process of random events:[46]

The mathematical solution [that is, the analytic solution] of the diffusion equation is an *approximation* of the distribution. The mathematical solution of the diffusion equation applies to the ideal situation of infinitely small steps. In setting up the diffusion equation, more assumptions were used than were put directly, in an elementary fashion, into the computing machines, and a solution has been reached by an entirely different computing scheme from any that would be used by hand.

The most concise formulation of King's view emerged in an animated exchange between King and the New York University mathematician Eugene Isaacson at one of the earliest meetings on the Monte Carlo method. King had just spoken on the problems of applying Monte Carlo methods to quantum mechanics:[47]

Mr. Isaacson: Isn't it true that when you start out on your analysis of the physical problem and you have a complicated finite difference process, you then look at a continuous differential equation and approximate that by a simpler finite difference process and so cut down some of your work?

Dr. King: I think one can dodge a good deal of differential equations by getting back of the physics of the problem by the stochastic method.

King's view—that the Monte Carlo method corresponded to nature (got "back of the physics of the problem") as no deterministic differential equation ever could—I will call *stochasticism*. It appears in myriad early uses of the Monte Carlo, and clearly contributed to its creation. In 1949, the physicist Robert Wilson took cosmic-ray physics as a perfect instantiation of the method: "The present application has exhibited how easy it is to apply the Monte Carlo method to a stochastic problem and to achieve without excessive labor an accuracy of about ten percent."[48] And elsewhere: "The shower problem is inherently a stochastic one and lends itself naturally to a straightforward treatment by the Monte Carlo method."[49] These two radically different metaphysical pictures of how simulations relate to nature are illustrated in Figure 1. On the Platonic view, "physical reality" was or ought to be captured by partial differential equations. In the particular case of diffusion, the physical reality was the spread of red dye in a thin capillary tube. It was "represented," as it rightfully should have been, by a differential equation. This equation would, in any particular case, have a

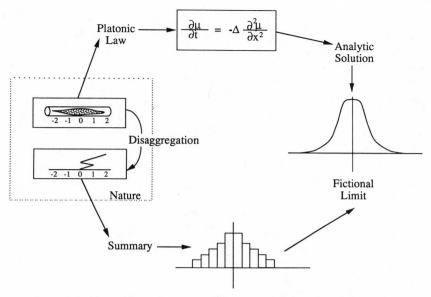

Fig. 1. Ontology of the Monte Carlo. In the Einstein–Dirac view, the capillary-tube diffusion process in nature (left) is modeled by a partial differential equation (top), and the solution for the diffusion equation is obtained analytically (right). By contrast, the King–Kahn view (stochasticism) has it that the capillary-tube diffusion process is *imitated* by the stochastic process of, for example, flipping a coin and moving to the right when a head falls and to the left when a tail falls. These "runs" are then tabulated in the bar graph (bottom), which summarizes the diffusion process without ever passing through the stage of an equation. In the case of diffusion, Einstein clearly saw the differential equation as an approximation; in the long run he believed differential equations would capture the underlying continuity of nature. King and Kahn saw the diffusion process as typical: nature is discontinuous and random—the computer Monte Carlo could imitate this sort of process without the approximation of differential equations.

solution, as illustrated at the right of the figure. On the stochasticist view, the walk of a counter right and left in the simulation, like the diffusion itself, was predicated on what at root was a fundamentally random process. Thus for the stochasticist, the simulation was, in a sense, of a piece with the natural phenomenon. The simulation was not just an alternative representation; it was a phenomenon of the same type as diffusion, but disaggregated into the motion of individual entities. Each individual particle ends at a particular point on the x axis; when we collect these ending points into a histogram, we get the summary indicated at the bottom of the figure. Smoothing the curve, at the right of the figure is a fictional limit, obtained when one (imaginatively) extrapolates to an infinite number of "runs" of the experiment. To the Platonist, the stochasticist has merely developed another approximative method. To the stochasticist, the Platonist has interposed an unnecessary conceptual entity (the equation) between our understanding and nature—stochasticism, he or she claims, offers a direct gaze into the face of nature.

As I have already stressed, the problem of neutron diffusion was at the heart of the creation of the method, and it continued to provide difficult arenas of application, mostly in the sphere of weapons design and effects, and in related civilian applications of reactor design and safety problems. Householder, writing on neutron age calculations in water, graphite, and tissue, contended that "the study of the diffusion of heavy particles through matter provides an ideal setting for the Monte Carlo approach. Here we are not forced to construct an artificial model to fit a given functional equation, but can go *directly* to the physical model and, in fact, need never think about the functional equations unless we choose."[50] This notion of directness is central; it underlines the deep philosophical commitment to the mimetic power of the Monte Carlo method. This same faith in the metaphysical replication of process and representation appears in "health physics," where one finds a prominent Oak Ridge contributor, Nancy Dismuke, writing of neutron propagation through tissue: "The type of problem I am concerned with is a *natural* Monte Carlo problem in the sense that the physical model suffices as the model for proceeding with the calculation. An experiment is carried out (on a computing machine) which at every stage resembles closely the true physical situation. Whenever a random selection must be made in our experiment, the corresponding physical situation seems to be a matter of random choice."[51] A. W. Marshall spoke of the Monte Carlo as most productive when it was applied to "nature's model" of a process such as diffusion, avoiding the differential equations.[52]

Unlike the authors just cited, some advocates for this "naturalness," "directness," or mimesis of "nature's model" were people unwilling to endorse the program on abstract grounds of correspondence; instead, they argued on pure pragmatic grounds that it was the set of stochastic processes that were most effectively modeled by Monte Carlos. Listen to the intervention of a programmer (a Dr. Howlett) at the British atomic weapons research center at Harwell during a conference in the 1950's:[53]

I am responsible for the computing service at Harwell and represent the user of the Monte Carlo technique who views it, professionally at any rate, as just another numerical weapon with which to attack the problems he is asked to solve. . . . I have to answer the questions, what are the problems to which Monte Carlo is best suited, how good a method is it—how much better than conventional numerical analysis—and how can it be improved?

In a sense this is the hardest-line pragmatic view possible: simulated reality would be accorded just that degree of credence that it earned in competition with other numerical processing packages. A slightly softer version of the pragmatic approach emerged from those who wanted to marginalize the Monte Carlo to a heuristic function, a kind of suggestive scaffolding to be kicked aside at the earliest possible moment. This emerged clearly at the December 1949 IBM Seminar in the following two interventions, the first by the Princeton mathematician John Tukey:[54]

One point of view for the use of Monte Carlo in the problem is to quit using Monte Carlo after a while. That, I think, was the conclusion that people came to. . . . After you play Monte Carlo a while, you find out what really goes on in the problem, and then you don't play Monte Carlo on that problem any more.

A bit later, the Cornell mathematician Mark Kac commented on Robert Wilson's use of the Monte Carlo to examine cosmic-ray showers in a similar way:

They [Wilson and his collaborators] found the Monte Carlo Method most valuable because it showed them what goes on. I mean the accuracy was relatively unimportant. The five per cent or the seven per cent accuracy they obtained could be considered low; but all of a sudden they got a certain analytic picture from which various guesses could be formulated, some of them of a purely analytical nature, which later on turned out to verify very well. . . . One of the purposes of Monte Carlo is to get some idea of what is going on, and then use bigger and better things.

In recent years this faith in the directness of the method has continued; even as late as 1987, a textbook on experimental particle physics could stress the special relation between stochastic objects and their Monte Carlo representations:[55]

Historically, the first large-scale calculations to make use of the Monte Carlo method were studies of neutron scattering and absorption, random processes for which it is quite natural to employ random numbers. Such calculations, a subset of Monte Carlo calculations, are known as direct simulation, since the "hypothetical population" . . . corresponds directly to the real population being studied.

This more contemporary view wanted it both ways—metaphysical justification for simulations of stochastic processes, and a pragmatic validation of deterministic ones. But not everyone was so sanguine. Some scientists responded with barely concealed fury toward the pervasive introduction of the Monte Carlo method into physics. Monte Carlos, as far as the applied mathematicians John Hammersley and Keith Morton were concerned, ought to be expurgated line by line from the discipline. In January 1954, they put it this way:[56]

We feel that the Monte Carlo method is a last-ditch resource, to be used only when all else has failed; and even then to be used as sparingly as possible by restricting the intrinsic random processes to bare essentials and diluting them wherever possible with analytic devices. . . . We cannot emphasize too strongly that the random element of Monte Carlo work is a nuisance and a necessary evil, to be controlled, restricted, excised, and annihilated whenever circumstances allow; and this applied mathematician's attitude, of seeing how much can be done without appealing to random processes, we contrast with the pure mathematician's attitude, of seeing how much can be done by appeal to random processes.

To these two authors, the battle line was drawn between the applied mathematicians and the pure mathematicians. Why? To these applied mathematicians, the extraction of physical laws from the world demanded analytic techniques because they led to reliable results. On this view of the world, pure mathematicians were merely playing with curiosities when they spent their efforts eval-

uating simple problems like linear differential and integral equations with only one or two variables, for these were easily tackled
with everyday numerical methods. Such frivolous mathematical
activities would be better replaced by serious inquiry into the
true challenges, such as nonlinear integral equations with several
unknowns.

Two distinct attacks on Monte Carlo enthusiasm had thus
emerged by the early 1950's. There were those who, on principle,
remained deeply suspicious of the artificiality of pseudorandom
numbers and the simulations themselves. And there were those,
like Hammersley and Morton, who for pragmatic reasons doubted
the new method's reliability. Either way, it was the status of the
enterprise itself that was in question. Not only was the broader
community split over whether these new simulations were legitimate; it was split even further within the camps of the defenders
and the attackers. Viewed "globally" the enterprise of simulating
nature was on the shakiest of grounds. In the absolute absence of
any agreed-upon interpretation of simulations, it might be thought
the whole enterprise would collapse. It did not. Practice proceeded
while interpretation collapsed.

The Pidgin of Monte Carlo

Pure mathematician, applied mathematician, physicist, bomb builder,
statistician, numerical analyst, industrial chemist, numerical mete-
orologist, and fluid dynamicist: each has a view about what the
Monte Carlo was. To the pure mathematician, the Monte Carlo
was a measure defined on the space of infinite graphs, for a coupled
set of Markovian and deterministic equations. For statisticians, the
method was another sampling technique, with special application
to physical processes; they considered such techniques well known,
and as a result were at first hesitant to join the plethora of postwar
conferences, discussions, and research efforts. To the numerical
analyst like Howlett at Harwell, the method was just one more
numerical tool for the solution of integrodifferential equations. To
the industrial chemist such as King, the stochasticist view made his
subject amenable to a form of direct representation never before
possible. Bomb builders saw a modeling technique for a physical
process so complicated by radiation transport and hydrodynamics

in hot media that all their usual means failed, both experimental and theoretical. The symbols and procedures of the method, therefore, sit variously in each domain, and connect differently to the terms, theorems, and style of each discipline.

Yet for all this diversity, it is clear from our discussion thus far that at meeting after meeting, beginning with the Institute for Numerical Analysis in Los Angeles of June 1949, and continuing through the IBM Scientific Seminars of 1949 and the 1954 Gainesville meeting, representatives of these various professions could and did find common cause. Moreover, individuals, including Ulam, King, Householder, John Curtiss, and others could alternate between problem domains without difficulty. Von Neumann passed as quickly to meteorology as he could to the Superbomb, to reactor problems, or to fluid-dynamical shock calculations. A chemical engineer could speak easily to a nuclear physicist in this shared arena of simulation; the diffusion equation for chemical mixing and the Schroedinger equation were, for purposes of this discussion, practically indistinguishable. Yet this is not to say that the nuclear physicist and the chemical engineer themselves became identical or interchangeable—far from it. If anything, with the growth of nuclear engineering and cyclotron design, and an expansion of meson theory, nuclear physicists were growing ever farther from their chemist colleagues.

In the heat of the moment, a kind of pidgin language emerged, in which procedures were abstracted from their broader signification. Everyone came to learn how to create and assess pseudorandom numbers—without the full mathematical apparatus for evaluating the "essence" of being random. It became a matter of common ability to transform these pseudorandom sequences into the particular weighted distributions needed for a particular problem. Everyone learned the techniques of variance reduction, such as importance sampling and splitting. Terms like "sampling" and "finite game" captured pieces of each of the broader and fuller fields that interacted with one another. What had been arcane knowledge of the mathematician—Courant's criterion—became a standard test to be applied to the difference equations as they were readied for loading onto computers.

Slowly, this pidgin language began to build into something more than a matter of provisional utility. Research in the Monte

Carlo, and into its inner structure, began to gather a momentum of its own: theorems were postulated and proved; rules of thumb were generated. The crude mediating language began to acquire its own journals, its own experts. By the 1960's, what had been a pidgin had become a full-fledged creole: the language of a self-supporting subculture with enough structure and interest to support a research life without being an annex of another discipline, without needing translation into a "mother tongue." The principal site of creation for this hybrid language was Los Alamos. For it was there, in the search for improved nuclear weapons, that a new mode of coordinating activities was built, where scientists from different disciplines (different practice and language groups) could form a trading zone.

Of course not everyone shared all the skills of this new "trading zone." Some focused on the game-theoretical aspect; others, more on variance reduction or convergence problems. There were formalists and practically minded researchers, workers interested in particular method problems such as inverting matrices and others content to exploit basic results. For a few years, between 1944 and 1948, the exploration of sampling and stochastic processes proceeded in the hothouse environment of the weapons laboratory, nurtured with the endless resources of the Atomic Energy Commission and the devoted efforts of von Neumann and Ulam. Already during this time, a shared proficiency developed that made the translation of problems into the pidgin of the Monte Carlo inviting, almost required. Perhaps because secrecy at first had both bound together and isolated the originators of the method, when the news broke outside the AEC enclave it had an even greater fascination among the great uncleared; it was a fascination so intense that it sparked warnings about overenthusiasm.

Fiction writers have recognized (more clearly than philosophers or historians) the powerful and problematic temptations of this artificial reality. In William Gibson's remarkable trilogy,[57] cyberspace—the unlocated computer-driven reality outside physical existence—contains as much significance as our daily three-dimensional space.

The seductiveness of cyberspace as an alternative to coping with the harsh edges of the everyday is, and was, apparent to those who work with simulations. Years after working on simulations of the first H–bomb, a physicist (then a young postdoc) told me: "I had a strange attitude toward the reality of hardware and the reality of explosions, which it's hard for me to explain now. But [it] was intense and real at the time. I didn't want to see the actual hardware of an atomic bomb in the laboratory . . . in the machine shop, in the metallurgical facility. And I didn't want to see a nuclear explosion."[58] The alternative world of simulation, even in its earliest days, held enough structure on its own to captivate its practitioners. And in their fascination they learned a new way of work and a new set of skills that marked them for a long time to come.

Elsewhere, one mathematician went so far as to identify an incident that occurred in the seventeenth century as the first recorded instance of what might be regarded as an application of the Monte Carlo method, while other statisticians, in some annoyance, presented genealogies of precedents and antecedents to show that sampling methods were nothing new to the statisticians. A few days after the IBM Gainesville conference, Herman Kahn, then writing a brief "historical" introduction that would set Ulam's place in this now–extended pantheon of Monte Carlo progenitors, queried Ulam on how to set the scene. Ulam responded this way:[59]

It seems to me that while it is true that cavemen have already used divination and the Roman priests have tried to prophesy the future from the interiors of birds, there was not anything in literature about solving differential and integral equations by means of suitable stochastic processes. In fact the idea of Monte Carlo seems to me to consist mainly in inverting the procedures used before, that is to solve problems in probability by reducing them to certain special differential equations. Of course, sampling processes were used in statistics but the idea of using probabilistic schemes to solve problems in physics or pure mathematics was not used before Johnny v. N. and myself were trying it out.

This seems to me about right: the Monte Carlo in some ways was the culmination of a profound shift in theoretical culture, from the empyrean European mathematicism that treasured the differential equation above all, to a pragmatic empiricized mathematics in which sampling had the final word.

When it came time to justify the new technique, the community's responses were varied. If the borrowed error-analysis methods of experimentation provided a first, epistemic argument for calling Monte Carlo runs "experiments," a second, metaphysical one came close on its heels: like nature itself, and unlike differential equations, the computer-based Monte Carlo proceeds by a series of (at least simulated) random occurrences. In this strong sense, the early Monte Carlo applications to the diffusion of gases, the scattering of neutrons, the production of cosmic-ray showers simulated nature. On this view (the view of many early practitioners of the Monte Carlo), the Monte Carlo offered a resemblance relation between sign and signifier in a pre-sixteenth-century semiotic sense. As the Monte Carlo became a standard tool for the resolution of problems with no stochastic elements, King's position became harder to defend, and the vision of simulations as offering a uniquely privileged vantage point began to dissolve. Nonetheless, the sense of direct access to a problem "as nature poses it," "behind" the equations never quite left the players of Monte Carlo.

Others challenged the legitimacy of knowledge based on simulations. For many theorists, an analytic solution continued to have a cachet forever inaccessible to the electrical engineers and their approximation methods. If any theoretical representations stood in Plato's heaven, they were the delicate hypersurfaces of differential equations—not batch-generated random numbers and the endless shuffle of magnetic tape. For experimenters, the Monte Carlo never came to occupy a position of "true" experimentation, as exemplified in debates that continued decades later over the legitimacy of according doctorates to students who had "only" simulated experiments. This left the engineers and their successors, the computer programmers, in a peculiar position. They spoke an intermediate language, a kind of formalized creole: the language of computer simulations understood both by theorists and by experimenters. (It was no accident that conferences flourished with names such as "Computing as a Language of Physics.") Accordingly, the simulators became indispensable as links between high theory and the gritty details of beam physics and particle collisions. But just as they occupied an essential role in this *delocalized trading zone*, they also found themselves marginalized at both the experimental and the theoretical end of particle physics.

I have argued here that the atomic bomb served as both the subject and the metaphorical generator of the Monte Carlo technique. Simulations were essential in enhanced-fission-weapon work and in the basic design of the thermonuclear bomb. Not only did the weapons projects provide the resources; they yielded the prototype problems on which virtually all early thinking about simulations were predicated: neutron transport, stability analysis, radiation diffusion, and hydrodynamics. Everywhere one turns—from von Neumann's early work on mounting the Monte Carlo on computers, to Ulam's original reflections on the method; from Herman Kahn's variance-reduction methods, to a myriad of applications in air bursts, tissue penetration, and the diffusion of gamma rays— one sees the hand of weapons design. There were, naturally, exceptions: Robert Wilson continued his work on cosmic-ray showers with the Monte Carlo, and Enrico Fermi used the computer for calculations on pion–proton resonances.

But the early world of simulations was steeped in weapons considerations; nuclear bombs saturated every aspect of these early discussions, from language to the self-representation of the simulators. In this respect, I close with a paraphrased summary of Jerzy Neyman, a statistician at the Statistical Laboratory (Berkeley) who opened a roundtable discussion on the Monte Carlo with an analysis of origins of the new method:[60]

Speaking first of the history of science, he [Neyman] observed that it seemed rather general that an idea begins to explode with sporadic, disconnected events. Each of these may trigger off further explosions in turn, as we imagine the events to occur in a chain reaction. In time these explosions of ideas occur more frequently and we eventually have what might be likened to a mass explosion in which the ideas blossom out and become common knowledge.

Perhaps. Looked at from a microscopic point of view, an atomic bomb does proceed in this way: disconnected fissions, scatterings; a Markovian universe plummeting into detonation. But from outside, we see a mission-oriented laboratory, a group of scientists allied with a military infrastructure struggling to create particular weapons and the intellectual superstructure to facilitate their design and implementation. Have we witnessed the cascade of the participants' narrative of history into the narrative of the physicists' simulations?

In baldest possible form: the computer began as a "tool"—an object for the manipulation of machines, objects, and equations.

But bit by bit (byte by byte), computer designers deconstructed the notion of a tool itself as the computer came to stand not for a tool, but for nature itself. In the process, discrete scientific fields were linked by strategies of practice that had previously been separated by object of inquiry. Scientists came together who previously would have lived lives apart, and a new subfield came to occupy the boundary area. Notions of simplicity were stood on their heads: where compact differential equations previously appeared as the essence of simplicity, and numerical approximations looked complex, now the machine-readable became simple, and differential equations complex. Where the partial differential equation had appeared as the exalted furniture decorating Plato's heaven, now Monte Carlo methods appeared to re-present truly the deeply acausal structure of the world.

Methodologically, the lesson—stated as provocatively as possible—is this. A contextualized, practice-based analysis serves nicely to underwrite a specific kind of social constructivism: categories of simplicity and conformity to the world will not survive as transcendent categories. But the picture of science in total rupture between frameworks strikes me as ever more chimerical. Three generations of framework relativists are enough.

RICHARD CREATH

The Unity of Science:
Carnap, Neurath, and Beyond

A great many different theses have been advanced under the banner of the unity of science, and these and others have been attacked when the unity of science has been disputed. It would be useful to sort out some of these before fixing on any opinion on the general topic. Indeed, there are more theses than there is space to discuss here, so I shall arbitrarily limit my remarks to certain classical discussions within the Vienna Circle, to one contemporary writer, Peter Galison, and to some very general remarks of a political nature. By way of preview, I shall argue that what our Viennese predecessors were really defending is far more sane and sensible than has been supposed, that Galison's view, however much it may seem to be anti-Viennese, is in harmony with that classical tradition, and that the broadly political hopes and fears expressed under the banner of the unity of science would be addressed more perspicuously in a different way.

Carnap

Classically, the unity of science movement was initiated in the mid-1930's by the left wing of the Vienna Circle as part of its controversies with the Circle's right wing. More specifically, the movement's organizational leader was Otto Neurath, and its intellectual leader was Rudolf Carnap.[1] Even these two often interpreted the unity of science in different ways, so I shall discuss them separately. Because he is the more systematic it will be easier to begin with Carnap. By the time that Carnap discussed and defended the unity of science, he was a conventionalist and pragmatist and also a physicalist, though not always in the same way that Neurath was. In addition, at this time he and Neurath and Carl Hempel favored coherentist over foundationalist approaches to justification. In his most extensive treatment of the unity of science, Carnap divided the issue into two parts: the unity of the language of science and the unity of the laws of science.[2]

In the discussion of the former there is a lot of talk of reduction. But Carnap does not mean what we would mean by this word. The issue is not ontological but epistemic. Once all the argle-bargle is stripped away, the central question that is bothering Carnap here is: What is sufficient as a source of news, as basic evidence, for us to weave the fabric of science around it? And his answer is a very tolerant empiricism, namely that observation at the level of middle-sized physical objects is *a* sufficient source. More technically this gets played out as the requirement that the applicability of all the concepts that we use in science must at least sometimes be empirically and publicly testable. This seems pretty benign.

Notice that Carnap claims only that observation at the level of middle-sized physical objects is *a* sufficient source. To be sure this rejects the British and Cartesian traditions in empiricism, according to which the only basic evidence is phenomenal or, more broadly, concerning the contents of our own minds. Since Carnap is a conventionalist and pragmatist, he does not deny that one could set up the conceptual framework in the British way. But it is not the only way, and Carnap is ready to urge the virtues of making the tests public rather than private. Note also that Carnap says nothing whatever about certainty. If asked, however, he would have said that on the physicalist basis he was proposing no claim in empirical science is certain, and that includes the observation reports themselves.

As a final point on the question of language, nowhere in any of Carnap's writings is there a prohibition on the development of technical vocabularies that would be quite different, even unrecognizable, from one special science to another. Thus, the unity of the language of science is not to be refuted by quoting seemingly unintelligible jargon from a technical scientific journal. What Carnap does care about in defending the unity of the language of science is public testability and the possibility of bringing different scientific specialties to bear on the same practical or scientific problem.

To the second part of the unity of science, that is, to the unity of laws, Carnap devotes only three paragraphs.[3] In that space, the only claims that Carnap chooses to emphasize are that there is no unity now and that whether there is ever to be any unity of scientific laws is not a matter for philosophic speculation but must, rather, wait on the evidence and on the further theoretical developments that this evidence makes possible. The unity of two sets of laws Carnap equates with the derivability of one set from the other, and he thinks that the more unity the better. While he is more or less indifferent as to which fields are to be derived from which, he apparently supposes that if ever all scientific laws are to be derived from the laws of a single science, that single science will probably be physics. The important point, however, is still that whether science is ever to be unified is not an *a priori* matter; it must wait on the evidence.

All this seems eminently sensible, but it would still be worthwhile to reflect briefly on Carnap's motives. First, these views regarding the unity of laws reflect the fact that Carnap is at this point in his career a coherentist rather than a standard foundationalist. Unifying the laws would show their relative coherence in a very strong way. Theories that explain much with very economical means are on the whole to be preferred to theories that assume much and explain little. This is not a very daring epistemological preference. Most of us share this preference. Most of us applaud Newton's unification of celestial and terrestrial mechanics, the unification of light, electricity, and magnetism in the nineteenth century, and the unification of thermodynamics with statistical mechanics in the molecular theory of gases. Though there are those who would insist that the unity of laws is of only pragmatic interest, it may well be that all epistemic preferences are ultimately pragmatic matters as well.

Carnap's remarks on unity were designed to do something more than express his coherentism. They were also designed to combat Cartesian dualism, not by denying the dualism, but by rejecting its apriority. This in turn allowed Carnap and Neurath to attack a brand of social science that had been inspired by dualism but that they viewed as moribund.[4] The radical separation of minds and bodies had led to a radical separation of social and natural sciences, and this had fostered within the social sciences the use of intuitive and introspective methods. Of these the former was dangerously antiempirical, and the latter was unproductive because the evidence was private rather than public.

Neurath

Even more than Carnap, the emotional force behind the unity of science movement was Otto Neurath. He wrote and talked endlessly about it. He organized its meetings and edited its encyclopedia; he was indefatigable. Yet it is curiously hard to say precisely what Neurath thought that the unity of science was or just why he supposed it was desirable. This is in part because his writings are polemical rather than expository. It is also partly due to something else. Let us call it the enormous fertility and suggestiveness of his ideas: there is no time for him to work them out; no one system can contain them all; and we must often fall back on our own interests in identifying his "real" views and "real" reasons. Still one can recognize themes to which Neurath repeatedly returns and enough of the broad outlines of a view to get some idea of what the doctrine was and was not, and what it was supposed to accomplish.

Carnap spoke of the unity of laws, by which he meant the derivability of one set of laws from another. Carnap thought this desirable where possible but warned against *a priori* pronouncements. Neurath, by contrast, is simply not interested in deriving some laws from others (in spite of his occasional friendly references to axiomatizing and systematizing), much less in finding a hierarchy of sciences.[5] Neurath never speaks of reducing one science to another but talks instead of symmetrical relations such as connecting, building bridges between, and filling gaps between various branches of science.

Carnap also spoke of reducing the language of science to a common physical–observational vocabulary, but as we saw what he meant was that all scientific claims should be publicly testable. Neurath too campaigned for the unity of the language of science, but it is far less clear what he meant. For example, he often took the distinction between an object language and its metalanguage as a threat to the desired unity, a distinction that was introduced in order to avoid the semantic paradoxes. Even if one thinks that the paradoxes can be dealt with in some other way (Neurath's way was to abandon the notion of truth or to twist it into something unrecognizable), the object/meta-distinction does not violate any argument that Neurath has for the unity of science, nor does it interfere with any utility that he claimed for the unity.

So what *does* Neurath have in mind? The answer can be given in two words, "holism" and "physicalism," but those will need to be unpacked in turn. The first of these arises in the contexts of confirming our scientific beliefs and of using them for practical purposes. In either context we make predictions. But doing this usually requires far more than a narrow theory drawn from a single discipline. Neurath uses the example of predicting the end of a forest fire. For this we would need news about trees from biology, news about fire from chemistry, news about fire-prevention institutions from sociology, and so on. In short, this is the Duhemian argument. Even when the required theory speaks only of, say, electric currents, the prediction will speak of a particular laboratory and the behavior of particular experimenters. While the Duhemian argument is sometimes ignored and sometimes exaggerated, we need not do either here, for Neurath's point is entirely sound. In confirming or in using our theories we frequently have to draw on many disciplines, which is hard to do if they are syntactically or semantically isolated from one another.

Actually, Neurath's holism is both less and more than this. It is more because he plainly intends to make it a full-blown coherentism, in which the unification of the various disciplines is itself the justification of them rather than merely a required step. Like most traditional coherence doctrines, this is extremely sketchy. It is also less than we might expect from holism, because the remark that predictions often draw on a multiplicity of disciplines imposes only an extremely minimal demand on the language of science. That there be some inferential connections between domains is a demand that even a reasonably enterprising Cartesian could meet. (After all, the point of the *Meditations* is to show how sometimes reasonably to conclude that there are objects out there corresponding to our sensations and judgments.) But Neurath intends to use his holism as a club with which to attack Cartesian dualism and also to attack the division between natural and social sciences that is built thereupon. Neurath's argument is just not enough for the purpose.

Turning to physicalism, let us preliminarily identify it with the claim that all scientific discourse can and should be "translated" into a language that speaks only of physical things. It is hard to say whether the physicalism or the holism is primary. Is it that physicalism is good because it helps facilitate holism? Or is it that holism is good because it is a natural outcome of physicalism? I suspect that the latter will turn out to be more charitable, but I also suspect that Neurath himself would take most comfort in the consilience of the two.

As is well known, Neurath complains unceasingly about metaphysics: in Tarski's truth definition, in Schlick's foundationalism, in Popper, in Russell, and so on. This may be thought odd, since physicalism certainly seems to be a metaphysical doctrine. Carnap too is an antimetaphysical physicalist, but he at least can claim that, while he prefers a physicalist language on practical grounds, he also allows other even phenomenalist languages, and that his principle of tolerance (and later his distinction between internal and external questions) saves him from anything like traditional ontological commitment. Neurath has no such rejoinder; he has neither the analytic/synthetic distinction nor any other resource that he would need. Instead, his best defense would be to admit quite frankly that for him the word "metaphysics" is a term of general philosophic abuse without any descriptive content at all. This may not sound very flattering, but it is better than an internal contradiction.

Actually, there is much in Neurath's physicalism to admire. Perhaps chief within this is his version of a private-language argument. Neurath contended that any phenomenalist conception of observation in which the observations were private and logically independent of one another (a characterization that would include not only Cartesianism, but also Russell's external world program and Carnap's *Aufbau*) resulted not only in a "methodological solipsism" but also in a solipsism of the now. Since there is no way of checking, there is no way of knowing whether one is employing the observational vocabulary in the same way one did previously, and there is, moreover, no independent ground (save for a pious hope) for thinking that one is always or even usually right in such judgments. By contrast, when one moves to the physical language one can check one's judgments against those of other observers. Also, because one's judgments can now genuinely disagree with each other, the fact that they generally do not is at least some evidence that they are somewhat reliable. All this is subtle and extremely interesting. Moreover, despite the fact that related arguments and related versions of physicalism are given, not only by Carnap and Wittgenstein, but even by Schlick and others, Neurath's version seems to be completely original and in many cases stronger and more satisfactory.[6]

Before we conclude, however, that Neurath is a physicalist in exactly the contemporary sense we need to look again. Perhaps shockingly, it can be argued that his physicalism is in important respects a species of idealism. I am fully aware that as Neurath saw the battle lines between materialism and idealism in nineteenth-century Germany his intent was to be unequivocally on the side of the former. Alas, we do not always do what we intend or fully know what we do. Consider for a moment the Kantian distinction between the world as we represent it and the world as a domain behind those appearances. This double world is just what Neurath would call a redoubling metaphysics. So like the German idealists he lops off the second of these two worlds. This is the inevitable result when one rejects the notion of truth or identifies it with the body of accepted statements, not even in the long run but now. Neurath is as candid as he can be that this is his approach to truth. Carnap avoids this by quickly distinguishing truth and confirmation. Quine also avoids this by distinguishing truth and current belief. Officially, Quine also rejects a Peircean theory of truth.

So why does Neurath call his view physicalism rather than idealism? The answer is straightforward. Neurath's notion of the physical differs from the sensory only in being how things appear to *us* rather than how things appear to *me*. To speak physicalese is to conceive of the discourse as public rather than private. Thus Neurath can be completely sincere and correct in thinking that he is rejecting the privacy claims often associated with idealism. What he achieves is not what we mean by physicalism but rather what might be called social idealism.

Remember all that was needed in the above private-language argument was that my basic observations could conflict (or agree) with one another, and there could be similar conflict or agreement between someone else's observations and my own. Nothing about that argument, which is the main intellectual support for Neurath's physicalism, requires us to go beyond how things appear to us. Moreover, there is nothing at all surprising in an idealist's using a language of physical objects. Berkeley himself insisted that the ordinary person's talk of tables and chairs is entirely correct. Even Berkeley is ambivalent as to whether ideas are in some sense shared.

When the idealists lopped off the world behind appearances, they had nothing left to compare appearances to. Thus, an observation can in no sense be a comparison of something with reality, and justification can involve only the comparison of one appearance with another. This last forces a coherence theory of justification, which in the absence of a separate theory of truth is often mistaken for a coherence theory of truth. Neurath follows this pattern almost exactly. He hotly denies that observation is a comparison of anything with reality; indeed, he insists that only statements can be compared. It is not at all clear whether Neurath has a theory of justification. It seems instead that he tries to replace it with an empirical theory describing the conditions under which the body of accepted statements changes. But this latter theory is cast entirely in terms of the "agreements" among the component statements. Agreement involves more than just logical consistency and hence is presumably equivalent to coherence, whatever that means. So Neurath has a coherence theory of (the successor to) justification.

So in every way Neurath does seem to be a social idealist. Though I do not share that view, I do not intend this description as a criticism. Idealism is neither silly nor obviously wrong. Of course, English-speaking analytic philosophy began with the wholesale rejection of idealism, so it may seem to the unwary that an attribution of social idealism is tantamount to a refutation. That is not my intent. Instead I want to emphasize that Neurath uses many of his terms in unexpected ways and that in particular his notion of physicalism is far from the contemporary one.

But Neurath's physicalism and holism are jointly indistinguishable from his conception of the unity of science. It is a very different conception from Carnap's. The two men seem to share much: both favor the unity of science; both make the unity of the language of science central to their conception of this; both favor unifying disciplines in still further ways; and both are physicalists and holists. But their conceptions of each of these component features are so very different that it hardly makes sense to speak of the unity of science as if it were some one doctrine. Even among the Viennese it is not. Nor are these Viennese views the same as those defended later by such writers as Oppenheim and Putnam.[7]

Thus, in attacking or otherwise appraising doctrines of the unity of science one must be very careful to articulate just what thesis is under discussion. Some who look favorably on the unity of science in fact favor only one or another version. Correspondingly, some who attack the unity of science aim at doctrines that were never held, and some confuse distinct views that were held by quite distinct authors. Finally, as we shall see, some people who are taken by others to be critics of the unity of science have views that are entirely compatible with one or another of the Viennese.

Galison

One person who has seemed to others to be a critic of the unity of science is Peter Galison. And because he is a knowledgeable reader of Carnap, it may seem to some that he is directly challenging Carnap's views. Galison's work that I have specifically in mind is his discussion of linguistic holism and also his development of a powerful and persuasive metaphor of brick walls.[8] What I understand Galison to be saying is both correct and important, but I also think it is entirely compatible with Carnap's presentation of the unity of science.

Consider holism. Language, of course, is a cultural phenomenon, and meaning in the broadest sense can be understood only by grasping the relevant features of the culture from which it emerges. This is true of English and Urdu, but it is also true of the more technical sublanguages of quantum electrodynamics and neurophysiology. The doctrine might be called linguistic holism, and in its milder forms is completely wholesome. Linguistic holism is not always mild and thus benign; some believe that cross-cultural understanding is not possible, whether the cultures in question are scientific cultures or more broadly political. For decades philosophers have worried about radical translation in the full knowledge that whatever its difficulties the same problems infect our attempts to understand our closest neighbors. The extreme form of linguistic holism has powerful consequences: only those who have adopted a theory can understand it; any changes in theory are reflected in changes in the vocabulary of experimentation; and changes in theory drive changes everywhere else.

Galison sees this extreme holism as a version of the unity of science; I see it as the complete Balkanization of science. But that is just the point; there is more than one sense of unity involved. The kind of unity that Galison opposes is not the kind of unity that Carnap was advancing. Both of them reject the extreme holism that turns various disciplines into island empires. Since, as Galison shows, experimentalists in fact have a culture that is not entirely dependent on theory, extreme holism would likewise pit experimentalists and theoreticians against one another as unintelligible aliens. Here again Galison wants the kind of unity that Carnap wants rather than that imposed by extreme holism. In fact, contextualism itself seems to be a milder form of holism, a form that might indeed be called framework relativism, where the cultural context is taken as the framework.

Galison introduces into his discussion of holism an extremely interesting metaphor. Now a metaphor is not an argument, but it can be far more persuasive than one, and Galison's is a very powerful metaphor indeed. The holist picture in which divides between theories are reproduced at the levels of language and of experimentation is likened to a brick wall in which all the bricks line up. Such a wall is weak, certainly weaker than a wall in which the bricks overlap. To vary the figure, a cable or rope does not derive its strength from the strength of the individual strands nor from the fact that the strands are continuous from one end to the other. Rather the strength arises from the fact that short, weak strands overlap and weave together.

There are several things to notice about these metaphors. First, they are not new; the cable is the same metaphor that C. S. Peirce, the American pragmatist, used in defending his coherentism. This is no complaint against Galison; it puts him in pretty exalted company. In fact, the same pragmatism and coherentism are just what get expressed in Carnap's discussion of the unity of science. It is also worth noting that the metaphor is not really directed against unity or order at all. Rather it suggests that some orders are better than others and that the strength of the whole can arise as the holistic result of the proper combination of parts that are not individually strong. A well-made brick wall is not a heap, and a heap is not strong. A cable is not a wad of steel wool, and a rope is not a ball of cotton. Indeed, for Peirce, and Carnap, and Galison coherence requires order—the right sort of order—and it also requires that the parts be in some sense independent. Bits of evidence that are separately quite weak can, if they are derived independently and they fit together in the right way, become quite strong. Evidence that is the mere by-product of theory has little probative value for that theory. But if that evidence is derived independently and it fits with the theory, then it is much more powerful support. The whole point of Galison's recent discussions has been to show that experimental evidence is derived with the requisite degree of independence and that it can be understood without the extreme linguistic holism that would threaten this. In all of Galison's discussion I see little that Carnap would not welcome.

The Imperialism of Physics

Finally, we need to consider a fear that for many people stands in the way of accepting the unity of science as a desirable goal. Baldly put, they fear and resist the imperialism of physics. Well, I do too, not because of the unity involved, but because of the imperialism involved. In a wholesome unified science biologists and sociologists would not become the lackey running dogs of physicists. Instead there should be a kind of disciplinary egalitarianism. Every field would still gather evidence, experimentally or otherwise. Every field would still try to make sense of that evidence in the locally most advantageous ways. But everyone would have to listen to everyone else, for no one's evidence could be discounted as utterly and permanently irrelevant. Relevance is after all a two-way street. Physical theory might explain biological evidence, but that biological evidence might just as easily bring down a physical theory or tip the balance between one physical theory and another. In any case the unity cannot be imposed *a priori* but would emerge only to the extent that it reinforces theories from the various fields.

As far as I can see this is the unity-as-cooperation view that was championed by Neurath. As such it is quite independent of his idealism.

While most of the opposition to the imperialism of physics seems to come just now from biology and the social sciences, the case of chemistry is instructive. It may well be that all the chemical regularities discovered so far are derivable from established laws of physics. It does not matter. So far it has not diminished the number of chemists, nor is it likely to. It has not diminished the scope and importance of chemical research, and it has not diminished the willingness of chemists to tell physicists to go jump off a cliff. Even within physics the derivation of thermodynamical laws from those of statistical mechanics has hardly meant that meteorologists have been replaced by elementary particle physicists. The unification of laws, even to the limited extent that it has been achieved, has not had the feared effect.

Nor, if we keep our heads, should it. What we should be aiming at, just as Neurath said, is the rational and equal cooperation of all the sciences rather than the domination of any field by another. Happily, this is where Galison's work is especially helpful, for it reminds us that theory need not dominate experiment nor vice versa. It is also just where Carnap's work is helpful. Carnap reminds us that the unity of science is a preference for coherence, not a fact about current science. It is thus up to us to describe more fully the kind of cooperative unity we want and to reject the imperialist unity we fear.

STEVE FULLER

Talking Metaphysical Turkey About Epistemological Chicken, and the Poop on Pidgins

❖ Science students all too often either believe they should shun philosophy or that they should borrow whatever philosophy there is off the shelf. It is my contention that . . . science students will have to take philosophy much more seriously; they even might have to redefine their own metaphysics in order to deal with the bizarre ontological puzzles revealed by their discoveries.
—Bruno Latour

A point that Harry Collins and Stephen Yearley repeatedly drive home in "Epistemological Chicken" is that methodological principles are not to be pursued as ends in themselves.[1] For Collins and Yearley, the most radical trends in Science and Technology Studies (STS)—reflexive constructivism and actor-network theory—result from a failure to appreciate this point. The principle in question, *alternation*, enables sociologists to understand an alien practice by adopting the standpoint of its practitioner. Depending on how alien the practice is, alternation may require that sociologists suspend many of their normal beliefs, including fundamental ones. However, Collins and Yearley stress that eventually sociologists "return home," so to speak, without any permanent damage done to their native epistemological powers. The sociologists' ability to return home intact then enables them to construct an explanation of the differences between the beliefs of their own culture and those of the one under study. Because STS radicals like Bruno Latour and Steve Woolgar refuse to return home, they end up adding nothing significant to understanding the social conditions of knowledge production. Either they take their ability to alternate as undermining the very possibility of knowledge (Woolgar), or they lapse into forms of hypothetical reasoning that range from banal to fanciful (Latour). Thus, instead of contributing to a storehouse of social-scientific knowledge, the radicals' promiscuous pursuit of method veers uncomfortably between the two philosophical foes of science, classical skepticism and speculative metaphysics.

Yet, as Collins and Yearley themselves document, both Latour and Woolgar explicitly embrace a return to philosophy in STS. Moreover, as someone generally sympathetic to the two radicals, I can report that "Epistemological Chicken" gives a relatively even-handed presentation of their positions. Clearly, Collins and Yearley have failed to pick up on something here. That something is what I would like to explore in this paper, namely the difference between thinking about intercultural relations in terms of bilingualism and in terms of trade languages; hence, the need not only to "talk metaphysical turkey" but also to deliver the "poop on pidgins."

Linguistics provides two models for thinking about the relationship between two cultures. One is *bilingualism*, the phenomenon that Kuhn said motivated his account of incommensurability in *The Structure of Scientific Revolutions*. The basic idea is that a bilingual speaker generally finds it easier to alternate between languages ("Gestalt switches") than to translate between them.[2] By analogy, then, one can imagine that Galileo could think about motion in Aristotelian and in his own non-Aristotelian terms without being able to construct one-to-one correspondences between concepts in the two theories. Bilingualism highlights the discrete and holistic character of languages, which, when tied to cultures, can place considerable practical demands on one's allegiances; hence, the Kuhnian talk of "conversion experiences" and "living in different worlds." Collins and Yearley clearly play to this model in their account of alternation. However, a rather different model of intercultural relations is implied by *trade languages*.[3] Their existence suggests that, as languages come into contact for specific reasons over an extended period of time, they tend to spawn a hybrid tongue—a "pidgin" or "creole"—that serves to break down the cultural barriers that were associated with the original languages. While trade languages typically have limited applicability, they have been known to develop into all-purpose, grammatically independent languages.[4] The fact that such a development can occur suggests that the incommensurability experienced by bilinguals may be more an artifact of the lack of communication between the two languages in question than a symptom of some deep ontological difference between them. This point, I maintain, is the key to what happens when Latour and Woolgar fail to return home from alternation. What happens is that the process of alternation effectively shrinks the distance between the native and the alien—between the sociologically grounded and the philosophically outlandish.

Talking Metaphysical Turkey

The radical side of STS is opening up metaphysical possibilities that are orthogonal to the ones that have been exploited in the Western philosophical tradition. First, in the spirit of our tradition, I will systematically distinguish the exploited from the unexploited possibility, and I will then show how the extremist methodologies of radical STS introduce us to the underutilized metaphysical horizon.

A metaphysics is, if nothing else, a scheme for organizing reality. To construct a metaphysics, one begins by relating two dimensions: an apparent (folk) dimension that distinguishes particular "individuals" from the "properties" they share with other individuals, and a real (scientific) dimension that distinguishes "abstract" terms from the "concrete" objects in which they can be realized. The fundamental question for any metaphysical scheme is how to map the first dimension onto the second: Which apparent objects are abstract, and which are concrete? Except in extreme forms of Platonism, abstract objects typically cannot exist unless materially realized in some concrete objects, which, in turn, place constraints on the ways in which the abstract objects can appear. By themselves, abstract objects are indeterminate. And so, while abstract objects enjoy the metaphysical virtue of primitiveness, they must pay the price of existing never in a pure form but only as part of a mixture with other such primitives. There are two general strategies for relating individuals and properties to abstract and concrete objects: a *standard* and a *nonstandard* strategy. The standard strategy connects Plato and Aristotle with Russell and Quine, while the nonstandard one connects Hegel and Durkheim with Latour and Woolgar.

The standard strategy takes properties to be abstract and individuals to be concrete. Individuals are said to be composed of a combination of underlying properties. In this vein, a medieval scholastic may speak of resolving a particular into its component universals, or a twentieth-century analytic philosopher may declare that a proper name is coextensive with, or its meaning is exhausted by, a "definite description," that is, a finite conjunction of predicates. Given the abstract nature of these properties, the individual is often portrayed as the spatiotemporal point at which the properties converge. Without spatiotemporal moorings, abstract objects would be "indeterminate," in the sense of "unbounded." Unsurprisingly, then, the standard strategy gives precedent to the problem of *the one and the many* as the classic metaphysical conundrum: How can the same property, say, "humanity," belong to an indefinite number of individuals? It is as if one would expect that the character of the property should diminish as it participates in the definition of each new individual, when in fact such participation only contributes to the fundamental status of the property. "Being," the only property that is shared by all individuals, turns out to be ultimate metaphysical foundation. By contrast, the nonstandard strategy takes individuals to be abstract and properties to be concrete. A property—sometimes called a *concrete universal*—is said to consist of individuals organized in distinct way. For example, political theorists have periodically spoken of each person as literally part of a "body politic" or a "social organism." To be a "German" would not be to have a property that every other citizen of Germany has, but rather a property that comes into being only by virtue of the lawful interaction of the citizenry. Without this interaction, the identity of each individual German would be "indeterminate," in the sense of "incomplete," or as Durkheim would say, "anomic." In a more linguistic vein, the nonstandard strategy switches from what structuralists call the *paradigmatic* to the *syntagmatic* dimension of definition. Instead of the standard strategy of defining by substituting synonymous expressions (i.e., a definite description), one defines by functional differentiation in a system.[5] Thus, a sentence determines what particular words mean, a paragraph what particular sentences mean, and so forth. This gives rise to the problem of *the part and the whole* as the classic metaphysical conundrum: How can the activities of spatiotemporally separate individuals be arranged so as to enable the emergence of some higher-order unity? It is as if one would expect that individuals, given their inherently partial nature, would interfere or simply ignore each other's activities, when in fact it is only through this interaction that they receive a sense of common purpose.

The standard and nonstandard strategies underlie two radically different conceptions of inquiry, which, in many respects, update the neo-Kantian distinction between nomothetic and idiographic approaches to scientific methodology. Without giving away too much of what follows, we can say that the standard strategy defends itself as a more efficient way of carrying out the nonstandard strategy, whereas the nonstandard strategy responds by revealing the hidden labor costs of the standard strategy. According to the standard strategy, inquiry is an intensive, perhaps even microscopic, search for the essential properties into which an individual can be analyzed without remainder, and which together can be used to synthesize the individual without cost. By contrast, the nonstandard strategy works by differentiating a whole into its proper parts—but at a cost, since it is not clear that the process can be reversed, so as to allow the parts to be reintegrated into the original whole. The methodological extremism that Collins and Yearley find so distasteful in Latour and Woolgar may be fruitfully regarded as two complementary procedures for converting the standard metaphysical strategy into the nonstandard strategy. Let us call these the *Woolgar Procedure*, associated with radical constructivism, and the *Latour Procedure*, associated with actor-network theory. The Woolgar Procedure (WP) goes like this:

WP1. Turn each individual into an instance of a property.

WP2. Make it an open question whether the individual really has the property.

WP3. Find situations when the property seems integral to what the individual is doing and situations when it does not.

WP4. Conclude that the individual has the property only in the identified situations.

In this way, an abstract individual is (de)constructed from a concrete one. Woolgar talks about this process as the establishment of *ethnographic distance.*[6] Consider the following simple example:

WP1. Turn a scientist into someone who happens to be a scientist (or better: who *is said to be* a scientist).

WP2. Make it an open question whether that person really is a scientist.

WP3. Find situations when being a scientist is integral to what the person is doing and situations when it is not.

WP4. Conclude that the person is a scientist only in the identified situations.

The Latour Procedure (LP), in turn, produces concrete universals from abstract ones in the following manner:

LP1. Turn each property of an individual into relations in which the individual stands to other individuals.

LP2. Turn all the aforementioned individuals into nodes of a common network that bears the name of the original property.

LP3. Distribute responsibility for maintaining the network equally across the nodes.

LP4. Conclude that the property exists only to the extent that the network is maintained.

According to Latour, this strategy involves *opening black boxes*.[7] Now, let us see how this works for a simple example:

LP1. Turn each scientific artifact into a relation in which the artifact stands to other artifacts and people.

LP2. Turn all the aforementioned objects into a common network that is identified as science.

LP3. Distribute responsibility for maintaining science equally across the network.

LP4. Conclude that science exists only to the extent that the network is maintained.

What makes Latour and Woolgar committed to the nonstandard strategy is their denial of the assumption on which the conundrum of the one and the many rests. That is, they reject the idea that a property's ontological status increases—or at least remains undiminished—by increased participation in the construction of individuals. In other words, Latour and Woolgar presume that properties do not have potentially infinite scope, but are rather subject to scarcity. Thus, "being a scientist" or "being scientific" is valorized just insofar as its predication is clearly restricted to certain situations and objects, and not to others. Once ambiguity is generated about the property's application in particular situations (as in WP) or the property is applied equally to all the nodes in a network (as in LP), then much of the property's original value is lost. This point is especially crucial for understanding Latour's recent attempt to dissolve the distinction between human and nonhuman by attributing agency to all the nodes in an actor-network.

At first, I and other commentators thought that Latour was trying to revive the Aristotelian doctrine of *hylomorphism*, the infusion of spiritual properties in material things, a mere metaphysical cut above primitive animism.[8] It was easy to reach this preposterous conclusion as a result of having interpreted Latour's move in terms of the standard strategy, whereby granting agency to objects such as computers and doors would increase the amount of agency in the world, thereby producing an image of the laboratory that Piagetian children in the first two stages of cognitive development might have, namely a world populated by objects moved by ghosts. However, this absurd conclusion can be avoided by imagining—as the nonstandard strategy would suggest—that there is a fixed amount of agency in the world, such that the more objects that possess it, the less of it each object possesses. And so, by spreading agency thin, the mystique of being an agent disappears.

In a turn of phrase that would appeal to Latour's thermodynamically oriented philosophical mentor, Michel Serres, the concept of agency is thereby "wasted," in that it can no longer be used as a principle of order that privileges certain objects at the expense of others.[9] While the allusion here is clearly to entropy, the import of Latour's usage is to dissipate the structures of activity/passivity that are needed to maintain any sort of power relation. Another important precedent for this line of thinking, which Latour himself fails to draw on (no doubt because of its teleological character), is Hegel's account of the progress of Reason dialectically unfolding itself in the world. As Hegel put it, originally only one person was free—the oriental despot—and everyone else was enslaved. Freedom under such circumstances corresponded to arbitrary will, as no one had the power to check the despot. An unconditional reasoner is thus able to behave irrationally—at least from the standpoint of subsequent stages, in which this despotic concentration of power is distributed to ever larger numbers of people until it is completely "democratized" (which, of course, took on a special meaning in the Prussian context—much to Marx's displeasure). In each stage, the character of reason and freedom is altered because one needs to act in the presence of others whose agency one had not recognized before.[10] In Freudian terms, one might say that democracy is "socially sublimated" tyranny. But be it expressed as the spread of freedom or the dissipation of energy, this nonstandard way of bringing about "the death of the subject" stands in interesting contrast to the more standard strategy of such recent French poststructuralist philosophers as Foucault and Derrida. They tend to presume that subjectivity will subsist unless explicitly eliminated or reduced to "effects" and "traces," while Latour follows Serres in supposing that a subtler but more effective way of erasing the significance of the subject is by flooding the world with subjectivity.[11]

This last point returns us to the metaphysical conundrum countenanced by the nonstandard strategy, namely how disparate parts are given a sense of wholeness. Woolgar and Latour characterize the disparateness in instructively different ways. Woolgar opens a free zone of possible ways an individual's identity may be extended into the future by holding that all properties of an individual are (linguistically) occurrent, never dispositional. Specifically, an individual has a property only in situations where the property is attributed to her or him. Latour operates on a larger scale, suggesting that individuals tend to scatter in a variety of directions, interfering with one another's actor–network building, until a node emerges through which all the individuals must pass before they can achieve their disparate ends. In both cases, closure and unity are articulated by a narrator who is elevated from the status of mere participant to that of representative. Indeed, Latour and Woolgar refer generically to this process as *representation*, which is meant to be taken in both its semiotic and political senses simultaneously (as, say, Hobbes did).[12] Thus, when one object, *X*, *represents* another, *Y*, three things happen:

1. *X* speaks in the name of *Y*.
2. *Y* no longer speaks for itself.
3. There is no longer any need to refer to *Y*.

Thus, representation is not only informative and economical, but it is also—for better or worse—repressive.

In keeping with the skeptical cast of his thinking, Woolgar doubts that representation can continue uninterrupted for very long. For just as, say, Woolgar may be trying to give closure to a narrative about scientists in the lab, the scientists are trying to do the very same thing themselves as they make sense of the sort of individual that their participant-observer, Woolgar, is—*especially* when he relates tales of the lab. Thus, Woolgar's own identity is constructed at the same time that he is trying to construct the identity of the scientists. The attempt by Woolgar and other ethnographers of science to "represent" these multiple acts of representation has led to the *reflexive turn* in STS, as epitomized in the cultivation of "New Literary Forms," which are stylistically modeled on the ironic and paradoxical fictions of Borges and Calvino.[13] It is easy to see the reflexive turn as another instance of dissipating a property—in this case, "being a representation"—that would otherwise issue in one set of objects' enjoying privileged status over another.

The actor-network theorist on whom Latour draws most heavily, Michel Callon, has further attenuated this property by granting voices of resistance to nonhuman objects in STS narratives.[14] And while Collins and Yearley find the methodology that informs Callon's literary practice obscure, semiotics provides many precedents, as it routinely translates technologies from instruments of control to media of communication, or "sign systems."[15] In the example from Callon that particularly incenses Collins and Yearley, trawling technologies enabled scallops to articulate an agenda that was at odds with the agenda proposed by the seafaring humans. What would ordinarily be taken as a failure to control nature is thus rendered as a disagreement between two interlocutors, humans and scallops, who communicate in "Trawling" over boats and nets, much as two humans might communicate in English over the telephone. The point that seems to elude Collins and Yearley here is that, by extending voices to scallops, Callon is not so much claiming that we can communicate with the sea creatures as well as with other humans, but rather that our ground for intraspecies communication is really no better than our interspecies grounds. That is, Callon is advising us not to presume too *much* of our ordinary communicative skills. In both intra- and interspecies cases, the two interlocutors judge communicative success by their own criteria, which, in turn, depend on judgments about the degree to which the communicative medium is perceived as faithful to one's own wishes and the extent to which the interlocutor is perceived to have complied with those wishes. In other words, scallop-talk is important less for what it makes of scallops than for what it makes of the criteria by which we identify communication. Put as a more positive point, if "communication" and "control" mark the two poles by which we distinguish our relations with humans and things, Callon is showing that the poles are separated by a continuum with no clear dividing point.

The above line of thought takes a particularly interesting turn in Latour's attempt to subvert what has traditionally been the most privileged form of representation, namely *explanation*.[16] In this context, it is important to see that Latour is referring to the image of explanation that has dominated Western science since the heyday of Newtonian mechanics, namely an image that unifies by reducing, that explains the most phenomena by the fewest principles. This has perhaps been the most intellectually compelling statement of the standard metaphysical strategy. It implies a radical asymmetry whereby one representative stands for as many representables as possible. It is also the conception of explanation—the so-called covering-law model—that is most familiar to contemporary philosophers of science who count the positivists among their intellectual ancestors. Latour's avowed "political" concern here is that various schools of STS, including the reflexivists just mentioned, tend to want to "explain" science exactly in this objectionable way. The principal symptom of this tendency is the development of protodisciplinary discourses within STS that promise to capture the complexity of the knowledge enterprise in a few simple terms—or not so simple terms that nevertheless reduce the complexity that is normally captured by the distinction between the natural and the social sciences (e.g., "social epistemology").[17] Before evaluating Latour's solution to this problem, it will pay to examine in more detail the sense in which that pinnacle of cognitive pursuits, explanation, is supposed to be an instance of political economy in action.

Consider explaining as a social practice. Its point is to bring into relief properties common to a set of individuals that had been previously occluded. A typical result of providing an explanation (its perlocutionary force, as Austin would say) is that these individuals are treated in a more uniform fashion than before, thereby reinforcing the appropriateness of the explanation to each new situation. However, Latour advances beyond a speech-act analysis of explanation by revealing the amount of hard work that needs to be done in order for an explanation to succeed. In particular, the relevant individuals must exchange their voices for that of the explainer. This may be done with or without resistance from those individuals, but, in either case, the idea is that, as a result of the exchange, the individuals must go through the explainer in order to have their own interests served. At the very least, the individuals must find that the cost of contesting the explanation outweighs the benefits. In that sense, the explainer becomes what Latour calls an *obligatory passage point*.[18]

The more ambitious the explanation, the wider the variety of obstacles that the explainer is likely to encounter. For example, if you try to put forth an explanation that is common to objects that move on earth and in the heavens, you will probably have to contend with specialists on terrestrial and celestial motions, the legitimacy of whose careers may depend, in large part, on the two types of motions' requiring separate expertises. You may thus need to make allowances to these specialists. Nevertheless, the successful explainer can keep these concessions hidden—in, say, *ceteris paribus* clauses that form the conditions under which the explanatory principle can be applied in particular cases. Ultimately, success in this arena is measured by the explainer's ability to get the locals to regard their respective clauses as *merely* local, in the sense of being relevant only for bringing to the surface principles that are already latently present in their locales. If the locals should combine to realize that what had been in their best interest not to oppose as individuals is now worth opposing as a group, then the explainer is faced with the amalgamation of anomalies that precipitates a Kuhnian "crisis." But failing the emergence of class consciousness among the anomalous, their "tacit consent" enables the explainer to rewrite the history of her or his travails so as to make it seem as though disparate phenomena were spontaneously attracted to his or her explanatory principles, when in fact she or he had to negotiate with each one separately in their own terms. Latour speaks of this special power as the conversion of contact motion into *action at a distance*. One suggestive way of characterizing this conversion is in terms of a Piagetian developmental transition from *spatiotemporal* to *logicomathematical* cognitive orientations.[19] In the former orientation, the child treats objects as having a found order, or "natural place," and thus will maneuver around them so as not to cause a disturbance. In the latter, the child realizes that the objects can be moved from their original locations to a central place where the child can rearrange the objects as he or she sees fit, which typically involves grouping some together and separating them from others. In this way, Piaget speculates, primitive set theoretic relations of inclusion and exclusion emerge, which are, of course, essential to any covering-law model of explanation. And because the material character of the objects offers little resistance to the child's manipulative efforts, she can play with them without concern for the consequences of her interaction with them. Piaget regards this sort of play as a primitive instance of pure theorizing, in that the child's manipulations are determined almost entirely by the rules of the game she has imposed on the field of objects. One obvious point that often goes unnoticed in discussions of Piaget's views, which nevertheless bears on our understanding of Latour, is that even when a child operates in the logicomathematical mode, he continues to arrange objects in the world of space and time, but simply denies to himself that the actual arrangements he comes up with represent the full range of possible arrangements. By analogy, then, once an explainer has overcome an initial array of opposition, she will tend to presume maximum generality for the scope of her explanation, until she is forced to do otherwise. At that point, the explainer perhaps begins to take her own rhetorical reconstruction too seriously, believing that she can "act at a distance" without any "hysteresis" effects, that is, without any interference in the transmission of explanatory power from fundamental principles to disparate cases.[20] This phenomenon is especially common in accounts of science that speak of, say, the historical trajectory of Newtonian mechanics as the application of Newton's principles to an ever wider domain of objects, as if the process were one of uninterrupted assimilation.

Given the way the nonstandard strategy works, Latour's proposed solution to the totalizing tendencies of explanation should perhaps come as no surprise. It involves flooding the world with explanations, thereby eliminating whatever economic advantage a set of principles might have over a set of phenomena. Each phenomenon is to be given its own unique explanation, presumably one making plain all the effort it takes in that case to have the *explanans* represent the *explanandum*. The curious feature of this move is that it retains a vestige of the standard strategy's approach to explanation, in that the explanatory principles still replace the things explained, though now on a one-to-one rather than a one-to-many basis. And so, while it is easy to see how Latour's recommendation would remove the mystique from the practice of explaining, it is not clear what fate would await the individuals on either side of the explanatory divide. Will *explanans* and *explanandum* come to resemble one another, so that, as it were, putative causes and effects will appear interchangeable? Apparently, this was a possibility that Nietzsche countenanced in his original forays into deconstruction.[21] Or, will other practices emerge to reinstate the distinction that explanations draw? After all, nominalists from Peter Abelard to John Stuart Mill have believed that the homogenizing—and, in that sense, generalizing—tendencies of abstraction were inescapable features of the human mind. In that case, we might expect to find a mutant strain of explanatory practice that appeals to the recurrent properties of "locality," "situatedness," "contexture," and other tools of social construction in order to account for why things appear as they do: in a word, ethnomethodology as science. At that point, we would have returned home— indeed, to Collins and Yearley's own methodological doorstep!

What Latour seems to have retained is the "correspondence" relation of the standard strategy, which ensures that nothing is lost in the translation from *explanandum* to *explanans*, except, of course, the uniqueness of the *explanandum* now that it has been reduced to yet another instance of the *explanans* in action. Latour thus envisages one-shot explainers eventually exhausting themselves by virtue of having to reproduce in their explanations that uniqueness as well. However, this conception of translation, which leaves no lasting impression on the explainer and the explained, but only in their relationship, is little more than a conceit of logical positivism. In fact, the theory of translation with which practicing translators implicitly operate sets a more instructive example about how the nonstandard strategy should proceed on matters of explanation. In particular, translators are aware that the consequence of any translation is to move the two languages either a little closer together or a little farther apart. The former, an instance of *dynamic equivalence*, reproduces the pragmatics of the source language in the target language (typically at the expense of conceptual nuance); whereas the latter, an instance of *formal equivalence*, reproduces the semantic distinctions of the source language in the target language (typically at the expense of practical relevance). The difference that biblical translators draw between "hermeneutics" and "exegesis" intuitively captures what is at stake in the distinction: texts for believers versus texts for scholars.[22] The point is that, as the process of translation is repeated between two languages, the languages themselves start to blend into each other in interesting ways, reflecting the balance of trade between dynamic and formal equivalents. Here, it is worth beginning to examine the recent flurry of appeals to *trading zones* and *trade languages* as ways of understanding how representatives of disparate disciplines (or phenomena) come to terms with one another. As the reader can already see, I favor an *interpenetrative* construal of linguistic trade.

The Poop on Pidgins

Let me start by distancing my own sense of interpenetration from the way in which the trading-zone metaphor has been deployed in histories of physics and economics. The metaphor is most closely associated with the historian of twentieth-century physics Peter Galison and the economic historian Donald McCloskey.[23] McCloskey offers the most succinct formulation of the idea, one that goes back to Adam Smith's *Wealth of Nations*. As a society becomes larger and more complex, people realize that they cannot produce everything they need. Consequently, each person specializes in producing a particular good that will attract a large number of customers, who will, in exchange, offer goods that the person needs. Thus, one specializes in order to trade. McCloskey believes that this principle applies just as much to the knowledge enterprise as it does to any other market-based activity. Galison's version of the trading zone draws more directly from the emergence of pidgins. His account has the virtue of being grounded in a highly informed analysis of the terms in which collaborative research has been done in Big Science–style physics. For example, determining the viability of the early nuclear bombs required a way of pooling the expertise of pure and applied mathematicians, physicists, industrial chemists, fluid dynamicists, and meteorologists. The pidgin that evolved from this joint effort was the Monte Carlo, a special random-number generator designed to simulate stochastic processes too complex to calculate, such as the processes involved in estimating the decay rate of various subatomic particles. Nowadays, the Monte Carlo is a body of research in its own right, to which practitioners of many disciplines contribute, now long detached from its early nuclear origins.[24] There are two questions to ask about the models that McCloskey and Galison propose: (1) Are they really the same? In other words, is Galison's history of the Monte Carlo trade language properly seen as a zone for "trading" in McCloskey's strict economic sense? (2) To what extent does the trading-zone idea capture what is or ought to be the case about the way the knowledge enterprise works?

The short answer to (1) is no. McCloskey is talking about an activity in which the goods do not change their identities as they change hands. The anticipated outcome of McCloskey's trading zone is that each person ends up with a greater number and variety of goods than when she or he began. The process is essentially one of redistribution, not transformation. In this way, Galison's trading zone is closer to the idea of interpenetration. The Monte Carlo simulation is an emergent property of a network of interdisciplinary transactions. It is not just that, say, the applied mathematicians learn something about industrial chemistry that they did not know before, but rather that the interaction itself produces a knowledge product to which neither discipline had access previously. By contrast, then, McCloskey's idea perhaps captures the eclecticism of the human sciences in the postmodern era, which, to answer (2), calls its desirability into question. Interestingly, another economist, Kenneth Boulding, has already offered some considerations that explain why "Specialize in order to trade!" is not likely to become a norm of today's knowledge enterprises—though it perhaps should be.[25] Boulding points out that in order to enforce Smith's imperative in the sciences, one would need two institutions: one functionally equivalent to a common currency (i.e., a methodological standard that enabled the practitioner of any discipline to judge the validity, reliability, and scope of a given knowledge claim), and one to an advertising agency (i.e., brokers whose job it would be to persuade the practitioners of different disciplines of the mutual relevance of each other's work). Short of these two institutions, the value of knowledge products would continue to accrue by producers' hoarding them (i.e., exerting tight control over their appropriate use) and making it difficult for new producers to enter their markets.

However, Galison's trading zone has its own problems from the standpoint of interpenetration that I promote. His deployment of the metaphor does a fine job of showing how a concrete project in a specific place and time can generate a domain of inquiry whose abstractness enables it to be pursued subsequently in a wide variety of disciplinary contexts. In this way, he partly overcomes a limitation in McCloskey's trading zone, in that he shows that the trade can have consequences—that is, costs and benefits—that go beyond the producers directly involved in a transaction. But Galison does not consider the *long-term* consequences of pursuing a particular trade language. Not only does a pidgin tend to evolve into an independent language, but it also tends to do so at the expense of at least one of the languages from which it is composed. This empirical point about the evolution of pidgins may carry some normative payoff, insofar as the mere invention of new languages does not clarify the knowledge enterprise if old ones are not at the same time being displaced. Since we are ultimately talking about scientists whose energies are distributed over finite space and time, cartographic metaphors for knowledge prove appropriate. You cannot carve out a new duchy without taking land away from neighboring realms—even if the populations of these realms are steadily growing. The strategy of interpenetrability that I support (and develop)[26] is, ultimately, a program for rearranging disciplinary boundaries. It presumes that the knowledge enterprise is most creative *not* when there are either rigid boundaries or no boundaries whatsoever; nor is creativity necessarily linked with the simple addition or elimination of boundaries; rather, creativity results from moving boundaries around as a result of constructive border engagements.

Let me conclude with a methodological image. I envisage the texts of, say, Marx or Freud as such border engagements, the conduct of cross-disciplinary communication by proxy. They implicitly represent the costs and benefits that members of the respective disciplines would incur from the revolutionary interpenetration proposed by the theorist. For example, in the case of *Capital*, the social epistemologist asks what would an economist have to gain by seeing commodity exchange as the means by which money is pursued rather than vice versa, as the classical political economists maintained. Under what circumstances would it be worth the cost? Such questions are answered by examining how the acceptance of Marx's viewpoint would enhance or restrict the economist's jurisdiction vis-à-vis other professional knowledge producers and the lay public. More specifically, it would require looking at the audiences that took the judgment of economists seriously (for whatever reason) at the outset, Marx's potential for affecting those audiences (i.e., his access to the relevant means of communication), and the probable consequences of audiences' acting on Marx's proposal. And while this configuration of *Capital*'s audience would undoubtedly further our understanding of the reception and evolution of Marxism, the social epistemologist is aiming at a larger goal, namely the generalizability of the judgments that Marx made about translating distinct bodies of knowledge into a common framework: What was his strategy for removing interdisciplinary barriers? How did he decide when a key concept in political economy was really bad metaphysics in disguise, and hence replaceable by some suitably Hegelized variant? How did he decide when a Hegelian abstraction failed to touch base with the conception of material reality put forth in classical political economy? Is there anything we can learn from Marx's decisions for future interdisciplinary interpenetrations? So often we marvel at the panoramic sweep of revolutionary thought, when in fact we would learn more about revolutionary thinking by examining what was left on the cutting-room floor.

An interpenetrative approach would thus differ from that of mainstream hermeneuticians and literary critics in emphasizing the *transferability* of Marx's implicit principles to other potentially revolutionary interdisciplinary settings. However, none of these possibilities can be realized without experimental intervention—specifically, the writing of new texts that will, in turn, forge new audiences, whose members will establish the new terms for negotiation that will convert current differences into strategies for productive collaboration.

PART II

Contexts

MARIO BIAGIOLI

From Relativism to Contingentism

Although routine bashing of whiggish history of science has been among the favorite sports of science historians since Butterfield, we may be now witnessing something of a return of the repressed. Reviewers of recent works in the history and sociology of scientific knowledge have called attention to the emergence of "whiggish social studies of science," while some historians of science display their presentist frame of reference in their printed work and discuss it more explicitly in informal settings.[1] In the last decade or so, other historians and philosophers of science had already discussed "presentism" or "present-centered history" as appropriate approaches to the study of past science.[2]

The relativist historians' "neo-whiggism" seems methodologically equivalent to the sociologists' frequent bracketing-off of what they take to be the dangers of so-called reflexivity. Many historians of science and sociologists of scientific knowledge who take a relativist stance in examining the belief systems and practices of the scientists they study end up writing their interpretations from a more or less unacknowledged nonrelativist frame of reference.

Differently from philosophical analyses of reflexivity that approach the issue in terms of its methodological implications, I want to consider the emergence of "neo-whiggism" and the sociologists' frequent bracketing-off of reflexivity as possible indicators of the current socioinstitutional predicament of science studies and of its practitioners' cultural and professional identities. I believe that, if properly contextualized, the issues raised by the reflexivity debate in sociology and by the limits of the historians' ability to understand the past "on its own terms" do not need to deepen the anxieties of relativist historians and sociologists about the cognitive status of their disciplines. Instead, the reflexivity debate may provide a useful starting point for avoiding the deadlock that has characterized the rationality/relativism debate in recent years.

Interpreters in Hiding?

The last few years have witnessed the emergence of a range of reflections about the interpretive limits of the social sciences—history and ethnography in particular.[3] However, these debates do not seem to have yet influenced mainstream historiographical or anthropological practices. Quite to the contrary, they have been often ignored or cordoned off by historians and anthropologists in an attempt to prevent the double-edged sword of relativism from being turned against their own work. With some notable exceptions, science studies too have avoided or tried to control similar methodological issues.[4] As Malcolm Ashmore shows, sociologists of scientific knowledge usually tend to dismiss, exorcise, or simply pay lip service to the implications of reflexivity.[5] Similarly, judging from the very limited debate on reflexivity within history of science, its practitioners (myself included) keep plying their craft without losing much sleep over that issue.[6] As we proudly repeat the methodological dictum of our discipline ("You shall interpret the past on its own terms") we do not seem to spend much time analyzing the major tensions barely hidden underneath such an optimistic professional credo.

Because of its methodological sophistication, some of the recent work on scientific discoveries offers a good example of the historians' routine effacing of their own frame of reference. These studies argue that the scientists' historical narratives about their discoveries are rooted in and help stabilize the very closure of the debates through which those discoveries had become legitimized (and then passed into history) as such. In short, those narratives are constitutive of (and constituted by) the scientific facts they are about. However, while the historians contextualize the scientists' own accounts of discoveries by relating them to the dynamics of the debates and of the structure of the communities or networks involved in them, the historians' narratives are not usually presented as being affected by comparable processes.[7]

In history of science, this routinized avoidance or boxing of reflexivity is particularly intriguing because of the higher epistemological status that history of science was thought to have gained at least since Kuhn's *Structure*. If science is now often seen as a process whose dynamics can be understood through its history (or, more generally, by looking at it in "action"), then one would have expected that the features and limitations of historical interpretations of science should have received extensive attention by historians. Instead, these issues have been addressed almost exclusively by philosophers like Kuhn and Feyerabend, who, because of the historical nature of their theories of scientific change, could not avoid confronting the problem.[8] Although reflexivity becomes an issue whenever one recognizes cultural differences between his or her own culture and that being studied, the problem emerges more explicitly when—as argued by Kuhn and Feyerabend—one may encounter incommensurability between the belief systems of the various historical actors or between those of the historical actors and the historian studying them.

Both Kuhn and Feyerabend have argued that it is possible to circumvent incommensurability and reconstruct alien worldviews either by becoming bilingual (Kuhn) or by enriching one's language (Feyerabend). However, I believe that their assessment of the implementability of their interpretive guidelines has been overly optimistic. For instance, it may be telling that Kuhn did not present a sustained analysis of the implications of incommensurability for the writing of the history of science, but introduced the issue by means of an autobiographical reminiscence.[9] The tale of young Kuhn coming to terms with the incommensurability between Aristotelian and Newtonian physics through a sudden "conversion" was offered as a living and true exemplar of the general process being discussed. Quite fitting with Kuhn's own view of scientific change, his conversion narrative did not present explicit rules about how to do history of science in an environment of incommensurability, but seemed to suggest that what one needs is an unverbalizable skill—the historian's "tacit knowledge." However, the vividness of Kuhn's exemplary conversion tale may hide some of the sociocultural conditions that frame the possibility of experiencing such a gestalt switch.[10]

Therefore, while the "neo-whig" relativist historians tend to gloss over their presentist point of view by making it "tacit," Kuhn invoked some kind of "tacit knowledge" to argue for the possibility of a nonpresentist and fully contextualized understanding of past science. As I hope to show, we may not need (or want) to rely on such opaque categories or stances.

Locating Historians of Science

In the 1960's, Kuhn's and Feyerabend's appreciation of the otherness and the legitimacy of old scientific worldviews reflected an innovative stance within history and philosophy of science. Today, especially in the wake of "multiculturalism" and its reception in academia, beliefs in the specificity and legitimacy of the "other's" culture and in the necessity of understanding it on its own terms are quite commonly held (or at least professed) by academics in the humanities and social sciences.[11] Such beliefs convey the comforting feeling that, if approached with a relativistic attitude, the interpretation of "the other" (historical, ethnic, cultural, etc.) is going to be "fair" and that dialogue (or at least consent) among different cultures may be achieved—at least in principle.

By avoiding any privileged point of view, academics suggest that relativism would keep clear from the dangers of hegemonic discourse. While relativism provides important tools to denaturalize claims and beliefs and to expose the role that power, domination, and other unpleasant dynamics may have played in their establishment and maintenance, relativist academics are often believed to be immune from these dangers because of the reflexivity they develop by practicing relativism. A relativist may expose the workings of power but should be able not to reproduce them in his or her historical interpretations. As epitomized by the transition from the "melting pot" to the "salad bowl" as metaphors for national identity in the United States, relativism's methodological and political "civility" has been a crucial tool for promoting respect for other cultures within academia and without.

However, the potential for cultural criticism is not an essential quality that belongs naturally to certain methodologies while being *a priori* alien to others. Methodologies are resources that yield different results when they are deployed in different cultural, political, and institutional environments. For instance, while relativism can be a powerful critical tool when deployed by a marginalized group against a hegemonic discourse, it may yield very different results when used by the status quo in a "democratic" environment as a way to delegitimize minority claims.[12] Symmetrically, while rationalism has been politically critical in the past—as when it was used in an attempt to delegitimize societies structured around birth-related privileges—it can, in other contexts, lend legitimacy to hegemonic master narratives. In short, both rationality and relativism may, when employed in suitable environments, lend legitimacy to existing forms of power. The crucial difference lies mostly in the different ways they do so.

As Donna Haraway has recently argued, rationalists speak as if they were "nowhere while claiming to see comprehensively."[13] Instead, she sees relativism as "a way of being nowhere while claiming to be everywhere equally." As she puts it, both stances are "god tricks": one claims not to be speaking from a specific, identifiable place and pretends, instead, to be able to evaluate the matter *globally*—either by being *everywhere* or by seeing *everything*.[14] Haraway's proposal for avoiding both "god tricks" is to think in terms of situated, partial knowledges. While her proposal is itself situated, as it reflects her perception of the predicament of feminist science studies, I believe that the proposed shift to situated (rather than simply *local*) knowledge is one that is relevant to other constituencies and that it may provide an important starting point to overcome the epistemological (and political) shortcomings of both relativism and rationalism.

As I understand it, the main difference between local and situated knowledge is that the latter is not presented as a knowledge that a given group simply "happened" to develop. Rather, situated knowledge is something that is produced through being located in a certain position that allows for a *specifically* partial perspective. Consequently, although such knowledge may well be *partial*, it would not be *arbitrary*. Moreover, being partial is no sin, because the belief that accurate knowledge can be produced only through global perspectives (seeing *every*thing or being *every*where) is shown to be maintained only through the *ad hoc* introduction of god tricks. These are aporias resulting from the assumption of the possibility of *global* knowledge (in either its logocentric or its relativistic brand). Therefore, the shift to partial knowledge proposed by Haraway is not a gesture of epistemological retreat but rather the result of the awareness of how knowledge is *necessarily* produced through partial perspectives.

Trying to apply some of these considerations to the current methodological practices of historians of science, I see the emergence of "neo-whiggism" in history of science (or the avoidance of reflexivity in sociology) as an important (though implicit) admission that, in the end, our god tricks do not work. Relativist historians of science cannot be "nowhere while claiming to be everywhere equally." Quite to the contrary: we are anchored in (or stuck with) our present, and our location imposes specific limits to our understanding of the historical "other." In short, despite frequent reassuring-sounding claims to the contrary, we cannot understand the past on its own terms. As indicated by the implicit assumption of a presentist point of view in some of the recent historiography, our allegedly global respect of the historical "other" is bound to break down.

This, I think, is no simple failure of "methodological nerve." By boxing the reflexivity question or by writing implicitly "neo-whig" history we are doing more than simply effacing what we may perceive as a problematic feature of the relativistic framework. Rather, such a defense of relativism might also reflect an attempt (though not necessarily a conscious one) to defend a discourse that, after a "subversive" past, has recently developed symbiotic ties with the university, which, especially in the United States, has become interested in representing itself as an institution producing pluralistic, fair, and nondogmatic culture. If in the past being a relativist put one at risk of being represented as a science hater anarchonihilist, now the same stance may allow one to fashion himself or herself as an interpreter whose relativistic impartiality legitimizes him or her as a reliable interpreter of all the many tensions, struggles, and negotiations that characterize the workings of science and society.[15]

In short, by being relativists, we may be also defending our identity as university-based social "scientists"—a new kind of expert for a new kind of socioinstitutional environment and cultural agenda. For example, on 3 June 1992, in a radio program about the Los Angeles "riots" on KCRW (a Santa Monica–based, NPR-affiliated radio station), a relativist academic was introduced as "an expert in cultural diversity."[16] When placed in the contemporary academic context, relativism provides a venue for representing oneself as an "objective" interpreter—that is, as somebody whose comprehensive perspective does not derive from one's panoptical (and therefore hegemonic) vision, but from "being everywhere" and therefore "understanding everybody." We are no longer after who is right or wrong: now we can understand science as a process or as a form of "action," something that is neither good nor bad but simply *is*.

When practiced by academics, relativism tends to transform the institutional locatedness of their knowledge into a "politically correct" method—one that gives credibility to the interpreter while legitimizing the host institution as a place where nonhegemonic, nondogmatic, other-friendly discourses are developed. Through these discourses, the university contributes to the mythologies of consent, fairness, and respect of difference on which modern participatory democracies are deemed to be based. By this I am not trying to suggest that relativism is politically problematic in a general sense and that, consequently, it does not have anything else to contribute to cultural and political debates.[17] As we know, the respect of other cultures is far from being something we can take for granted. Moreover, here I am not talking about the uses of relativism by marginalized or oppressed groups, but rather those produced by constituencies operating in allegedly legitimate and legitimizing institutions such as universities.

To summarize, I suggest that by "locating" the historians and sociologists of science (and making explicit the partiality of their perspective on scientific change and practices) we may be able to address three important and related issues. One is that the hesitations about acknowledging "neo-whiggism" and the boxing of reflexivity are strategies aimed at covering a problem that does not exist—a problem that is caused only by our insistence at playing god tricks. The second is that, by dropping the relativists' god trick of "being nowhere while claiming to be everywhere equally" we may also be able to avoid many of the epistemological problems for which relativism is attacked by rationalists. Third, playing relativistic god tricks is not just a harmless self-deception. When played by relativist academic historians and social scientists, god tricks help legitimize the university as the institution where "scientific" social knowledge is produced. In short, making explicit the historians' and sociologists' location is both epistemologically rewarding and politically critical.

From Relativism to Contingentism

I will now try to develop a localizing critique of relativism in science studies—a view that I would call "contingentism." Paradoxically, such a critique may be developed from an analysis of the role of incommensurability—a notion that is usually taken as the very emblem of relativism.

I have recently proposed a genealogical interpretation of the phenomenon of incommensurability based on an analysis of a debate between Galileo and a group of Aristotelian philosophers.[18] Instead of taking a *synchronic* view of the incommensurability sometimes existing between competing scientific paradigms, I have suggested a genealogical perspective aimed at identifying some of the processes responsible for the *diachronic* emergence of incommensurability. While a synchronic approach is bound to present a scenario in which different groups and cultures hold different and possibly untranslatable belief systems, a genealogical view of scientific change gives visibility to the *mechanisms* linking the production and maintenance of groups' identities to the knowledge they produce over time. The adoption of a diachronic view makes it possible to perceive the phenomenon of incommensurability not simply as a problem but rather as a key to understanding how new paradigms or worldviews develop out of or away from old ones during a nondirected process of scientific change in which different groups or cultures fashion new identities for themselves.

Moving from a preliminary analogy between Darwin's population-based notion of species and Kuhn's community-based view of paradigm, I suggested an analogy between incommensurability and sterility. Just as a variety's inability to breed back with the original species marks the beginning of a new species, the inability to communicate between an emerging paradigm and the previous one (that is, of "breeding cognitively") may be seen as the sign of the establishment of a new "scientific species." However, as I hope will become clear later, my position is quite distinct from what is commonly known as evolutionary epistemology.[19]

It is through this perspective that incommensurability becomes more than just an obstacle to communication among competing groups of scientists and begins to appear also as a necessary component of the process of scientific change. In a sense, incommensurability represents a cognitive cost that sometimes cannot be avoided for scientific change to take place. If scientific knowledge is the collective product of a group of interacting scientists, such a group needs to remain cohesive and committed to the articulation of its paradigm.[20] The maintenance of the group's cohesion depends also on the members' sharing a professional and disciplinary identity, and having an identity is connected to representing oneself as different from others—a process that involves various ways of "keeping the other at a distance." In the case of science, the maintenance of group cohesion is often connected to scientists of competing groups' talking past each other or boxing adversaries' claims within one's linguistic and conceptual categories—a behavior that is very likely to produce misreadings. Consequently, the possible breakdown of communication between competing scientists cannot be seen as merely unfruitful. The hypothetical scenario resulting from everybody's willingness to learn every other group's worldview might not be characterized by a perfectly ecumenical and rational science, but rather by the absence of different groups, disciplines, and paradigms, and, consequently, by the termination of cognitive activity itself.

Seen from this angle, incommensurability ceases to be the unfortunate result of cohesion-keeping processes that happen to reduce one group's willingness to dialogue with another. Instead, these noncommunicative behaviors function as a sort of containing belt that makes cognition possible by keeping its actors together and committed to the articulation of their lexical structure. In the long run, such an articulation may produce a new lexical structure that may be incommensurable with that of the original group.[21]

An interesting reframing of the process of theory-choice results from this genealogical interpretation of incommensurability. By shifting our attention away from the process of choosing between competing theories to the processes that allow a group to maintain cohesion, build a common sociodisciplinary identity, and articulate a worldview, this view suggests that competing groups do not need to engage in a fully constructive dialogue in order to produce science. To be dropped, a claim does not need to be falsified (nor a research program superseded). Like species that die off not because they are directly eliminated by others but because they no longer fit the environment, paradigms can come to an end not because they are replaced or refuted by others but because they no longer fit that ecological niche—that is, the reward system of science and the socioinstitutional context in which they are located. However, as I will argue in a minute, the "extinction" of a given worldview does not mean that it is unsatisfactory by any absolute standard. Local contingencies (rather than the hidden hand of rationality) have a lot to do with it.

Different groups develop and hold different representations of the world, but it is not at all clear why they would always need to convince the competitors that theirs is best. The success of a representation is not necessarily achieved by having it chosen and adopted by all competing groups. Rather, it may simply be that a paradigm or set of practices appears to have been adopted by everybody simply because the groups that did not adopt it became professionally extinct. In short, intergroup justification of beliefs is not generally necessary. Somebody belonging to a given scientific group does not necessarily need to justify his or her beliefs to members of other groups. Simply, those beliefs are the only ones he or she has. In a sense, to ask people why they believe what they believe is a bit like asking them why they look the way they look.

This genealogical view of scientific change helps reframe relativistic pictures of science both by stressing the diachronic (and localizing) processes that limit a group's knowledge options and by acknowledging the role of the "out-there" in the cognitive process. In fact, a scientific tribe ontologizes its worldview not simply because it does not have access to alternatives, but also because such a worldview embodies (by the very fact of having survived) the result of the successful interaction between that tribe and the environment (both natural and social) with which it happened to interact. However, there is still some degree of arbitrariness (for lack of a better term) in the tribe's knowledge, in the sense that the tribe's cultural genealogy could have taken different paths and could have led to much different cognitive and sociocultural scenarios.

Consequently, although the knowledge of that tribe reflects a "success story" in the sense that it indicates that the tribe has survived a more or less long interaction with its environment, the quality of this knowledge cannot be assessed according to any general and external parameter. Although one may say that a tribe's knowledge is "absolutely good," that statement has to be understood as meaning that the tribe that held that knowledge managed to survive, and that we have no external point of reference by which we may evaluate the degree of its quality. The way in which it is "absolute" is by default only. Although the dynamics of this genealogical process indicate that other scenarios may have come about, those are scenarios that, in practice, we cannot reconstruct.[22] In short, this perspective suggests that the "out-there" had and has an essential input in the knowledge of the tribe, but that, at the same time, one cannot apply categories such as truth or progress to this process except in a very specific sense.

The notion that is central to this view of scientific change is the fit between a given group and its worldview. However, the notion of fit is not meant to refer to the closeness between the physical world and the group's representation of it. Once viewed in a genealogical framework, the problem of evaluating closeness (or hazy notions such as simplicity or elegance) becomes something of a red herring. A worldview's being good (in the sense of having contributed to the group's socioprofessional survival) does not imply that one's representation of the world was necessarily close to it. Although a good representation is one that fits the environment, fit does not need to be thought of in a *mimetic* sense. We do not need to think of representations of the world as good or bad copies of it, but simply as contingently effective or ineffective—that is, as making it possible (or impossible) for a given group or culture to survive as such.

Evidently, this does not suggest that, as far as representations of the world are concerned, "anything goes" in any given context, but rather that, in different contexts, worldviews may "go" for different reasons. That a tribe holding a certain view survived socioprofessionally implies that something "went" at some point. As we know, not all worldviews would have been effective in allowing a group to survive in a given socionatural context, or in a given sequence of such contexts. However, because both tribes and socionatural contexts change both historically and geographically, I see no point in assuming that the interaction between representations and environments must lead to their survival or extinction *always for the same reasons.*[23]

This is something similar to saying that at one time and place one species may have become extinct because it could not survive a change in the environment's temperature, whereas in a different place or period another species may have been exterminated by the arrival or development of a disease previously unknown in that ecological niche. However, had they survived those unexpected changes, these species might have turned out to be doing quite well in today's environment. In short, representations *do* interact with the "out-there" (as shown by the fact that some of them survive though others become extinct), but such effects cannot necessarily be traced back to the same cause. Nor can we say that those causes were rational. At the same time, the locality of the parameters of fit should not be read as implying that science is arbitrary. The contingency framing scientific change is crucial to its "adaptation" and eventual possible survival.

Consequently, I do not think that the fit between the environment and its representation can be conclusively evaluated through *a priori* rules such as "rationality."[24] "Fit" can be detected only *a posteriori*—accordingly as a tribe did or did not survive. Nor do I think it is correct to talk about degrees of fit. Fit is, so to speak, a binary category. Either the culture fit (and survived) or it did not fit (and became extinct). For similar reasons, I think it improper to say that a representation *fits* a given socionatural environment.[25] True, a paradigm's being still around suggests that, in *some* way, it does *not not fit* the environment. However, the only certainty about fit is necessarily an *a posteriori* and *negative* one. All we can say with certainty about fit is that the cultures or groups that are no longer around happened *not* to have fit the environment at *some* point and in *some* way.

Finally, it is worth pointing to the fact that, by considering scientific change in terms of paradigm survival rather than paradigm choice, we do not need to draw the line between society and nature. I am not saying that the natural should be subsumed under the social or vice versa, but simply that such a distinction is not useful once one looks at scientific change with a genealogical perspective. What makes a worldview extinct is the socionatural environment, and not nature or society taken separately.[26]

This genealogical view of scientific change avoids, I think, the charges of arbitrariness usually leveled against relativistic views of science by showing the processes through which a given group may adopt certain beliefs because of the impossibility (at that location and time) of holding alternative views and yet remaining members of that same tribe. At the same time, it offers a picture of the development of knowledge as a process of full interaction with the contingencies (and the play of chance) that a scientific tribe may encounter along the way. In short, the "out-there" does enter the picture, but in ways that cannot be reduced to the rules of Rationality.

Understanding these mechanisms helps a shift of focus from *local* to *located* (or situated) knowledge in ways that are, I think, congruent with Haraway's proposal. As I hope to show, this shift is useful not only in order to bypass a deadlocked debate on rationality and relativism, but also in order to locate (and acknowledge the limits of) our possible interpretations of past (or other) scientific cultures given our cultural and institutional position.

Once we consider the historiographical implications of this genealogical perspective we may realize that neo-whig history of science becomes a quite appropriate methodological option. Fortunately, what I have kept referring to as neo-whig history is really not so whiggish after all. Rather, it can be best described as presentist history. Differently from whig historians, presentists do not write history as leading to the present in order to legitimize it. Although it is written from the present, presentist history acknowledges that the present from which it is written might have turned out quite different from what it happens to be.

As historians of science, we are not only located in the present but we are also tied to the current state of scientific knowledge and to our institutional predicament and culture as our ultimate frames of reference for historical interpretations. However, as discussed before, today's scientific knowledge cannot be said to be the best possible in any general sense. It is good only in the (important) sense that it made it to the present. Therefore, while the present state of science is the only system of reference we have, this system might have turned out quite different. This present is a *fact* simply because it is the *only artifact* we happened to get.

Consequently, it is not whiggish to say So-and-So was right if all we mean by that is that So-and-So's scientific tribe happened to be successful and that *we happen to belong to a culture that is genealogically connected to that of that tribe.* True, So-and-So survived also because his or her worldview was "good." However, as mentioned before, the meaning of "good" needs to be understood within this genealogical framework. Had the socionatural environment been different, things might have gone differently, and we might have been writing not only histories of a different science, but also from a very different point of view.

However, we cannot (re)construct how things might have gone—we do not have alternative points of reference beyond *this* present. Given these epistemological limits, all we can do is not to look for other ways in which science might have evolved (that would be another god trick), but to study the *mechanisms regulating the process* through which it became what it is now, and to understand how this very process frames our interpretation of the evidence about the genealogy of modern science.[27] While it is appropriate and rewarding to study science as a process, we need to understand that we do not have any external point from which to view that process.

Of course, the present is not one but many (depending on who we are and where we are situated in it). Nor do we need to be happy with or accept the present as we have it. As exemplified by Haraway's mixing of history and philosophy of science with science fiction, it is quite possible to write in this present while proposing categories, images, and metaphors that may be useful in changing it. My goal here was not to present a paralyzing picture of our predicament but rather to point at the processes through which we academic social scientists may contribute to normalization (through contextualizing and historicizing) by overlooking the processes through which we ourselves are historicized and situated.

Contingentism and Canon Formation

This takes me to Simon Schaffer's "Contextualizing the Canon"— a response to "contingentism" he presents in this volume. Schaffer too finds it problematic that relativist historians of science do not treat their claims as being produced through the same processes of knowledge construction they describe in the natural sciences. He too wants to correct this asymmetry *and* avoid reflexive regressus by focusing on the genealogical processes that locate the historians and their claims. However, his genealogical approach is different from mine.

Schaffer sees the canon of a discipline as "the corpus of exemplary texts that provide a standard of that discipline. The canon of the human sciences provides resources for any currently possible historiography of the natural sciences." I read Schaffer as presenting the formation of the canon in a given social or human science as a process analogous to what sociologists of science have called the "closure" of scientific controversies. In the natural sciences people canonize experiments and instruments, whereas in the human sciences people canonize books. In both cases, the closure or canonization provides the practitioners of a discipline with a temporarily stable frame of reference against which to evaluate their claims. According to Schaffer, reference to canonical texts puts an end to the historians' "regress"—at least temporarily. Canonizing is like black-boxing.

However, as Schaffer argues through a number of examples, among the things relativist historians are best at doing are precisely the contextualization, historicization, and relativization of canonical texts. In these, they typically move backward from the time at which the canon is established and reconstruct the precanonical context—the array of contending candidates that were eventually silenced (at least temporarily) by the text that happened to be successfully canonized. This he calls the "teleological" (and I would say "presentist") dimension of relativist history. Therefore, while the documentation of a canon is a move toward removing the asymmetry between the historians' views of the construction of past scientific knowledge and their own historical narrative about it, that step is not sufficient to solve the problem. True, the canon may provide historians with a relatively stable frame of reference, but such a stability is not immune to the relativist historians' contextualizing skills. That relativist historians routinely contextualize other canons *but not their own* shows that, by itself, a canon is not sufficient to put the historians' regressus fully under control.

What we need to do, Schaffer argues, is to supplement the documentation of a canon with "a social history of canonization." It is not enough to contextualize, historicize, or relativize a canon;

one needs to show the processes through which this takes place—processes that should be applicable to one's own canon as well. The process of canon formation plays in Schaffer's scheme a role comparable to the process of genealogy of cultures in the framework I propose.

My first query is about how a social history of canon formation would look. While such history may provide us with a number of examples, it does not necessarily follow that we would be able to *abstract* the *process of canonization* from those discrete instances. To put it differently, what is the difference between documenting the canon and writing the social history of canonization proposed by Schaffer? While this distinction is crucial to the workings of his two-piece approach (because it is this distinction that would stop the historians' regressus), there seems to be a tension in the separation of these two components—a tension that has all the features of an "aporia." In fact, works like *Leviathan and the Air-Pump* or McKeon's and Ashcraft's contextualization of the canonization of the novel or of Locke's *Two Treatises* are *both* case studies *and* analyses of the process of canon formation, and yet their analyses do not seem to be free from regressus. Schaffer assumes that we can actually do something more than this, that we can have access to some sort of "meta-" point of view about canon formation. But is that really possible? And, if not, what would be left of his proposal?

I also have a corollary question about Schaffer's exclusion of science studies from the range of examples he provides about canon formations and contextualizations. This, I assume, results from Schaffer's belief that a discipline cannot undo its canon without undoing itself as a discipline. It seems that Schaffer would be right in not being too disturbed by this local asymmetry if the historians could actually gain sight of the *process* of canon formation through its social history. Had they access to that meta-level, the historians not only would know that canons are only contingently stable but would also understand how canons (including their own) emerge and disappear. They would be aware of operating in a domain of canon change but would also understand the dynamics of that change. However, as mentioned above, it is not clear whether canon formation would actually be graspable through its social history, or whether the understanding of that process we gain from studies like *Leviathan and the Air-Pump* would end the regressus. In short, is Schaffer's proposal really adequate to avoid the limits of contextualism and historicism he is addressing?

By contrast, the genealogical process of situating I have sketched above tries to locate the practitioners of science studies in ways that are both more rigid and more flexible than Schaffer's proposal. The localization I propose is more rigid because it points to a few constitutive (and therefore unavoidable) constraints of the historians' knowledge. My central claims are that presentism is inescapable and that the necessary character of presentism is sufficient to stop the historians' regressus. But while my claim about the necessity of presentism is categorical, the view I propose does not impose constraints on whatever canons (including none) people may want to adopt.

That my model does not talk about canons is not accidental. Being located is not reducible to having a canon, and I do not believe that, as Schaffer puts it, "the canon of the human sciences provides resources for any currently possible historiography of the natural sciences." Being situated in a certain position frames one's perception of the resources one can use to legitimize his or her interpretation of science, and the canon of the human sciences (whatever that may be) may (or may not) be seen as a resource. What about interpretations of science that mobilize science fiction as much as more "canonical" science-studies literature? Further, while I do agree with Schaffer that (in certain disciplines and in certain historicoinstitutional contexts) canons did and do play the role he attributes to them, I am much less convinced about both the viability and the desirability of a canon in science studies today. Not only does science studies seem to have no unified canon, but—to continue to play on the analogy between canon and closure—we may not even have (or want to have) a "core set" (to use Collins's terminology) that would make closure or canonization possible. Given the interdisciplinarity of our field (history, philosophy, sociology, anthropology, gender and cultural studies) and the porousness of many of these communities and discourses, it seems very unlikely that we will ever be able to get to a canon even if we wish to do so.[28] More important, is "canon" something that is really worth defending (though in a "reflexive" manner) or—as I have tried to propose—should we rather try to think about alternative ways in which regressus within relativist science studies can be avoided?

Conclusions

Contingentism argues that there is no need to be apologetic about relativism, to defend it, and to pretend that we can be everywhere. Contingentism provides, I think, a better link between the process of scientific change and its historical or socioanthropological interpretation. In doing so, it bypasses some of the deadlocks of the rationality/relativism debate, and, by locating the actors (both scientists and science-studies practitioners) and framing their interpretive options, it avoids the pitfalls of relativism and its possible problematic political uses resulting from playing experts—though of a "politically correct" brand.

In fact, the exposure of the aporias of relativism provides tools for criticizing the ways academic discourse uses those aporias as opaque spots in which the power dimensions of that discourse can remain hidden. Contingentism indicates that, despite academic mythologies of politeness, we cannot interpret the past (or the "other") on its own terms. Interpretations of the "other" are bound to be partial; that is, partiality is not a failure but a necessity. All we can do is be aware of the ways in which they are *necessarily* partial— that is, of how these partial understandings of the "other" (not unlike incommensurability) allow for the development of different cultures (scientific or not).

Contingentism does not present a critique of relativism in an attempt to go back to rationalism. Rather, it indicates that some of the problems of relativism derive from confronting cultural difference with an incomplete understanding of the processes framing the genealogy of those differences. Understanding those processes supports the relativists' critique of rationalism while avoiding some of its epistemological and political problems. In fact, by pointing at the genealogical limitations on our ability to understand other scientific cultures, it helps us understand the processes that frame scientific change and the ways in which we and our discourses about science are situated by those same processes.

SIMON SCHAFFER

Contextualizing the Canon

❖ In acquiring one's conception of the world one
always belongs to a particular grouping which is
that of all the social elements which share the
same mode of thinking and acting. We are all
conformists of some conformism or other. . . .
The question is this: Of what historical type is
the conformism?

—Antonio Gramsci

Mario Biagioli's "From Relativism to Contingentism" argues two
theses about the contemporary condition of relativist studies of the
sciences. First, it is claimed that historians of science who use
sociological resources implicitly ignore the fact that their historiography is affected by the same social processes that they document in the natural sciences; second, that to correct this asymmetry
it is necessary to render explicit the location of the historian and to
understand the processes that produce the position currently occupied by historians of science. The uniqueness of this position,
Biagioli urges, is a result of the genealogical process through which
our current standards and conventions have emerged. Biagioli calls
his program "contingentism." Historians of science are ineluctably
tied to their own context. But, Biagioli recommends, they should
recognize that the situation from which they work could have
turned out differently—their milieu is contingent. The study of its
emergence does not legitimate it. The essay sketched out in what
follows proffers a possible response to contingentism. Biagioli is
very interested in the way our own conventions have emerged and
been canonized. The canon of a discipline is the corpus of exemplary texts that provide a standard of that discipline. The canon of
the human sciences provides resources for any currently possible
historiography of the natural sciences. Canonization is part of the
process that Biagioli identifies in group solidarity: "The mainte-

nance of the group's cohesion depends . . . on the members' sharing a professional and disciplinary identity, and having an identity is connected to representing oneself as different from others."[1]

To document the canon is to address the first of Biagioli's theses. It exemplifies the conventions by which historians of science ply their trade. But, typically, historians of science moved by relativist sensibilities have set out to break this canon, to show that classics had very different meanings when placed in their local context of production. What would happen if historians of science engaged in the enterprise of explaining how canonical texts became classics, rather than simply contextualizing those texts? The purpose of what follows is to show that a social history of canonization is required as a complement to the social history of knowledge production, and that it can answer demands that the social historiography of science be properly reflexive. In so doing, it addresses a fundamental concern of the current interest in *disunity*. Much of this concern centers on two observations about the way scientists work. They operate with very heterogeneous resources, and their operations have a distinctive temporality. Science is not to be seen as a single form of knowledge made in a single way, nor is it to be treated as a synchronic network of logical implications and empirical verifications. Canonization matters here because canons are made and used to formalize traditions, to differentiate cultures, and retrospectively to make disciplines' histories into effective resources. It is a process that helps manage diversity and define temporality. Biagioli agrees that even such an apparently general category as "rationality" can be "perceived as a set of guidelines (rather than prescriptions) developed *a posteriori*."[2] The construction of such guidelines demands that scientific communities do their own historical work. They become devotees of a form of common law. The language of judgment and precedent is pervasive when we talk about guidelines and customs. Past judgments are used to make traditions and precedents, then developed to help communities look robust and their behavior seem self-evident.

Contingentism joins other recipes for dealing with the "methodological horrors" of reflexivity. No doubt their interest depends on the supposed failings of the sociology of scientific knowledge, which stands accused of insufficient attention to its own interpretive stance. It is alleged that sociologists have shied away from such attention because they supposedly hold that such self-analysis would undermine their own authority. According to Steve Woolgar, for example, the methodological horrors should not be managed by erecting a hierarchy of knowledges that are variably infected by problems of representation, nor managed by supposing that the horrors only affect other inquirers' work. So we should not suppose that some sciences escape the problems of representation and reflexivity; nor should we suppose that our own work is invulnerable. There is nothing very startling in these worries. Any interpretive enterprise will confront this version of the hermeneutic circle. Any inquiry concerned with representation will have to deal with representation's political constitution as a problem at once of depiction and of delegation. Woolgar proposes "new ways of *interrogating* representation."[3] In what follows, I recall some familiar ways of interrogating representation's power. I stress the language of custom, precedent, and judgment because I am interested in the way in which communities' common sense is constituted as such, including our own's.

A political agenda is implied by the history of canonization and self-evidence. Here I appeal to the canons observed by materialist historians who have explored the ways in which the practices and norms of rebellious traditional cultures contested the new ways of capitalist life. These explorations show how self-evidence is set up politically. For example, E. P. Thompson's account of plebeian custom in early modern England provides good resources for our own study of the ways in which taken-for-granted forms of life get established and help affirm or subvert the most apparently "natural" values. Moral and political economies involved rival means of representing behavior, needs, and interests. In a justly celebrated analysis, Thompson displaced a condescending view of the irrationalities of eighteenth-century food rioters, the "rebellions of the belly," with a careful recreation of the protesters' moral economy, a network of customary expectations to which they appealed as the standards of right conduct. His work suggests why the language of common law works so well in analyzing the production of canons and their application.[4]

I also appeal to canons of the sociology of scientific knowledge, which cannot be convicted of a willful suppression of these concerns with self-analysis and the origins of self-evidence. To defend the claim that "the application of a term is a judgment," a key tenet of recent sociology of knowledge, Barry Barnes points out that "it can always be said that previous usage was wrong . . . and that it must be revised." He denies that "usage is determined in advance," and he argues that reflection on human institutions is required to account for such behavior. "To understand concept application we must understand ourselves." The social history of canonization is part of this process of self-understanding. It explores the contingencies through which inquirers' present conduct becomes, for them, anything but contingent. It addresses the processes through which communities define themselves as distinct and restricted. It also helps explain how members of such communities judge their own work universally compelling. It supposes that we belong to such a community.[5]

This essay outlines a program for such a social history in four stages. In the section that follows, "The Rules of Life," the role of the canon of science studies is introduced by considering two exemplary figures, Hobbes and Durkheim. Both have been subjected to rather detailed contextual analysis. Debates about these analyses, significantly, often explicitly engage in reflection on the sacred quality of canonical texts. I use the example of the conflict between David Bloor and Joseph Ben-David about the proper reading of Durkheimian sociology. The next section, "The History of Pregivenness," extends this reflection to show that the canons of cognate disciplines, such as philosophy, in the case of John Locke, and literature, in the case of the Augustan novel, deploy the sacred categories of blasphemy and heresy as they develop notions of membership and loyalty to disciplinary structures. Thus an analysis of canonization requires both a contextual account of the origins of canons and of their beatification. In the third section, "Moments of Danger," it is suggested that science studies itself should move beyond what Barthes called "mythoclasm," the display of the original or local purposes served by our conventions, to a careful analysis of the process by which these conventions are produced and reproduced. The political agenda of this analysis is clarified by consideration of the role of "national heritage" and other themes of contemporary conservative culture recently analyzed by critics such as Raymond Williams and Patrick Wright. Finally, in a section entitled "Canonization and Apotheosis," the process of canon formation at the very limits of a cultural field is explored. I consider recent critical and historical work by Greenblatt, Bernal, and Said, together with a series of anthropological accounts of the death of Captain Cook. It is stressed that the difference between the allegedly original senses of practices and those to which they are subsequently subjected is not to be treated as an unfortunate and incorrigible misfortune, nor does there seem any reason to limit our own attention to these original senses. There is an illusion of familiarity here. Just because our own canons are the products of long histories of commentary and manipulation, their original sense, which might seem easily accessible, is hard for us to render. Hence there is some virtue in the juxtaposition of canonization within our own cultures and canonizations at moments of cross-cultural contact.

The Rules of Life

> By the Books of Holy SCRIPTURE, are understood those which ought to be the Canon, that is to say, the Rules of Christian Life. And because all Rules of life, which men are in conscience bound to observe, are Laws; the question of Scripture, is the question of what is Law. . . . Seeing therefore I have already proved, that Soveraigns in their own Dominions are the sole Legislators; those Books only are Canonicall, that is, Law, in every nation, which are established for such by Soveraign Authority.
>
> —Thomas Hobbes

In *Leviathan*, Hobbes recalled the connection between the construction of authoritative texts and the "question of what is Law." Canonization, the process of making holy, is concerned with sovereign power. So there is an obvious problem in speaking of the context of a canon. A canonical set of exemplary texts provides a discipline with an ancestry. Canonization is, precisely, a process in which the contingent trappings of mundane context are sloughed off, and essential, semipiternal nature revealed. No doubt this is why recent intellectual historians have found it so worthwhile, and so comparatively easy, polemically to contrast the anachronistic, disciplinary meanings given to canonical texts with the meanings they attained in the settings of their original production. But this ease is slightly deceptive. It is not simply a question of pointing out our saints' feet of clay. Richard Watson has urged the importance of serious study of what he calls "shadow histories," narratives of philosophy's past that, whether historically accurate or not, provide working philosophers with material for their own arguments and that reinforce rival versions of the canon. His suggestion quickly prompted an energetic discussion of the power and validity of myth and history in such disciplines as analytic philosophy. Historicism has its interests too. They become apparent when we consider material that has become canonical for our own work. Here I consider both Hobbes and Durkheim, whose work has been canonized and also provides resources for understanding that process of canonization.[6]

In our recent study *Leviathan and the Air-Pump*, Steven Shapin and I used resources drawn from this historicism to read *Leviathan* as a text in natural philosophy and epistemology so as to estrange some important aspects of the experimental canon. Bruno Latour has subsequently read Hobbes's work, and ours, as a contribution to the modern constitution, a means through which nonhuman agents came to be representative of human subjects.[7] Hobbes was also concerned with the fate of his own *Leviathan* as a canonical academic text. Hobbesian interests pointed toward the *institutional* and *practical* settings in which discipline would be maintained and discipline's rule of life established. This book, he reckoned, would play a crucial role: "I think it may be profitably printed, and more profitably taught in the Universities, in case they also think so, to whom the judgment of the same belongeth. For seeing the Universities are the Fountains of Civill, and Morall Doctrine, from whence the Preachers, and the Gentry, drawing such water as they find, use to sprinkle the same (both from their Pulpit, and in their Conversation) upon the People, there ought certainly to be great care taken, to have it pure, both from the Venime of Heathen Politicians and from the Incantation of Deceiving Spirits." Thus Hobbes rightly reckoned canonization would be contested. His antagonists, such as the Oxford professors Seth Ward and John Wallis, agreed with him that the formation of an academic discipline was polemical. But they disagreed with him as to the profit to be garnered from canonizing *Leviathan*.[8] In contrast, and for some of our own contemporaries, this text is now the canonical work of a visionary political philosophy. How did this transformation take place?

Disciplinary politics helps provide an answer. In his celebrated essay "The Activity of Being an Historian," Michael Oakeshott set out a philosophical defense of the autonomy of the discipline of history. The autonomy in question was, specifically, an autonomy from natural science. In this sense, Oakeshott replayed an older German debate about the place of *Geisteswissenschaften*. He reckoned that historians were scientific to the extent that they had no practical interest in the past. He noted that historicism and scientism appeared contemporaneously in modern Europe. But Oakeshott complained that historians had wrongly and slavishly aped the natural sciences, that historical events were not like scientific objects, and that the search for necessary and sufficient reasons infected the historical disciplines. This emancipation proclamation helped make historical work look independent of the problems of scientific representation. Our own concerns with reflexivity owe a great deal to the resources that Oakeshott helped produce. He needed a reworked canon to substantiate his tradition. It needed a canonical text that accounted for the unity of historicism in the face of the individualism of historical events. Oakeshott found it in *Leviathan*. "The context of the masterpiece itself, the setting in which its meaning is revealed, can in the nature of things be nothing narrower than the history of political philosophy." This orotund sentiment became relevant to Oakeshott's reform of intellectual history through his combination of localist interpretation and universalist morals. Among Hobbes scholars, Oakeshott approached Quentin Skinner in his willingness to concede the contingencies of *Leviathan*'s formation. But Oakeshott insisted that this "particularism," as he called it, should not "mislead us into supposing that nothing more is required to make a political philosopher than an impressionable political consciousness; for the masterpiece, at least, is always the revelation of the universal predicament in the local and transitory mischief."[9] The theological language is very marked. Without a history of the emergence of the curriculum of political philosophy after 1800, in which several patriarchs of our own disciplines, such as William Whewell and Ferdinand Toennies, were deeply involved, we will not understand how *Leviathan* became this kind of masterpiece, how it became sacred.

By situating the canonical text at the intersection of the universal and the local, Oakeshott and his fellows remind us that canonization is the making of a saint. A common resource that sociologists have used to deal with this process is the work of Emile Durkheim. Here the reflexivity of the social analysis of knowledge production becomes obvious. Durkheim is doubtless a sacred figure, an ancestor whose texts provide the canons of a discipline. At the same time, the principal means for grasping this beatification as a socially accountable process are provided by Durkheim's own work. It is possible to provide a contextualization of that work. In 1887, aged twenty-nine, Emile Durkheim was appointed a lecturer in social science at Bordeaux. For the next fifteen years he worked hard to establish a discipline he called sociology, write its rule book, and find an authentic Francophone ancestry for it. "To determine the part played by France in the development of sociology during the nineteenth century is, in large measure, to write the history of that science; for it is among us and during that century that it was born, and it has remained an essentially French science." The canon was simple and scriptural. Montesquieu and Condorcet prophesied the scientific future but could not realize it themselves. Saint-Simon, Comte, and Cournot succeeded in scientizing the study of society. The inauguration of this project to make an autochthonous and Girondin canon was, significantly, a doctoral thesis composed in Latin, presented at the Sorbonne in 1893, and devoted to the work of Montesquieu. Steven Lukes notes that Durkheim's Sorbonne defense of his thesis, and of the accompanying text *The Division of Labour*, was greeted with rapture back in Bordeaux. Thus *La petite Gironde*: "We are happy to state that, thanks to M. Durkheim, sociology has finally won the right to be mentioned at the Sorbonne. . . . It was indeed an event of great importance. It could not fail to concern both those interested in the progress of social science and those who are concerned for the good name of our University of Bordeaux." The universal project of social science intersected with the local purposes of Girondin culture.[10] Durkheim reckoned that the early eighteenth century lacked the technology of social science. It possessed no historicism, a faith in travelers' tales, no statistics, no psychology. Durkheim built his definition of a social science into his specification of its tools. Thus he wrote that "since society is a large living organism with a characteristic mind comparable to our own, a knowledge of the human mind and its laws helps us to perceive the laws of society more accurately."[11] The thesis is a good case of canonization. It occurs in a specifically pedagogical setting (a defense at the Sorbonne); its aim is disciplinary (the installation of sociology within the human sciences); finally, it established a sacred succession for the enterprise.

There are important relationships between this story about the social origins of Durkheim's program for a scientific study of society and recent uses of that program in the study of science. It is striking that the debates about the relativist implications of the sociology of scientific knowledge have positioned themselves within this canon. When David Bloor set out to explain the sources of resistance to the sociological analysis of scientific knowledge, he explicitly used Durkheim's model of the sacred to show that scientific knowledge is often deemed to be profaned when it is itself subjected to scrutiny. Bloor argued that "science and knowledge are typically afforded the same treatment as believers give to the sacred." Then he appealed to Durkheim's claim that the sacred is a means of protecting and authorizing the basic principles of social order. Social models are deployed when subjects reflect on the principles of their knowledge and their conduct. This will be especially so for the most privileged areas of culture. "We become perplexed about knowledge in the first place . . . when rival claims to provide knowledge are offered by different social groups such as the church and the laity, . . . the powerful and the weak."[12]

Bloor offered a Durkheimian explanation of the position of critics of his program. Any sociologist among those critics must either reject the discipline's canon or become a sociologist of scientific knowledge. In his response to Bloor's proposal, Joseph Ben-David worked hard to sanctify the canon. Durkheim, he claimed, provided no warrant for sociological relativism. "Durkheim's stress on the importance of the social context of all human thought does not imply any social determinism." Bloor read Durkheim wrongly by concentrating on "some misleading terminology in his early writings." Ben-David explained that Durkheim expected close linkage between social structure and knowledge forms only at early stages of history. The progress of society was marked by an increasing objectivity of collective knowledge. The strong program "is not a proper application of Durkheim's theory." Furthermore, by "becoming interpreters of particularistic, context-dependent epistemologies," sociologists of knowledge were acting like philosophers rather than social scientists. Ben-David endorsed a sociology of science that would "explain the relative continuity in the normative structure of science and in the goals of the scientists." Since Ben-David also represented himself as a scientist, he found it important to show that Bloor's sociology of scientific knowledge disrupted the continuity in the normative structure of the discipline. So Ben-David charged sociologists of knowledge with blasphemy, since they abused the discipline's canon, and with apostasy, since they had abandoned the proper tasks of the discipline.[13] Such religious imagery is simultaneously an example of the sociological analysis of knowledge production and an attempt to defend its legitimacy against its critics. Reflection on the authority of this process must necessarily draw on an articulated picture of the construction of authority itself.

The History of Pre-givenness

The enterprise that seeks to understand the formation of a discipline's canon will set out to map the purposes and processes through which texts are set up as authoritative. The case of Hobbesian texts reminds us that the early modern period is peculiarly susceptible to this approach. The appearance of a range of new literary technologies within the sciences and the republic of letters has irresistible appeal to scholars engaged in the desacralization of disciplines. Michael McKeon's *Origins of the English Novel* and Richard Ashcraft's *Revolutionary Politics* are recent examples of this strategy of simultaneous engagement with, and criticism of, the literary canon of key disciplines. McKeon situates the work of Defoe, Swift, Fielding, and Richardson in the contexts of radical transformation in social order, especially patterns of inheritance; in credit, especially in real estate and stock jobbery; and in natural knowledge, especially travelers' tales and natural philosophical reportage. But, of course, his work does not damage the exemplary status of *Tom Jones* or *Robinson Crusoe*. Indeed, these canonical texts serve to warrant the structures of change that his argument charts. McKeon puts his concern with the explication and contextualization of the canon like this: "By the middle of the eighteenth century, the stabilising of terminology—the increasing acceptance of 'the novel' as a canonic term, so that contemporaries can 'speak of it *as such*'—signals the stability of the conceptual category and of the class of literary products that it encloses." McKeon's work is therefore inevitably presentist: "My procedure in this study will be to work back from that point of origin to disclose the immediate history of its 'pre-giveness.' "[14]

Similarly, Ashcraft's remarkable study of the radical clubland agitation that spawned John Locke's *Two Treatises* announces its purpose as a challenge to existing studies of the canon of political philosophy. Ashcraft explicitly sets out to juxtapose the canonical (Locke's text) with the uncanonical ("manuscripts, newspapers, correspondence, and hundreds of tracts and sermons"). He reckons that the neglect of this other literature "simply reinforces the propensity to identify political theory with philosophy—or, rather, with a few works written by philosophers."[15] In his response to Ashcraft's claims, the historian of political philosophy J. G. A. Pocock distinguished a reading of the *Treatises* as part of the Lockean corpus from a reading in context: "This may be a case in which to return texts to their immediate practical context is to restrict their significance rather than expand it." The result of Ashcraft's contextual analysis, unlike that of Pocock's, is to make Locke's work canonical for the protagonists of the Glorious Revolution. "Across the political spectrum," Ashcraft points out, "the work was perceived as a radical political statement, clearly relevant and popular in the context of the political events of the Glorious Revolution. . . . What happened to the *Two Treatises* after it appeared in print is another story." That other story should also be our concern. The critiques of McKeon and Ashcraft show that contextualizing the canon is a key element in their own disciplinary strategies: between sociology and literary criticism or between political science and philosophy. Disciplines tell stories of their formation that depend on the construction of a set of exemplary texts. Their rivals show that these texts were originally solutions to different, extra-disciplinary problems. The methods these rivals use also represent rival disciplinary strategies.[16]

Alongside the histories that simply recontextualize the canonical texts, we also need a contextual history of the work of canonization. After learning of the radical conspirators among whom Locke worked, we can begin working on that "other story" of the way

Locke's *Two Treatises* became a founding document for the discipline of political philosophy. This other story can help itself to the strictures of historicism. In a recent study of the formation of the American philosophical canon, Bruce Kuklick has attempted this task. He sketches the contrasts and conflicts between the tradition that stretched from Plato via Ramus to the New England Puritans, and that which was dominant until the third quarter of the last century, increasingly antiskeptical in tenor and Scottish in accent: Locke, Reid, Stewart, and William Hamilton. From within the history of the American philosophy curriculum, Mill's *Examination of Hamilton* (1865) emerges as a significant moment for the reorganization of the canon. Kuklick's representation of the canonization of Common Sense fits interestingly with Paul Wood's account of Reid's transformation into Smilesian saint through the work of Dugald Stewart. Viewed from the milieu of the Edinburgh professoriate, Wood shows how Stewart made Reid's philosophy "unrelated and indeed vastly superior to that of Kant." Hamilton developed this move into a systematic comparison between Common Sense and German idealism.[17] Kuklick explains that "nineteenth-century Americans believed that in the work of Sir William Hamilton, the Scottish position had overcome Kant's criticism of Reid. . . . Hamilton was recognized as having refined Scottish ideas to take account of whatever was valuable in Kant." These stories suggest the need for a model of theoretical work that links it with pedagogy and institutions, with the proliferation of curricula and of the resources with which reputations are made, used, and broken.[18]

Some very significant aspects of the organization of inquiry are illuminated by this attention to the social processes through which disciplines reflect on their own formation. Kuklick's story is characteristic of the pattern of philosophy's own sense of self. As Jonathan Rée has observed, disciplinary histories of philosophies construct an anticanon, an account of past texts whose errors have now been corrected, and against which the achievements of contemporary philosophical work must be judged. German idealism functioned as part of this anticanon for much modern Anglo-American philosophy, the Grendel whom the philosophical Beowulfs had slain. The formation of such anticanons is but one aspect of the cultural work that disciplinary histories perform. Legitimation and calibration through historical reflection are not epiphenomenal to the organization of the sciences, but fundamental for their social order. Scientific work is also the production of historical accounts that warrant achievement, define and exclude alternatives, and carefully position the author in the scientific milieu. These histories possess a history, a succession of genres from the heroic accounts of early modern doxographies through the progressive conjectural histories of the Enlightenment to the celebrations of cunning and genius so marked in Romantic institutions.[19] These canons and anticanons of the sciences are now used to sustain distinctions between the social, source of error and pollution, situated in science's past, and nature, source of truth, now grasped by the present. To describe these myths is not to deprive them of power. By proclaiming the end of historicity in the making of knowledge, they now position the inquirer in a narrative of what Mary Louise Pratt calls "anticonquest," the careful presentation of the inquirer as innocent with respect to the world that is represented and controlled. The interest of these stories of anticonquest has simultaneously given them immense importance in the production of authority, and yet made them seem marginal to the work of inquiry.[20]

Moments of Danger

> Historicism contents itself with establishing a causal connection between various moments in history. But no fact that is a cause is for that very reason historical. . . . A historian who takes this as his point of departure stops telling the sequence of events like the beads of a rosary. Instead, he grasps the constellation which his own era has formed with a definite earlier one.
>
> —Walter Benjamin

The move beyond mythbreaking is a familiar one. In his 1971 essay "La mythologie aujourd'hui," Roland Barthes catalogued the tenets of what he called "mythoclasm," the exposure of those processes by which "the quite contingent foundations of the utterance become Common Sense, Right Reason, the Norm, General Opinion, in short the *doxa.*" Historians who have used resources from the human sciences to contextualize the canon of the natural sciences have joined in this mythoclasm.[21] But Barthes also appealed for a new program, one that would "no longer simply *upend* (or *right*) the mythical message" but also reconstitute the processes by which these messages worked. Traditions and memories are invented, and acquire power.[22] Important clues about this empowerment are to be found in the work of cultural historians, notably E. P. Thompson, Raymond Williams, and Eric Hobsbawm, who have explored the significance of invented and customary tradition in moments of cultural crisis. In the early 1970's, Williams began to examine the notion of an "old England," green, rural, tranquil, which had allegedly been recently displaced by an urban, rationalized, violent culture. The contrast was drawn between eternity and history, in which, in Marx's evocative formula, "patriarchal, idyllic relations" seemed to be torn asunder, "all that is solid melts into air."[23] What Williams found, notoriously, was what he named a "problem of perspective." At every critical historical conjuncture, a form of rural past was nostalgically invoked. Just as historians might ironize the canon by noting the contingency of its local, contextualized senses, so Williams encountered Leavis, Hardy, George Eliot, and John Clare all harking back to an Eden that had only just vanished. Williams convincingly argued that analysis should not halt at irony. It must specify the changes of the allegedly permanent dichotomy between country and city: "We have to be able to explain in related terms both the persistence and the historicity of concepts."[24]

Williams's periodization is suggestive. He identified periods of rural protest when the city/country doublet figured most prominently: around 1800, when the insurrectionary masses seemed challenging to an incipient capitalist regime; and around 1900, when the mobility and isolation of the urban poor were crucial threats.[25] Milieus of intense cultural and political crisis bred talk of a vanished past and of some irreversible breach of tradition. The tradition in question was *not* itself an unchanging image. Each conjuncture allowed the refashioning of its past to suit its purposes. In a challenging collection of studies of "invented tradition," Eric Hobsbawm points to the relationship between novel cultural practices and the "aura" of immense antiquity that they are designed to acquire. The term "aura," due to Walter Benjamin, is intended to highlight the strange location of tradition in space and time, "the unique appearance or semblance of distance," as Benjamin put it, "no matter how close the object may be."[26] Hobsbawm's cultural-materialist perspective usefully indicates the relations of power, authority, and community that the invention of tradition involves. Forming canons and traditions requires labor so that the contingent and local may be made ancient and universal. Hobsbawm argues that such forms of labor link power and authority with the sense of particularism and moral economy. The systems of tradition developed in the Revolutionary epoch around 1800 and the *fin de siècle* of 1880–1910 were notable precisely because they embodied cultural senses of the particular in contrast to the individual, the loyalty to locale rather than anomic alienation. Thus developed the culture of division and loyalty that characterized industrializing societies. Though they may have been vague in intellectual content, such inventions involved very rich codes of practice. Especially crucial, Hobsbawm suggests, was the emergence of a set of traditions in dress, leisure, and iconography that soon came to specify the type of authority that nation-states might wield and to which loyalty should be devoted.[27]

The political implications of this refashioning and its crisis are very apparent to the subjects of the Thatcherite regime, in which talk of "heritage" and "Victorian values" has had a powerful ideological effect. Apologists for this capitalist revanchism have blamed British economic decline on ruralist, anti-industrial nostalgia; in the same breath, they have celebrated the national and paternalist heritage as a unique source of cultural value. In 1981, the year zero of the enterprise culture, Martin Wiener argued that by the end of the Victorian period a ruralist "collective self-image" had been secured: "Social integration and stability were ensured but at a price—the waning of the industrial spirit." He predicted that "Margaret Thatcher will find her most fundamental challenge . . . in changing this frame of mind." The Penguin Business Library helpfully brought out a revised—expurgated—edition of Samuel Smiles's *Self-Help*: a "book of *our* times," as it was described in the introduction by Thatcher's guru Sir Keith Joseph.[28] During the 1980's, the battle over the meaning and the canon of Victorian values was joined in earnest. James Walvin pointed out that the very multiplicity and perversity of historical interpretation could function as a weapon against the new Tory monolith. Patrick Wright produced a remarkable analysis of the heritage industry, of the

Falklands War and the raising of the Mary Rose, of the National Trust and the sale of Mentmore Towers, to break the myths of national identity and show their effects. He chose this passage from Walter Benjamin's critique of Ranke as his epigraph: "To articulate the past historically does not mean to recognize it the way it really was. It means to seize hold of a memory as it flashed up at a moment of danger."[29] Works such as those of Williams, Hobsbawm, and Wright can be helpful for historians of the sciences in search of ways in which the canon is made and remade in moments of danger. They help us see the productive work that the invention of tradition requires and how malleable that tradition can be. They also help us see that canons are not merely literary: they are embedded in the practices, rituals, and gestures that give their values meaning. Just as George Marcus has rightly challenged "the assumption that the canon is embodied in texts and that it is primarily realised through the activity of reading," so Steven Lukes reminds us that ritual practices often "invoke loyalties towards a certain, powerfully evoked representation of the social and political order" and that the same role is played by "other institutionalized activities that are *not* primarily identified by their participants as rituals."[30] In the final section of this essay, I treat canonizations embodied in ritual practices such as death and transfiguration, and institutionalized activities such as trade and empire. Canonizations of disciplines carry with them all these features of culture: an implicit system of authority, a defined relationship between the subservient and the masters, and the proclamation of an irreversible advance since an immemorial past. Most significant, these processes matter most when disciplinary systems are under threat.

Canonization and Apotheosis

> Faustus, these books, thy wit and our experience
> Shall make all nations to canonize us,
> As Indian moors obey their Spanish lords.
>
> —Christopher Marlowe, *Doctor Faustus*

Recent sociology of scientific knowledge studies work performed at the tense boundaries of fields of inquiry because there the tacit presuppositions of a culture become explicit. An emphasis on the disunity of scientific fields, and on the historical construction of incommensurability between these fields, indicates that much scientific work requires the management of these liminal encounters. Canons matter most at the edges of the world. Systems of classification are secured by making ambiguous zones sacred or taboo. In Marlowe's exposition of the ways of scientific ambition, the young Wittenberg professor John Faustus rapidly surveys the canons of learning: Aristotle's logic, Galen's medicine, Justinian's law code, Jerome's Bible, and Agrippa's magic. Each save the last is found wanting. Faustus reckons that magic can unify all other disciplines, and master them. The phrase with which Faustus's friends encourage his ambition brilliantly links the two senses of canonization. "These books" will secure the professor's authority. And this authority will be as effective as that exercised over the natives of the New World by their conquerors. But just as the "Indian moors" are dupes, since there is nothing truly holy about the Spaniards, so too would be anyone taken in by this form of scholarly magic. The play shows how canonization works most powerfully at the limits of licit knowledge, morality, and culture. It works as a form of deception: Faustus is wrong to suppose that he can master the world. Those whom he masters are themselves deceived by diabolical enchantment. The natives are everywhere gulled by aliens whom they take for saints. As Lévi-Strauss points out, the Spaniards were killed to see if they really were gods, even as the Spaniards slaughtered the natives because they took them to be less than human. The posers invested in this process were surely not symmetrical.[31]

A major concern of sociology of scientific knowledge has been to show this process at work in the realm of science. Those who take science to be a transcendent enterprise whose authority is secured by its distance from the messiness of cultural politics are to be cautioned that scientists are using common cultural resources that can be charted by conventional sociological analysis. David Bloor's use of Durkheimian accounts of the social function of the sacred is one version of this cautionary project. Here historicist sensibilities are used to enable the decanonization of the sciences. The project has a great deal in common with new historicist strategies in literary criticism. In a survey of these strategies, Stephen Greenblatt contrasts the tenets of new historicism with those of alternative senses of the term "historicism." Instead of historical determinism, he emphasizes that "it takes labor to produce, sustain, reproduce and transmit the way things are," so they are in principle mutable. Against the ideal of value-free history, Greenblatt argues that the analyst's values will be pervasive, though he doubts the virtue of rendering these explicit and of limiting study to areas where values and objects of study perfectly match. Last, he rejects a historicism that venerates the past, "the assumption that great works of art were triumphs of resolution." Much work in science studies also points to the labor needed to maintain order, the virtue of rendering one's own values strange, and the multiple senses of scientific canons.[32] Greenblatt's program notoriously works through the radical juxtaposition of episodes from violent encounters between the Renaissance English and the inhabitants of colonial and non-European worlds, and episodes from the classical canon of Renaissance literature. His reading of *Doctor Faustus*, for example, is placed alongside a report of the arbitrary destruction of a city in Sierra Leone by English mariners. The aim is to show how Marlowe's play reveals the basic contingency and duplicity of his culture's own canons of right conduct and right learning. In rather similar manner, sociologists of scientific knowledge have often profited from the juxtaposition of canonical episodes of scientific culture with less familiar episodes of encounter with other cultures.[33]

The power of this technique has been amply demonstrated in much recent work on the ethnocentric structures with which European scholarship represents other cultures. Notable examples are the works of Edward Said and Martin Bernal, both of whom have attacked the canons of major disciplines by writing compelling social histories of those canons' formation. Said's analysis of Orientalism insists on the productive work performed by representations of the Orient made in Europe in the nineteenth century. Indeed, his analysis relies on the "exteriority" of these representations: "Orientalism responded more to the culture that produced it than to its putative object, which was also produced by the West." Thus he works with a form of historical anthropology, in which "canonical texts" and cognate social structures are given great weight and the "hermetic history of ideas" is eschewed. Said's self-reflection is also striking. It obeys Gramsci's demand for the compilation of an inventory of the traces of the historical process left in the analyst's own concerns.[34] For example, concerns with anti-Semitism shadow both Said's analysis of the constitution of Orientalism and Bernal's historicization of *Altertumswissenschaft*. In his attempt to displace the canonical "Aryan model" of the origin of Greek culture with a version of the "ancient model," which recognized and documented the role of Egypt in ancient Greece, Bernal deliberately foregrounds his own marginality to the specialist disciplines of classical archaeology and Hellenistics. He traces his own experiences in the period of imperialist aggression and of renascent anti-Semitism. He notes the genetic fallacy: the "Aryan Model's . . . conception in sin, or even error, does not necessarily invalidate it." Yet the force of his argument notably depends on the detailed historical analysis of the cultural and political resources upon which the Aryan model drew when it was produced in the late eighteenth century at Göttingen and reproduced in nineteenth-century racist scholarship. Bernal writes a sociogenesis of the classics' canon. Both Said and Bernal fashion themselves as outsiders to the disciplines they criticize, and they both deal with disciplines whose purpose was to define outsiders. They both attack these enterprises by writing social history, by tracing genealogy, by displaying the corrupt ancestry of apparently pure scholarship. And they both criticize the ideologies of purity that Orientalism and *Altertumswissenschaft* so signally peddled. In a brilliant essay reviewing *Black Athena*, Iain Boal shows how the political history of courses in the canonical texts of Western Civilization, developed as anti-German and anti-Bolshevik programs in the first decades of this century, illuminates the milieus in which Bernal's work has now been appropriated and interpreted. These strategic choices show why a social history of canonization is so central a political task for contemporary disciplines. They also show that the issues of reflexivity and heterogeneity are by no means the unique burden of hapless science studies.[35]

Our own enterprise shares much with the strategies of these other historians of canonization. In particular, it shares an interest in the misprision of alien activity. Misprision of cultural practices' sense happens when considerations that clash with those in which these practices were developed are imposed upon them. It is certainly possible to analyze those processes that prompt such misprision. As Nick Jardine has pointed out, misprision "is virtually inevitable in the reading of canonical works of our own cultural tradition." This is because "continuous histories of interpretation and assimilation" must affect our encounter with these works. So the most apparently familiar features of our own disciplines are at least as subject to misprision as the strangest products of others' worlds.[36] Greenblatt, Bernal, and Said attend to encounters with foreign worlds and to the canons of entrenched disciplines, to the strangest and the most familiar aspects of culture. They make these canons strange; they represent the alien. They do not suppose that they are themselves delocalized, nor that they can travel effortlessly across the field of knowledge production. On the contrary, they make much of their own positions vis-à-vis these cultures and these disciplines. They show the enormous labor invested in making a representation of any cultural practice, not only those that seem distant but also, and especially, those that seem homely. Thus the misprision associated with canonization is highly productive, not to be treated as a sterile error. Indeed, the constructive manipulation of the canonical helps make communities and boundaries; it does not simply depend on them. The process of making the canonical involves processes both of estrangement and of socialization.

Canonization, the construction of norms and of saints, is especially important at the boundaries of cultures, where self-definition is established through differentiation. This helps explain why, as George Marcus has pointed out, the canon is not so described by the defenders of traditional authority but by those who seek to attack this authority's legitimacy. To point to the canon's existence is part of the process of subverting it.[37] Such gestures of delegitimation are peculiarly common at moments of cross-cultural encounter, where the tacit presumptions of an established culture are made visible and transformed. To illustrate the relationship between the sacred, the canonical, and the encounter with other cultures, I choose an episode that has itself become canonical for the study of such encounters. This episode is the death of Captain James Cook in Hawaii early in the European year 1779. Rival interpretations of this event's meaning show how the process of canonization (in this case, the deification of Cook) is ineluctably connected with cross-cultural negotiation (in this case, the relationship between various Hawaiians and Britons). The canonical explanation of Cook's death is that the Hawaiians took him to be the god Lono, and they

killed him as part of the sacrifice called for in the traditional Makahiki ceremony, the timing and route of which coincided with the
arrival and journey of Cook and his ships around Hawaii between
November 1778 and February 1779. Parallels have been drawn, for
example, between this encounter and that between Spaniards and
Aztecs, in which the myth of the return of Quetzalcoatl allegedly
dominated responses to Cortez and his men.[38]

Consider the following varieties of interpretation of Cook's
death. In a lecture delivered at the Kroeber Anthropological Society in 1977, the anthropologist Marshall Sahlins argued that the
fatal encounter was an example of a "working misunderstanding."
He challenged much of the standard account in order to explore the
symmetrical structures of such encounters. The Hawaiians took
Cook as a god merely because all great chiefs were deities. For
example, the British astronomical observatory, whence the sun
was observed, was treated as a sacred house; and the instruments,
as divinities. Sahlins pointed out that the British *also* took Cook to
be a god; and he reprinted a celebrated picture, Loutherbourg's
Apotheosis of Captain Cook: "The faith in which Cook held such an
exalted position was Imperialism." The significance of the coinci-
dences between Cook's path and the regime of the Makahiki festival had to be reinterpreted. Sahlins claimed that the Makahiki cult
was later redesigned, well after 1779, as an iconic representation of
Cook's voyage, in order to legitimate the regime of the powerful
Hawaiian chief Kamehameha, whose principal advisor took the
significant name Billy Pitt. The British cult outlasted the Hawaiian, which was deliberately transformed by the Hawaiian chief to
help represent this decisive historical moment. So the canonical
version was developed retrospectively to account for relationships
between Hawaiian political leadership and imperial power.[39]

However, in prestigious talks delivered at the Association of
Social Anthropology in 1979 and as the Frazer Lecture in Liverpool
in 1982, Sahlins significantly revised this argument. He continued
to accept that the myth of the return of Lono was developed after
Cook's death as "the explication of Cook's advent and integration
in the [Makahiki] ceremony." But now computer calculation and
detailed study of the Hawaiians' records showed that there was
indeed a striking coincidence between the ceremony's timetable
and that of Cook's movements. The Hawaiian priests did take

Cook to be Lono. Soon after Cook's death two of them asked the British when Lono would return. Sahlins now made this question crucial, for it indicated that an "opposition of practical interest" existed between the priestly identification of Lono and Cook and the chiefs' sense that this identity spelled trouble. Cook died because at a key point he disrupted the order of the ceremony and threatened chiefly power. Cook's unscheduled return precipitated a catastrophe resolved by his canonization by all parties. "Harbinger of the *Pax Britannica*, Cook was also a bourgeois Lono."[40] These shifting versions disaggregate and politicize Hawaiian responses. The standard version credits the Hawaiians solely with a mythopoeic misprision of Cook's character. Sahlins's early understanding was that the myth was engineered *post hoc* by interested chiefs. His later model describes political conflicts inside Hawaii of 1779. The shift also draws attention to the symmetry of explanations. Cook's apotheosis is equally recognizable in British culture, and is increasingly understood as a myth of imperialist commerce.

Sahlins's accounts have been widely debated and criticized by other anthropologists. In a paper "Mythopraxis and History," a group of Copenhagen anthropologists deny that the Makahiki festival existed as such in the late eighteenth century and that Cook's voyages ever corresponded to its pattern. They point out that our own information about this festival is derived from accounts gathered by missionaries from Hawaiian aristocrats in a European school in the 1830's. One of these aristocrats, David Malo, is the principal source on the Makahiki. They reckon that he helped portray the festival just so that Cook's death could be explicated as part of it. This legitimation, they claim, was part of the 1830's crisis among Hawaiian chiefs who tried to enlist British help against their American creditors. So these critics return to, and expand upon, Sahlins's original sense of the interested retrospection with which Cook's death was interpreted.[41] In a similarly revisionist account, Gananath Obeyesekere notes that the Hawaiian priests asked about Lono's return as part of a conversation about cannibalism. He reckons that the Hawaiians judged that the emaciated British had come to their islands to eat. The British mariners' repeated inquiries about cannibalism prompted the Hawaiian's fear that the British were cannibals. The gathering of native heads for such collections as those of the London anatomist John Hunter only reinforced these views. "Cannibalism is what the English reading public wanted to hear." In New Zealand in November 1773 one of Cook's lieutenants deliberately proffered human flesh to a Maori visitor in order to demonstrate native cannibalism. Here a new symmetrical misprision is uncovered. Both British and Hawaiians reckoned the others were cannibals. The development of Polynesian cannibalism was an indigenous strategy that matched practices of human sacrifice to what were taken to be British expectations. The return of Cook as Lono was then epiphenomenal to rational Hawaiian fears of the evil that these anthropophagous mariners would visit upon them.[42]

A range of representations of Cook's death has been provided by various Hawaiian chiefs and priests, British mariners, and modern anthropologists. The very flexibility of representation provides the resource to which these anthropologists appeal. They show stories and practices being manipulated and reworked to serve local purposes and to negotiate with inquisitive interlocutors. They describe what Mary Louise Pratt calls "autoethnographies," accounts produced in engagement with those of the colonizers. These accounts, she suggests, should be treated neither as the authentic expression of indigenous subjects nor as signs of those subjects' inauthentic and coerced absorption of alien codes.[43] The stories of cannibalism produced in the 1770's and of Lono's return produced by Hawaiians in the 1830's might be treated as autoethnographies. Their production and interpretation carry several lessons for our own study of canonization. Cook was canonized by the Hawaiians. A range of factors, such as coincidence with the Makahiki festival, the ambitions of Kamehameha, or the dangers of American creditors, is used to unfold the production of this canonization. Cook was canonized by the British. A range of different factors, notably the culture of imperialist capitalism and Christian doctrine, is used to highlight this process. Using their own canonical principles of symmetry, the anthropologists treat these canonizations not as misprisions, but as productive representations. The Hawaiians had reasons—their visitors' emaciation and conversation—to suppose the British cannibals. The British had reasons—metropolitan audiences and wealthy collectors—to represent Polynesians as cannibals. The Hawaiian chiefs had reason to suppose that Cook's identification with Lono and his unscheduled return threatened their rule. The Hawaiian informants had reason to suppose that the British would support them if their own myths prompted Cook's death. There seems no way of delimiting, in advance, the relevant explanatory factors used in these anthropologists' stories. Hawaiian chiefs and priests, American missionaries and traders, London journalists and virtuosi—all turn out to play a part. There is a close relationship here with the problem of defining the "evidential context" of some experiment.[44] What is taken to be evidentially relevant is itself dependent on skillful reflection on the canons of anthropology. Sahlins explicitly reflected on the relationship between the structure of his narratives and the sanctified milieus in which he presented them. Cook's death, Sir James Frazer's own stories, and the Frazer Lecture series held in his memory, for example, all involve the repetition of a performance of homage to a sainted ancestor followed by an attack upon him.[45]

This is what the process of canonization means for its culture. If we believe, naively enough, that misprision is a disease that affects our relation with the canon, to be cured or prevented, then its productive effects will escape us, or be devoid of interest. Perhaps I should perform my own Frazerian ritual. I take Biagioli's purpose in the development of contingentism to be the critique of this naive belief that we might ever attain a purified, universal grasp of any culture's values. Hence I endorse his very important emphasis on the production of group identity through differentiation, his account of the fruitful and productive consequences of incommensurability, his argument that representation must be seen as contingently effective rather than imitative, and his understanding of the importance of the genealogy of currently received standards. But his evolutionist analogies of sterility, competition, survival, and extinction still seem to distract attention from the labor required to maintain, and to negotiate, the boundaries between these differentiated cultural fields. The productive role performed by these appropriations and representations of the canon is crucial. They are resources that make and change the conventions and the communities of inquiry.

ARTHUR FINE

Science Made Up: Constructivist Sociology of Scientific Knowledge

❖❖❖ It must be acknowledged, that rational assent may be founded upon proofs, that reach not to rigid demonstrations, it being sufficient that they are strong enough to deserve a wise man's acquiescence in them.

—Robert Boyle

The idea of the social construction of knowledge belongs to a tradition in sociology that includes such seminal figures as Marx, Mannheim, and Durkheim, as well as George Herbert Mead, whose social constructionism also had a very broad range.[1] Until recently social-constructivist ideas had not been central to investigations in the sociology of science, which was dominated instead by the institutional approach of the Merton school.[2] This situation has changed over the past decade or so with the rise of a movement, referred to by its practitioners as the "sociology of scientific knowledge,"[3] that incorporates the idea of social construction in an especially striking way. Witness such titles as *The Manufacture of Knowledge: An Essay on the Constructivist and Contextual Nature of Science*, *Laboratory Life: The Social Construction of Scientific Facts*, and *Constructing Quarks: A Sociological History of Particle Physics*.[4] For brevity, I will refer to this movement simply as constructivism.

Turned now to scientific knowledge, constructivism has begun to capture the attention—although not always the admiration—of historians, philosophers, legal scholars, literary theorists, and "postmodern" culture critics.[5] Major university presses (e.g., Chicago, Harvard, Princeton) have begun to publish work in this genre. In philosophy proper, its impact so far has been largely among those versed in the Continental literature, where it seems especially intriguing to sympathetic readers of Foucault, and to some others who, in Steven Shapin's exasperated words, "wish to ape the modern French manner of academic posing."[6] In the philosophy of science the scope of the constructivist program fits in with the "big methodologists," like Popper and Lakatos and Kuhn, each of whom, not surprisingly, rejects it—but perhaps rather too strenuously. Larry Laudan's exchange with David Bloor is typical of the generally unsympathetic reaction of analytic philosophers to constructivist ideas,[7] where Laudan castigates the "strong" version of the program as the "pseudo-science of science."

Two doctrines conspicuously embraced by constructivism seem to generate much of the philosophical interest and reaction: its antirealism and its relativism. Although these doctrines are certainly much discussed in the constructivist case studies and methodological writings, their treatment there is philosophically careless and naive. To put it bluntly, constructivists write a great deal of nonsense on these topics. This expository sin is compounded by a dialectical one, for when it comes to defending their doctrines, constructivists tend to rely more on polemics than on careful argument. Their rhetorical style, moreover, is at once romantic and apocalyptic. They portray themselves as in the vanguard of a new dawn in understanding science, a profound awakening that sweeps away oppressive philosophical categories—truth, reality, rationality, universality. But although broad in understanding, their party is also depicted as small in numbers, an embattled minority pitted against the huge Scientific Establishment and their academic apologists (philosophers, historians, other sociologists, and so on). Of course this antiestablishment pose is increasingly *ex*posed by growing support for constructivism from the academy, in terms of positions, publications, and grants—not to mention admiring followers. So, increasingly, their polemics seem disingenuous. For philosophers, moreover, with their special regard for clarity of exposition and force of argument, the expository and dialectical sins of constructivism pose special obstacles to treating its doctrines with much sympathy. Still, that is what I should like to attempt here with regard to the antirealism issue, since I believe that one can use constructivist ideas to sketch an interesting and defensible program that presents a serious challenge to realism.[8]

A Constructivist Platform

I should like to begin with a description of the constructivist ideology and its background. However, despite its short tenure there are already several different constructivist schools, each with its more and less radical wings. It is, therefore, no easier to characterize constructivism than it is to characterize realism or antirealism. Still, as in these other -isms, I think there is a cluster of important doctrines that distinguish the constructivist party. Because their own rhetoric tends to the political, and they are clearly conducting a campaign for party members, I will refer to the cluster as a constructivist platform and to the doctrines that fall under it as the planks.[9]

Plank 1. Beliefs on a topic can (and do) vary. Prevailing beliefs are relative to particular prevailing social circumstances.

Plank 2. For any belief (whether true or false, rational or irrational) the question of why it is held (or not) is appropriate, and the answer is to be an explanation framed in terms of locally operating causes, and not in terms of the character of the belief (e.g., whether true or false) or in terms of rationality conditions (e.g., "It would be irrational not to hold").

Plank 3. Contingent sociological factors are (must be) relevant to explaining beliefs and judgments. In particular, beliefs are produced and judged to further local, collectively sustained goals and interests. The scientists' role in belief formation is active. They are agents doing things: making choices, forming alliances, pursuing local goals, advancing interests, and so on. All these are done in a rich field of social, cultural, institutional, and political forces; that is, they are all done together with other agents behaving similarly.

I have framed these planks in the language of "beliefs" and "judgments." Typically, however, constructivists do not distinguish between belief formation and the "making" ("construction," "manufacture," "production") of facts, using one idiom more or less interchangeably with the other, even when the result is literal nonsense. If we were to follow their lead in formulating the platform, and speak of facts rather than beliefs, then the first plank would issue in a rather striking relativism, and the third plank would amount to a strong sort of constructivism. Surely the conflation of the languages of beliefs and facts by the constructivists is not just ignorant usage, but a particularly forceful (and, to some, annoying) way of expressing a central doctrine—namely, that what makes a belief true (if I may use oldspeak) is not its "correspondence with an element of reality" (i.e., a "fact," realist-style) but its adoption and authentication by the relevant community of inquirers. This amounts to a loose consensus theory of truth and constitutes a special sort of semantic antirealism. If we adopt the customary semantic convention according to which facts are identified with what makes beliefs true, and if we also subscribe to such a consensus theory of truth, then facts turn out (literally) to be constituted by processes of belief formation. Moreover, whatever drives these processes then (literally) makes the facts. Of course, according to Plank 3, what drives belief formation is the activity of the scientist-actors. Hence, facts are made by scientists (literally!). Thus an antirealist, consensus theory of truth binds the three planks together to form a specifically constructivist platform (and one that is also relativist). Although the details of the consensus theory are not too important for constructivism, it is necessary for them that beliefs do get fixed, so that facts do get made. Hence the consensus is not, as with Peirce, in the sweet by-and-by. Constructivists require truth to be made by actual consensus, and not by some long-run idealization.[10]

The sociological turn, the emphasis on the social determinants of belief, is evident in these planks and in the underlying consensus theory of truth as well. In both instances social behavior is afforded not just a dominant but actually an exclusive role. Between the second and third planks, anything other than social behavior is actually excluded from playing an explanatory role in the fixation of belief. In a consensus theory, the social fact of the fixation of belief is promoted from being one of the marks of truth to being the whole of it. This is behaviorism with a vengeance, and, as we shall see, it is one of the places where constructivism comes undone. But apart from behaviorism, which is not in philosophical fashion, there are other, more fashionable doctrines to which the platform also owes some debt.

Central ideas in postwar philosophy of science, in particular the doctrines of the theory-ladenness of observation and of the underdetermination of theory by data, lend support to the planks. The former suggests that the observational data of science depend on the theories in the field. This makes for the ever-present possibility of variation emphasized in the first plank, and, when combined with underdetermination, it opens the door to the move to sociological explanation of both theory and data made in the third plank. A similar role is played by the Poincaré-Duhem thesis, according to which falsification in science is necessarily inconclusive. For this raises the question of how, then, is the target and conclusion of a falsifying experiment actually fixed? Again sociological explanation, as in the third plank, presents itself to fill the gap. In addition to these recent fashions in the philosophy of science, ones whose cogency not everyone acknowledges (and among whose dangers some might well include support of irrationalist platforms like constructivism!), there are also certain central disciplinary programs whose shortcomings make additional room for the constructivist moves.

I have in mind, in the first instance, various aspects of the confirmation industry. To begin with, the well-known paradoxes of confirmation, along with the "grue" problem, certainly warrant some skepticism about the viability of any general notion of "evidential support." That skepticism is reinforced by the wide range of inadequacies of all general programs in confirmation theory: hypotheticodeductive, Bayesian, and others.[11] Indeed it seems time to acknowledge that the idea of a general, explanatory theory of confirmation has turned out to be a philosophical dead end. The variety of evidential practices seems to have a "situatedness" that the philosophical search for a general theory has obscured. Therefore it is certainly not far-fetched to look at how "evidence" works *in situ*, in an open-minded way, without demanding that its operation in any one place must have a set of explanatorily relevant features in common with its operation absolutely everywhere else. It is also not so far-fetched to think that local social groupings and social factors may indeed determine in context how evidence does work, and how it "compels." Thus the failure of philosophical theories in this area brings us round to the constructivist platform. Of course, not inevitably: that is, it does not follow from the fact that "confirmation" fails of a general analysis and theory, nor even from the fact that the practices are context-bound and not generalizable (if that is a fact), that the contextual features that make for confirmation are exclusively (or even primarily) social. That latter is a guess, a hunch, a programmatic suggestion—but not an irrational one, given the presently moribund circumstances of the confirmation game, nor one especially far-fetched or implausible, either. It is, in fact, a rather interesting and intriguing suggestion for research. I believe that the constructivist platform as a whole can be looked at similarly.

What constructivism needs to take from philosophy of science are not specific, established doctrines. It needs only to take the programs of investigation that sought after general accounts of the structure and dynamics of theories, of observation, of confirmation, of hypothesis testing, of explanation, of discovery, and so on, and to note their widely acknowledged inadequacy.[12] The failure in every case is of the general theory to give an illuminating account of the known range of scientific practices. All the posited structures, rules, and maxims still leave unanswered why particular things occurred in particular contexts. Constructivism then adopts a sort of principle of sufficient reason (in the second plank) according to which there are indeed causes for what happens. Finally, it turns to the social to suggest that the causes are to be found in social interactions. I emphasize that this is not an argument for the constructivist platform but only an attempt to set it intelligibly in relation to philosophical practice, and thereby I hope to make it at least interesting to consider.

So set, constructivism shows up as social particularism. It does not see the operation of general algorithms in science, but only particular sets of local practices. These are associations of people, doing a large variety of different things no particular one of which, at any time, is forced upon them. Thus whether one is gathering data, deciding when to end an experiment (or how to interpret it), judging the relevance of a new item of information, revising a theoretical model, setting up standards, plotting a curve, responding to or making a criticism, reading a meter, and on and on without end, options are available and choices are made. In this sense, at every level and in every kind of endeavor (observational, experimental, classificatory, calculational, theoretical, methodological, and normative) science is open, and its judgments unforced. To do science is to make judgments and decisions that always outstrip any set rules of the game. The constructivist program, most simply put, is to try to explain what goes on in this open arena by reference to social factors: networks, interests, or whatever. So understood, constructivism offers an interesting contrast to other programs for treating science, in particular to realism and instrumentalism.

Realism, Instrumentalism, and Constructivism

To compare these programs it is useful to adapt a scheme that I have used elsewhere to contrast realism with instrumentalism.[13] The adaptation here involves comparison along five dimensions: (1) general valence with respect to science (pro or con), (2) reductionist attitude (or not) to scientific concepts, (3) treatment of truth in science, (4) hermeneutical orientation (or lack thereof), and (5) teleological stance with respect to science.

General valence. Realism is pro-science, advertising itself as progressive in this regard, by contrast with antirealist programs, which it labels as anti-science. But despite the realist polemic, instrumentalism (certainly in this century after being so baptized by John Dewey) is also pro-science, and it counts itself as no less progressive than realism.[14] As for constructivism, I'd better let a constructivist speak for himself: "There is no obligation upon anyone framing a view of the world to take account of what twentieth-century science has to say. . . . World views are cultural products; there is no need to be intimidated by them."[15] This reference to "intimidation," and the general debunking attitude expressed by Pickering, is typical of the romantic antiestablishment rhetoric of constructivist texts. Despite occasional disclaimers, the tenor of their preaching is against science.

Reductionism. Realism has no special interest in reductionist programs. Unlike some of its antirealist companions (e.g., idealism, phenomenalism, empiricism), neither does instrumentalism. They are both inclined to take scientific concepts as they come. In part the particularism of the constructivists supports the same inclination. In their relativist and ethnomethodological moods they are inclined to accept the conceptual framework of the natives at face value. Their social behaviorism, however, provokes different moods. "For the argument is not just that social networks mediate between the object and observational work done by participants.

Rather the social network constitutes the object (or lack of it). . . . There is no object beyond discourse, . . . the organization of discourse is the object. Facts and objects in the world are inescapably textual constructions."[16] By an "object" here they mean both concrete particulars, like individual hormones (or proteins), as well as general things, like mental illness. Thus not only is behaviorism allowed to drive the consensus theory of truth, with its implications for facticity; it also drives an ontological reductionism as sweeping in scope as old-fashioned idealism ever was. That scope includes science itself. "It is not that science has its 'social aspects,' thus implying that a residual (hard core) kernel of science proceeds untainted by extraneous non-scientific (i.e., 'social') factors, but that science is itself constitutively social."[17]

Truth. Realism is associated with a correspondence theory of truth, where descriptive terms in the scientific vocabulary are supposed to correspond to mind-independent objects in the world (at least for the nonhuman sciences). Instrumentalism is often represented as withholding predication of truth from the theoretical (= nonobservational) components of science. This is a poor way of expressing instrumentalism, however, since it makes instrumentalism vulnerable to questioning the divide between the theoretical and the observational, an issue over which it need not take a stand. It also makes instrumentalism seem arbitrary in focusing on just this divide. Better to think of instrumentalism as Dewey did, which is to subscribe it to the pragmatic account of truth as general reliability (or utility), right across the board. This amounts to taking " 'P' is true" and " 'P' is generally reliable" as synonymous, or near enough. Thus high-level laws in physics, for example those involving commitment to quarks, could be counted as true both by realists and by instrumentalists, although they would each understand something different thereby. Constructivists follow an earlier pragmatic route, which is to identify truth with fixed belief. In this enterprise they take William James's road, rather than that of C. S. Peirce, insofar as they opt for actual community acceptance rather than the acceptance of ideal agents (or acceptance at the end of inquiry).[18] Consistently with the general behaviorist reductionism discussed in the preceding paragraph, they take community acceptance not merely as a criterion of truth but as constitutive of it.

Hermeneutics. Realism and instrumentalism share a common attitude toward science. They see it as an enterprise in need of further understanding and interpretation. Not only do they look at specific forms of scientific practice—say, inference to the best explanation or the use of correlational data to support causal models—and seek a good way to interpret, to explain, and to assess the validity of what is going on. They also shape that good way to accord with, perhaps even lend support to, their (realist or instrumentalist) programs. For those programs themselves embody a general interpretation of scientific discourse and practice, the general interpretation that accords with the semantics embedded in their different conceptions of truth.[19] Constructivism joins realism and instrumentalism in treating science as fair hermeneutical game. This is very striking in the investigation of Latour and Woolgar at the Salk Institute, and their adoption of the anthropologist's pose of "strangeness" in order to make room for the interpretations they propose of laboratory practice. More fundamentally, however, it is built into the sociological part of constructivism, which seeks to expose the social character of what is ordinarily taken for granted in scientific activity. In this regard constructivism is closest to realism. They have in common an unmasking impulse, a basic inclination to peel away the conventional surface in order to see what is "really" going on. Instrumentalism shares with them exactly the opposite impulse: what you get is what you see. These impulses are the two sides of a common hermeneutic orientation, one that sets a perfectly general interpretive agenda for which all instances of practice are candidates (recall that the social explanations of Plank 3 apply to any belief) and into which they must all be made to fit. I would just note that this hermeneuticism is at cross-purposes with particularism, which ought to leave open whether episodes are candidates for interpretation at all, and if so what kind of understanding applies. Thus although hermeneuticism is built into constructivism, it is not a comfortable fit.

Teleology. Realism and instrumentalism each propose goals for the scientific enterprise as a whole. For realism, the fundamental imperative for science is to seek the truth: that is, the realist, correspondence-to-the-external-world type of truth. Instrumentalism is said to go for less: in particular, to be content with positing utility (read "empirical adequacy") as the aim of science. But this utility is nothing other than the general reliability that occurs in the instrumentalist's pragmatic account of truth. (Recall *Truth*, above.) Hence, understood in its own terms, instrumentalism too sets up truth as the goal of the scientific enterprise. By contrast, for constructivism, truth is just fixed belief, which is not a terrific goal for me as an interested scientist unless it is my belief that gets fixed. So constructivism parts with realism and instrumentalism, demoting truth as of secondary importance. Indeed the particularist strand in constructivism recognizes the variety of goals and aims of the many different scientific activities and groupings, and how they change over time, and so resists the impulse to collect them all up in one global telos. But the social-interest strand, nevertheless, does require one overriding aim. Like all social institutions, according to the constructivists, science seeks to perpetuate itself. When all is said and done, that is the name of the game. For constructivism that goal functions as a framework into which are set the more specific analyses of interests, influences, reward structure, expectations, training protocols, and so on, that make up the social net. Despite the animadversions about scientific rationality, for constructivism, no less than for realism or instrumentalism, positing an overriding goal for science provides a vehicle for seeing the functional rationality of scientific practice: that is, for seeing that practice as an appropriate means for achieving its end.

These comparisons of constructivism with realism and instrumentalism should help identify areas where constructivism, like these other programs, is vulnerable to criticism—a task to which I now turn.

Deconstructing Constructivism

Elsewhere in criticizing realism and instrumentalism I have focused on their treatment of truth, and on their hermeneutic and teleological stands.[20] These are the areas where I see problems for constructivism as well. To bring out the problem with truth let me attend to the central thesis in the second, explanatory plank of constructivism and contrast two versions of constructivism, respectively, methodological and metaphysical, in terms of their different rationales for the thesis.

THESIS (PLANK 2): *In explaining the fixation of belief, one should not bring in the truth (or falsity) of the belief itself.*

Methodological rationale. The scientist, considered as an agent in the process of investigation, has no access to the truth of the posits he is investigating independently of the process itself that fixes belief. So, given that the belief does get fixed, there is some way that it happens. The thesis of Plank 2 can be thought of as a reasonable methodological rule that directs us to search for that way, which we presume will involve a causal story, but not the inaccessible-at-the-time truth value.

Metaphysical rationale. There is no truth or falsity of the matter until beliefs about it get fixed. Hence (logically) one cannot bring truth values into the explanation of belief.

The methodological rationale involves a certain ambiguity, for one might agree that the truth value of a belief is not known until it is known, without agreeing that the truth (let us say) of the matter does not influence what comes to be believed. I might, after all, not know what is in a package until I unwrap it. But the contents do, nevertheless, influence what I come to believe as I proceed with the unwrapping. We have to be a little careful, however, because the contents are not truth values. And although the contents may affect what I come to believe, the truth of the proposition that the package has such-and-so contents is (indeed) not accessible to me along the way, and so (indeed) it does not influence the final belief. That is an elementary point about truth values. But still, one might object, it is because there was (let us say) an apple in the box that I came to believe that there was an apple in the box. If we follow the methodological prescription and bracket the truth of that, then how am I expected to account for the belief?

It is important to take up this challenge because the point at issue, although distinct from underdetermination and theory-ladenness, is liable to be confused with them. Those doctrines would suggest (underdetermination) that the fact of the matter does not determine our beliefs (or "theories" about it), and (theory-ladenness) that the particular representation of the situation (e.g., as an apple that is in the package) depends on a learned, prior set of possible representations, ones that might have been other than they are. To be sure, accepting these suggestions would open the way to supplementing the fact of the matter (i.e., that there is an apple in the package) by a sociological account to explain why this fact contributed to the particularly represented belief that it did. But the methodological rationale above is more radical than this. That line asks us to bracket the fact of the matter entirely, not merely to supplement it with some sociology.

To see whether this radical proposal is feasible, suppose we try the opposite. How might we bring the fact of the matter into an explanation of belief? Presumably the way would be something like this. We explain our belief that the package contains an apple by saying that at the end of the unwrapping activity (with all the socialization that certainly involves) I saw an apple. That is why I came to believe there was an apple in the package. This explanation sketch relies on a basic (perhaps even primitive) relation between seeing (in context) and believing (in that context). The issue raised by methodological constructivism is whether that relation is, so to speak, apple-specific. After all, we know that we might have formed the same belief about the apple without its actually being an apple that we saw. So even though in the case at hand we agree that there was an apple in the package, the question is whether in this very case it is the apple (*qua* apple) that makes for the connection between seeing and believing. Perhaps we do not need to refer to the apple specifically in order to explain the belief? Indeed we do not, for we can simply say this: it is because of what I saw when I unwrapped the package that I came to believe that there was an apple in the package. The phrase "what I saw" is noncommittal as to the character and qualities of its referent. This formulation, nevertheless, maintains the basic relation between seeing and believing (in context) that explains the formation of belief. I think this minimal account is what the methodological constructivist is after. He need not be committed to more—that is, to responding to the further question as to what "really" is the referent. (After all, his project is to explain how we answer this question.) In particular he need not (and in my judgment should not) give the phenomenalist answer, that it is a sense datum, or the like. The minimal formulation is sufficient. It succeeds in bracketing the truth of the belief, and thereby it opens the way to showing (provided it can be shown) "why particular accounts were produced and why particular evaluations were rendered . . . by displaying the historically contingent connections between knowledge and the concerns of various social groups in their intellectual and social setting."[21]

I have taken up the methodological rationale in the case of individually held perceptual beliefs, which is, I should think, the hardest case. For theory-and-evidence-based beliefs, especially those of a community, there would seem to be even more slack between the truth of the matter and acceptance of beliefs about it. Hence for more theoretical beliefs the methodological rationale for bracketing seems quite plausible, and sensible. The same cannot be said for the metaphysical rationale, whatever the character of the belief.

In maintaining that there is no truth of the matter until the belief gets fixed, the metaphysical rationale simply applies the consensus theory of truth. As Woolgar puts it, "Truth or falsity is perceived (and achieved) rather than inherent."[22] This account of truth, with its implicit appeal to specific communities of inquirers and contexts of inquiry, is relativist, and subject to the familiar array of philosophical objections to relativism. Similar objections apply to the theory of truth itself. Among these are the peculiarities forced on the grammar of "truth," if we adopt a consensus theory. For community opinion shifts, and, as the constructivists are fond of pointing out, this occurs even in the hardest of sciences and over the most central principles. Are we then to say that such principles were once neither true nor false, then became true, then false, and may become true again? The language of opinion, or shared belief, is structurally different from the language of truth. Most of us take this as a sign that there are jobs to do in tracking consensus and dissensus different from the jobs to do in tracking truth and falsity. Most of us, that is, understand that truth is different from consensus. This difference shows up strikingly if we consider the redundancy property of truth: that such-and-so is true holds in the case and only in the case that such-and-so (at least for nice such-and-sos). But the language of opinion is not redundant; it is neither necessary nor sufficient for a community holding the opinion that such-and-so for it to be the case that such-and-so. Precisely the point to having a language of opinions, it would seem, is to be able to deal with the acceptance of beliefs when the beliefs are not true. The failure of redundancy distorts that function, and has even worse consequences.

If the consensus theory has any virtue at all, one would think, it is at least definite about what is or is not true, even if the truth does vary with opinion. For when the community has settled on a view (perhaps firmly enough to last for a while) then it is true (according to the theory), and that is definite. But is it? That is, is it definite for us? For us to be able to judge that something is true is for us to be able to judge that the relevant community has settled on that opinion. But how do we judge the truth (or falsity) of whether the relevant community has settled on an opinion, on the consensus theory? Well, we have to judge whether the relevant community has settled on the opinion that they have settled on an opinion about the original item.[23] But to be able to judge whether that is true, given the consensus theory, we have to add yet another level of community judgment, and to be able to judge that. There is no end to this process, which is to say that there is no way to collapse the infinite tower of judgments required by the consensus theory so as to enable us to make any definite determination of truth in any given case whatsoever. Appearances to the contrary, as the unmasking constructivist might say, on the consensus theory, truth turns out to be just as transcendent and just as inaccessible as on the realist correspondence theory, or if picked out by Peirce's retrospective judgments at the end of inquiry. Moreover, on standards of intelligibility that the constructivists insist on elsewhere—that is, according to the idea that what makes sense has to make sense for us as social creatures—the failure of the consensus theory with regard to the determinability of truth judgments for us shows that this attempt to frame an intelligible notion of truth fails. The attempt is idle; it frames no notion at all.

The preceding considerations are arguments against the consensus theory of truth that underwrites the metaphysical constructivist project, arguments built on principles many of which the constructivist accepts. But the constructivist is a virtuoso at the Poincaré-Duhem defense: expert, that is, at dodging refuting arguments. So I expect that he could wriggle around these too, even if that means renouncing some of his other principles. Where that price is too high, he can always opt for accepting the counterintuitive consequences and announcing those as part of the new dawn in understanding science. "We used to think that truth was stable, and not fickle. We were wrong." "From Aristotle to Tarski we were persuaded that truth had the redundancy feature. No doubt that suited interests then, but it does not suit ours now. We were wrong." "As for accessibility, why trust these regress arguments? They make one dizzy. If on first glance it seems accessible, that is good enough for us." No doubt I exaggerate. The point is correct, however. Not even the best-looking arguments need persuade the committed believer. What is required, as Duhem understood, is good judgment (*bon sens*), and at best that can only be shaped by argument but not compelled. This is a feature of the openness of scientific discourse that constructivism itself (correctly) highlights.

That very openness, however, ought to make one sensitive to a different sort of consideration. The consensus theory of truth, like other attempts to frame a substantive account, hopes to find an informative formula to fill out the right-hand side of

'P' is true if and only if————.

The difficulty with all such efforts is this: if they were successful, they would yield a finished notion of truth; truths would be those things, and only those things, that satisfied the slogan on the right. The supposition behind such attempts, therefore, is that the concept of truth is closed: that is, that truth is a concept with determinate boundaries, and hence one amenable to such necessary and sufficient conditions. In particular the theory favored by metaphysical constructivism holds that all truths, in all historical eras, in all cultures, in all contexts, in all sciences, in all communities, with regard to every subject whatsoever, always in the past and always in the future, always have had, do now, and always will have something in common with each and every other truth, namely, consent by the relevant community. Never will anything be true that does not have the brand of consent. In the move to the consensus theory we must ask: What happened to particularism, and to the openness of science? Those key ideas of constructivism depend on the perception that science is a social activity, and so every scientific endeavor is as malleable and subject to change and revision as it is possible for human affairs to be. To protect that perception it is necessary that constructivism not close off the central concepts that underwrite scientific life, not by means of necessary and sufficient conditions, or anything else (since it is not logical gimmicks that are at issue here). Those central concepts certainly include the concept of truth, which is basic to the textual and representational aspects of science with which constructivism has been especially concerned. To say that the practices of truth-saying and truth-judging are open, however, is to say that one cannot project future practice from past practice. That means one cannot hope to fill out the right-hand side of the truth schema above with the description of some set of practices, for the "set" is not a well-defined entity. Thus the consensus theory of truth espoused by the metaphysical constructivists is at cross-purposes with the entire ethos of constructivism. The consensus theory is not just logically flawed and otiose; it runs counter to the insights over the openness of science that motivate the constructivist program. To be true to those insights constructivism must let truth be whatever it is and will be, and not consensus (or anything else) necessarily.

These considerations are perfectly general. That is, they apply not only to the constructivist attempt to fasten onto a reductive theory of truth; they apply as well to all the areas where constructivism is reductionist, closing things off rather than leaving them open to growth and change (see the preceding section, especially the paragraph *Reductionism*). Behind this reductionism is a leaning to behaviorism that manifests itself in the tendency to take the social in science as absolutely all of science. There is a sense in which this may well be true, just as there is a sense in which it is true that people are nothing but material objects. That sense might tempt one to think that the mental and the social and the political, to mention just a few handy realms, are reducible to the motions and interactions of material objects. That would be a mistake. It is no less a mistake to move from the sense in which science is nothing but the activities of human beings to think that all the realms of science can be reduced to the social. The mistake is to move from the fact that scientific concepts and activities are embodied in human practices, to conclude that this constitutes their essence and exhausts their content. That conclusion depends on (among other things) the assumptions that there is an essence (i.e., something like defining conditions, as in the consensus theory of truth), and that there is some definite thing that counts as content (that could be exhausted). Science is not like that, however, as constructivists well know. Scientific concepts and activities have no essence guiding their development (like Adam Smith's hidden hand). That is a central constructivist point. Because scientific activity is social activity, its practice is always liable to change. Past activity does not determine the shape of future development. Science is open. The constructivists see very well that for this reason the demarcation games and the old projects for methodology in philosophy of science are defunct. What they ought to see equally well is the futility of all reductionist games, including their own.

The social character of science makes it open. The openness blocks any "deep" characterizations of the constitution of scientific concepts, activities, and products. That includes realist characterizations as well as the characterization of them as essentially constituted by the social. So, ironically, the behaviorism to which constructivism is inclined is actually incompatible with the social turn that it has taken.

So far we have seen that metaphysical constructivism, with its consensus theory of truth, is an idle doctrine. The metaphysical part is at cross-purposes with the openness of the social-constructivist part. More fundamentally, however, that openness constitutes a wedge separating the social from the constructivist—that is, keeping the social from degenerating into a behaviorist reductionism. The hermeneutic and teleological dimensions of constructivism show a similar tendency for constructivism to be divided against itself. For if scientific activity is open, with no one's hand forced at any point, then how does it happen that one and only one mode of explanation and interpretation applies uniformly, right across the board? Since scientific activity is not generated by the uniform application of a general algorithm, why must all its varied activities be understood and explicable in terms of any single theoretical scheme, like a causal, interest-dominated model of social practice? As with any such scheme we can, of course, ask whether it is adequate, and, in this case, we can reasonably suspect that it will not turn out to be so. To go to basics, however, we can ask more fundamentally why the constructivists feel the need to approach science with any global, ready-made framework at all. That idea is central to the hypotheticodeductive model of theory testing that constructivism rejects. It is no part of the sociological tradition of field studies from which the constructivist program is drawn. To the contrary, the methodology of field studies resists the conception that what they are about is testing preset hypotheses in the field. That tradition is after a different kind of understanding, the kind that comes from subject-specific ideas arising in the course of close association, and examined in context to see whether they stand the test of time. The knowledge so acquired is framed in terms of concepts shaped piecemeal to suit the particulars of local circumstances, and not (or, at any rate, not necessarily) involving universal social mechanisms. The way in which the sociology of scientific knowledge breaks faith with its sociological roots suggests that constructivism has special concerns that call for a more universalistic approach. Those concerns, I believe, show up in the romanticism of constructivist rhetoric, and in its debunking attitude toward science. (Recall the reference to myth and intimidation in the quote from Pickering in the preceding section, and catch the sneer in Woolgar's *Science: The Very Idea!*) That has the stamp of antiestablishment psychology, suspicious of authority and needing to cut culture heroes down to size. So if one can see SCIENCE, overall, as just another bunch of people doing their own things (especially, pursuing their own interests, ultimately to insure the perpetuation of their kind), then we pull the sting of science's authority. The central metaphors of "making" and "producing" lead the same way. They suggest the image of scientists as laborers: no-hat construction workers, or laboratory-coated "proles" on the science-factory line.[24]

If this diagnosis is correct, then imposing a universal scheme for the interpretation and explanation of scientific activity serves an ideological function. Constructivists need it to cut the establishment down to size. Addressing that need, however, creates a rift in their program. It pulls them away from the piecemeal understanding characteristic of their own sociological tradition, and likewise, since no universal scenario can be preset for a truly open activity, it prevents them from conceiving of the scientific enterprise as truly open. Thus the unmasking hermeneutics that, as we have seen, characterizes both realism and constructivism derives from a related source. Just as the constructivists show realism as needed to foster regard for the scientific enterprise, so, similarly, the hermeneutics of constructivism is needed to tear science down. The same is true for teleology. The incongruity of inventing a single goal for an enterprise conceived of as emphatically multiform, plastic, and variable is striking. But that makes perfect sense if the purpose is to reduce our estimation of the institution of science by having us see it, like a political institution, as merely engaged in an effort to aggrandize and perpetuate itself. This is the constructivist downside of the realists' uplifting slogan about seeking the truth. It is the other side of the coin that connects realism with constructivism as opposing global programs. That currency is badly in need of deflation.

Methodological Constructivism and NOA

Realist and positivist presentations of science project an objectification that is false to scientific practice. That picture colors the public perception of science and makes it difficult for public policy to develop in an informed way. That false picture is also the image frequently picked up and elaborated by philosophers. (Not to embarrass my friends, let me just point to Husserl and the objectifying attitude that he imputes to Galileo and his more modern successors.)[25] It is the official line of many scientists, even if privately they know and speak better. Emphasizing the humanness of science is surely a needed and important antidote to these realist and positivist distortions. Thus it seems to me we can be sympathetic to the fact that the ideological and metaphysical needs of constructivism promote an important corrective to the standard objectification of science. We should recognize, however, that this corrective function is only contingently tied to those needs. Moreover, the needs themselves are subject to change and variation, and under competing pressures they can be even separated from the programs as such. This is especially true for constructivism, because those needs lead to the reductive, hermeneutic, and teleological excesses demonstrated above. These excesses fracture the program. I believe that if we hang on to the piece-by-piece approach to science as an open, social activity we can counter the misleading philosophical images of science, and salvage the best parts of constructivism as well. The result is a program for methodological constructivism that runs as follows:

1. Bracket truth as an explanatory concept.

2. Recognize the openness of science at every level, especially the pervasive activities of choice and judgment.

3. Concentrate on local practices without any presupposition as to how they fit together globally, or even as to whether they do fit together.

4. Remember that science is a human activity, so that its understanding involves frameworks and modalities for social action.

5. Finally, on the basis of all the above, try to understand the phenomena of opinion formation and dissolution in science in all its particularity.

This methodological program retains most of the original constructivist platform. Its most controversial feature is probably the bracketing of truth in (1), for which I would offer the methodological rationale already much discussed in the preceding section. The program dispenses with any theory of truth, and so, unless we tack one on, it will not lend itself to the reductionism of metaphysical constructivism (nor to its relativism, either). Similarly there is no global hermeneutic orientation to this program, although in (4) it places a clear emphasis on local, social factors. It does not enter into the game of teleology, inventing overriding goals or ends for science as a whole, either, a game antithetical to the openness theme of (2) and the particularism of (3). By stripping constructivism of its metaphysical attachments, this program does not prejudge the constitution of the scientific world—that is, whether the scientific facts and objects are essentially social, or essentially mind-independent, or whatever. Its attitude is to let the chips fall where they may. Thus blanket social constructivism, which derives from the metaphysics, is let go in order to retain the openness, particularity, and social orientation of the program consistently. I believe that a great deal of existing constructivist work, especially the detailed case studies, can easily be stripped of constructivist metaphysics (and rhetoric) with no loss in the contribution it makes to understanding science, and read instead as exemplifying this methodological program. Of course without the constructivist covering the theoreticians of constructivism may well feel exposed and insecure.

That feeling of exposure is a normal part of growth, and I think constructivism is already growing beyond science-bashing and postmodern bad faith (i.e., making one's way into the establishment while pretending otherwise). As a sign of further growth (and good faith) constructivists will have to let go of some of their romantic slogans and labels, and stop playing "Let's apply constructivist makeup to the face of science." As for a better general label, the "sociology of scientific knowledge" is a rather ponderous and establishment-sounding fallback that they seem already to have prepared for the occasion. In "SSK," however, it has a quite nondescript nickname. They might, instead, want to adopt NOA (pronounced "Noah"), which has a comfortable feel and is a philosophical attitude already made to accompany the methodological (or epistemological) program laid out above. For NOA (the "natural ontological attitude") is an open, particularist, and nonessentialist attitude to science.[26] It promotes a no-theory attitude toward truth, and thus avoids the metaphysics of realism or metaphysical constructivism. It places science squarely among other human activities, and so invites the social orientation of methodological constructivism to fill the ubiquitous gaps in understanding scientific practice left by overly rationalist methodological programs.

The Challenge to Realism

Despite the attitude of letting the ontological chips fall where they may, an attitude that NOA shares with methodological constructivism, these positions still offer a serious challenge to realism. For realism, when it pretends to be a system of beliefs supported by science, rather than a metaphysics simply imposed on science, relies on two forms of argument to give it support. The first is explanationist: roughly, that realism is to be believed because it is the best explanation for why science is successful. Since the issue between realism and instrumentalism, for example, is whether even the best explanatory hypotheses are to be believed (literally) at all, rather than just to be pragmatically used, it ought to be clear that the explanationist argument is question begging, and cannot offer reasonable grounds for the truth of realism.[27] One can also challenge the claim that realism actually does well in explaining the success of science, and even the presupposition that science is successful, when that is filled out and hedged in the requisite way.[28] Constructivism contributes a different dimension to the discussion. It offers the prospect for "explaining" much of the success of science, insofar as that obtains and is reasonable to inquire about, by showing the extent to which what counts as scientific success is tailor-made by the various scientific communities to fit just what those communities have the skills to do. This is, for example, Pickering's theme in his study of high-energy physics in the 1970's, especially with respect to the acceptance of the neutral current in 1973–74, and its connection with the flowering of gauge theory during that period.[29] We will not know what "success" this leaves for realism to address until the constructivists have had time enough to show what they can do.[30]

The second line of realist argument is an attempt to read realism off the details of scientific activity. Given its case-study orientation, and attention to detail, this is the sort of enterprise where constructivist sociology ought to have something to contribute. The realist idea is to see realism as integral to particular features of experimental practice. The implementation of this idea can take different forms. One is Ian Hacking's "experimental argument" for realism, with its injunction that when we can build a successful scientific instrument using an entity (an instrument like an electron gun), then the entity is real (i.e., really exists).[31] The challenge from constructivism is the prospect it holds out of giving a detailed accounting of the formation of belief while bracketing the truth of the belief. It holds out the prospect, for example, of accounting for the belief that we have built a reliable electron gun, without commitment to the truth of the description—that is, without presupposing the truth of the claim that what the "gun" does is shoot little electron bullets. If constructivism can do this, then the existential conclusion enjoined by Hacking would be seen as an ontological gloss not required in order to make sense of experimental practice.

Another way of trying to read realism off scientific practice is the interesting contextual approach of Richard Miller.[32] Miller restructures the debate over realism by deflating realism in two important respects. He eschews any global enterprise, so that not all of science nor even all of any one scientific theory (or whatever) need be realist. (Sensibly, he insists that we get down to particulars.) He also drops out the specifically metaphysical component of realism associated with the correspondence theory, which amounts to making no general assumption about the nature of the things referred to by science (e.g., that they are mind-independent, or even real). For Miller the only issue for realism is an epistemological one over the existence of unobservables: In science are we often in a position to claim that descriptions of unobservables are approximately true?[33] The realist says yes; the antirealist, no. Without assessing his complex account of approximate truth (which involves explanatory goals, historical consequences, and even extra-scientific interests), and of what grounds such a claim to approximate truth, we cannot really judge whether this epistemological reconstruction defines any sort of realism at all. It is not yet clear to me, for instance, whether historically well-defined individuals even get put in the right camp. Do serious instrumentalists, for example, come out as antirealists? Nevertheless the argument for the position that Miller calls "realism" is interesting. It consists of the many individual scientific arguments for the existence of the various unobservables in the scientific zoo. Miller is committed to the view that most of these (the "often" in the realism slogan) are good enough to support belief. We see here, I think, the vestige of the generalizing passion. For the universal "all" has got deflated to "most of" (or "often"), but the individual cases (of which Miller recognizes that there are indefinitely many) are still presupposed generally to stack up on the realist side. (How do we quantify the stack of an indefinite number?) I say "presupposed" because of course Miller himself can examine only a few cases (actually he treats just two: microbes and molecules), and he must then fall back on general considerations to try to persuade the reader either that other cases will be relevantly like these, or that in general the opposite supposition is hard to credit. The opposite supposition is that for the most part things turn out the antirealist way. It seems to me, however, that a particularist can make neither of Miller's suppositions. We cannot declare any general ontological faith, no matter how watered-down; for we are already committed to take things just as they come, judging individual cases on individual merit. That is how the scientific community does it, anyhow, and it is NOA's attitude as well.

Methodological constructivism suggests something a little stronger. To the extent to which that explanatory program works, it would subvert all the little arguments for realism that Miller anticipates. The constructivist program hopes to show how to account for the formation of the relevant beliefs without relying on the scientist's own account of what was compelling in the evidence and argument with respect to the truth of those beliefs, since were the scientist's account credited we would not be able to bracket the truth of the beliefs. Where this is successful, it undercuts Miller's idea that we can just follow the scientist's rationale for how they came to believe in one unobservable or another, supposing that rationale to give us good grounds for the approximate truth of the belief. If constructivism is correct about the openness of scientific rationality, close examination of scientists' accounts will show gaps. Those beliefs are grounded, rather, in contingent social factors. But that grounding in a historical social network does not lead to the truth of the beliefs. It could not generally lead to their approximate truth either, even on Miller's liberal understanding of that concept, unless Miller has built a consensus theory (or the like) of approximate truth into his account, which I do not think he has.

Unlike metaphysical constructivism, the methodological version is not an *-ism* that competes with realism or instrumentalism as a general philosophy of science. In this respect, too, it is like NOA. Thus methodological constructivism does not challenge the truth of realism. It challenges whether support for realism can be found in the practices of science. Most simply, if the explanatory tasks set for methodological constructivism can successfully be carried out, the case for realism will have lost even its apparent grounding in science. This prospect helps to show these other positions for what they are: namely, appendages to science that neither are supported by it nor contribute to its understanding. This profile of idleness is the one NOA has cast them in all along.

The realism/antirealism debate largely sidesteps science. The debate over a constructive reshaping of constructivism may be more important. For the hope is to liberate constructivism from its own global ideology, with its overblown rhetoric and poor philosophical understanding. The aim is to urge constructivism in the direction of an open, social particularism. That seems to me the heart of the program, the right corrective to philosophical (especially realist) distortions of science, and the place where lots of good work can be done too. Among the work to be done is to achieve some understanding of what is actually involved in rational acceptance and proof in science, of what, in Boyle's words, deserves "a wise man's acquiescence." (Recall the epigraph.) This job involves exploring the diverse range of contexts, historical and contemporary, in which inquiry is carried out. In that endeavor, and others, NOA joins hands with its constructivist allies, and wishes them well!

DAVID J. STUMP

From Epistemology and Metaphysics to Concrete Connections

✧✧ My account of truth is realistic, and follows the
epistemological dualism of common sense.
—William James

Truth kills—it even kills itself (insofar as it
realizes that error is its foundation).
—Friedrich Nietzsche

Ever since philosophy of science took a historical turn in the 1960's,
philosophers of science have paid close attention to the history of
science and to contemporary scientific practice, and the use of
evidence from case studies is often taken for granted in philosophical studies of science. Recent interest in experimentation, in the
cultural setting of scientific work, and in anthropological studies of
scientific laboratories by historians and sociologists has led to ever
more detailed studies of the history of science and current scientific
practice. Since historians, philosophers, and sociologists of science
all base their claims on case studies and include the detailed practice
in a concrete cultural setting, everyone in science studies can be
held to the standard of being true to science as practiced. I am not
claiming that science studies is trying to get science finally right,
but rather that there is a new historiographic standard in place,
which includes a demand for the inclusion of factors influencing
scientific practice that would have been parceled out to various
disciplines in the past. Science studies now must integrate cognitive
and social studies of science precisely because science is a heterogeneous set of practices that cannot be demarcated by a general principle. However, the new historiographical standards raise

questions: How must historical, philosophical, and sociological studies change with the new focus on the heterogeneous practices of science that are embedded in particular historical and cultural settings? How can general claims about science be made if we take case studies of single instances of scientific practice as our evidence? What follows from studies of scientific practice that integrate theoretical content, laboratory practice, materials, and instrumentation, and the social and political context as having an equal role in shaping science?

To get a handle on the issue of what follows from the new historiographical standards, I suggest that we look at arguments for realism and for relativism. I will first quickly review and criticize recent philosophical attempts to defend scientific realism by studying particular cases from the history of science, then delineate and criticize different strands of arguments for the social construction of scientific practice. I will then clarify the issues in debates over realism and relativism in order to show how some recent philosophical work can help us overcome the old dichotomies—scientific realism and instrumentalist, metaphysical realism and antirealism, and epistemological relativism and absolutism. These dichotomies all require global arguments, general requirements for knowledge, and interpretation of all scientific practice according to a single stance, so it will turn out that taking science as a set of situated practices that are influenced by heterogeneous factors can get us past the traditional dichotomies, as long as we remain faithful to the new historiography of science studies. Defending a positive philosophical position that makes realism and relativism impossible to formulate is part of this task, and purging the general epistemological and metaphysical stands that lead to realism or to relativism is the first step in allowing a new philosophy of science to develop. Philosophers, historians, and sociologists all have something to add to the new science-studies philosophy, and what has been a hostile debate is being replaced by signs of cooperation. I will end with some reflections on how to make general normative claims about science when starting from a local analysis of scientific practice and suggest that an extension of methods already in place in science studies will provide us with a framework that allows such a critical stance.

A Case Study of Case Studies for Scientific Realism

Scientific realism is the view that the theoretical entities postulated in scientific theories do exist and that science aims at finding true theories, not merely empirically adequate ones. Some may argue that I have picked a particularly bad example of local knowledge by starting with scientific realism, and they are right. Indeed, I have done so consciously because I want to present a clear case where claims about science fail to be local at all in order to raise methodological issues about studies of concrete scientific practice. Philosophers have attempted to apply a local analysis to the issue of scientific realism, and the attempt has been a dismal failure. Local arguments for the existence of theoretical entities were initiated by Dudley Shapere when he argued that the issue of whether or not theoretical terms refer must be decided term by term.[1] Michael Gardner applied Shapere's idea first to a study of the nineteenth-century debates on the existence of atoms and then to a study of the debate over geocentric and heliocentric theories in astronomy.[2] These pioneering case studies by Gardner accomplish the goal of formulating criteria for the existence of theoretical entities that were used in a particular scientific and philosophical context. However, Gardner tries to establish general principles for when scientists should accept theories as literal, a prescriptive generalization of the principles found in the local context.[3]

In his case study of Copernicus, Gardner develops rules for what count as "factors" in the acceptance of theoretical entities, and he decides on seven, the first three relating to ontological commitment, the next four relating to the truth of theory.[4] But Gardner's rules have been shown to apply in at most two areas, and there is no evidence that they should be generalized to apply in other cases in all their complexity. To give but one example, Gardner claims that a requirement in the debate over a realist interpretation of Copernican theory was the demand that the theory be "consistent with all observational data."[5] But clearly many theories have been accepted as true when they were not consistent with all observational data. Anomalies persisted in Newtonian theory long after it was well established and accepted. The attempt to formulate universal rules for justification of belief in theoretical entities from historical case studies puts an enormous burden of proof on the scientific realist. There are mountains of case studies to be written if we are to prove scientific realism with a sort of inductive generalization that a rule gleaned from a particular study can be applied to all types of sciences and through all history. The scientific realist will also have to deal with the many examples of good science that proceeded with entities that we now claim do not exist.[6] Inductive generalization from case studies seems a hopeless strategy if the aim is to provide a general criterion for the acceptance of theoretical entities.

Another important example of the case-study approach to scientific realism is Michael Friedman's *Foundations of Space-Time Theories*.[7] The development of the Special Theory of Relativity and of the General Theory of Relativity was read by the Logical Empiricists as an example of how science is itself instrumentalist. Friedman argues that the issue is more complex than they thought, though Einstein did use epistemological arguments that sound instrumentalist. Friedman expresses the issue of scientific realism in terms of how much of the mathematical structure used in physics is to be interpreted realistically, and he makes sense of distinctions scientists themselves make regarding parts of the structure of space-time theories.[8] Sorting out the reasons for the distinction between real entities and mathematical representations in the context of space-time theories is extremely valuable, and Friedman usually stays very close to foundational issues in space-time theories. He even says that the issue of scientific realism is secondary to the main purpose of his book.[9] Unlike Gardner, Friedman does not extend his argument beyond the range of his case study to give a general answer to the question of when we should believe in the existence of theoretical entities. However, his argument for scientific realism about space-time is based on general principles that he claims are "typical of the explanations found in mathematical physics."[10] In particular, Friedman claims that the unifying power of theoretical structure shows what is to be interpreted literally.[11] One must accept the general principles in advance, plus the specific details of the case at hand, in order to accept the reality of space-time.

These two attempts to argue for scientific realism exemplify two strategies for the use of case studies. Gardner's procedure is ground-up, since he starts from specific cases and argues to general principles. Friedman's procedure is top-down, since he argues for a general principle and applies it to a specific case. Neither procedure is successful as a local analysis of scientific practice, because the ground-up approach never justifies general claims (there simply are not enough historical cases to make generalizations possible)[12] and the top-down procedure is not really local analysis at all. It is simply an application of a general rule that is arrived at by some other means.

Richard Miller provides a third example of philosophical attempts to make a local argument for scientific realism. Miller again uses general principles, despite their being called "context-specific truisms," and seems to miss central issues in the realism debate. Miller claims that "any version of the question of scientific realism that is worthy of the label must be concerned with whether we are in a position to claim that descriptions of unobservables are approximately true. Since this is a question about justification, it is not surprising that different versions of this question turn out to be at issue as the consensus about the nature of scientific justification changes."[13] Arthur Fine points out that this judgment is not a question of justification; it is a question of interpretation.[14] Bas van Fraassen has argued persuasively that the constructive empiricist is always in a position to interpret any justification of the existence of a theoretical entity as merely a justification of the empirical adequacy of the theory, and to claim that only acceptance of the theory as empirically adequate is justified, while belief in the existence of the theoretical entities postulated never is.[15] Van Fraassen does not question the justification of the empirical adequacy of theories. His question is whether we should accept the theory (commit ourselves only to the view that the theory is empirically adequate) or believe the theory (commit ourselves not only to the empirical adequacy of the theory but also to the existence of the theoretical entities that the theory postulates). The scientific realist cannot win this debate with van Fraassen by simply adding more evidence, because any evidence can be recast as evidence for the lesser claims made by the constructive empiricist. Scientific realism requires a principle of abduction that is surely global and, I will argue below, unjustified.

It is important to distinguish the debate over scientific realism, which occurs within the philosophy of science community, from the debate over relativism that started with Kuhn's claim that scientific theories are only justified relative to a special social setting (disciplinary matrix or paradigm). Relativists do deny that theories are justified (in any non-question-begging sense) even as empirically adequate, and it is relativists whom Miller has in mind: "For typical anti-realists today, the acceptance of theories is always relative to a framework of beliefs, and the actual shared framework of current theoretical science is no more reasonable as a guide to truth than rival frameworks, dictating contrary conclusions."[16] It is now a commonplace idea that there are several forms of realism, and that one must be careful about which set of issues is at stake in any particular discussion. In this case there are good reasons to keep issues concerning acceptance of results and issues concerning scientific realism separate. There are two different epistemological arguments used as the basis for these skeptical positions. The first is a general argument about justification of results, in which it is claimed that results are always justified by circular reasoning or on the basis of unargued presuppositions (the framework); the second (van Fraassen's) relies on a special empiricist principle that is used to argue that theoretical entities have a different epistemic status from observables.

Both these epistemological arguments are global, since they apply to any scientific context. Indeed, the skeptical argument concerning circularity has a long history, and it applies everywhere, even outside science.[17] However, the issue of the acceptance of results, whether theoretical or experimental, and the development and establishment of scientific practices are amenable to local analysis, whereas scientific realism is not. Shifting the focus from scientific realism may get us closer to a truly local analysis of scientific practice that makes genuine use of case studies. By emphasizing the local situation instead of global realist interpretations and global rationality principles, the dichotomy between internal and external accounts of the history of science can be overcome. The local situation can be portrayed as influenced by multiple pragmatic and theoretical constraints and opportunities. Social influences can be a natural part of inquiry and can be seen as objective or even rational, though I do not want to put the debate in these terms. Indeed, I need to distance myself from traditional philosophical discussions of the acceptance of scientific results, because philosophical discussions of the issue are often put in terms of general rationality principles and universal scientific method. I adopt quite a different strategy for blocking skeptical arguments, one that undercuts the traditional standards for theory justification that form the basis for both relativist and foundationalist arguments about science. The key to overcoming the traditional standards and overcoming the skeptical argument about circularity is recognizing the disunity of science—the heterogeneous influences on scientists in actual practice.

Disunity and Epistemology

Traditionalists and their relativist doubles both assume that there is a large area of human knowledge called science and that it is meaningful to ask whether science is justified absolutely, as universally valid.[18] The common argument of both traditionalists and relativists is that the reasons that nonfoundationalists argue can be offered for acceptance of theories and experimental results beg the question, because these are only reasons according to local standards, and these standards can be called into question. Both claim that we need transcendental (universal and/or fixed) standards by which to judge our beliefs if we are going to say that we know. The traditionalists then say that there are such standards (or that there must be, if they are having trouble finding any), whereas the relativists proclaim that the fact that there are no such standards establishes their position.

The way out of this false dilemma is to recognize that one area of diverse scientific practice can be used to justify another without begging the question. As Ian Hacking and Peter Galison have emphasized, experimentation has "a life of its own" and often follows constraints that are independent of the theories under consideration. Peter Kosso points out that even in cases of very indirect observations, which are heavily theory-laden, it is quite possible for the theory under test to be epistemically independent of the auxiliary theories that are used to construct the observing instrument.[19] Thus, there is no automatic way to charge that theory-laden evidence begs the question. Once we reject the view that all standards are internal to a theory (thought style, disciplinary matrix, etc.), we see that it is an open question whether or not a given theory is related to another epistemically. One must look to the details of each particular case to find out whether experimental constraints are independent from the theory under test or whether they beg the question. I want to argue here that the notion that the results of science are independent from the theory and even from that material practices under which the experiment was produced (in the sense that results cannot be predicted in advance) is a much stronger notion than might be expected. It is all the realism and objectivity we need, and undoubtedly all that we can get, as well. There is no justification of science from the outside (from, say, an autonomous epistemology), but scientific practice is disunified enough to render non-question-begging knowledge possible.

The picture of science that emerges from a local analysis of scientific practices is very different from the realist picture, but very different from Kuhnian relativism and its latter-day cousins as well. Sociologists of scientific knowledge have often called themselves relativists, but have also complained that their position has been misread. The sociologists have been charged with claiming that "only social conditions matter," and that "scientists do not pay attention to evidence, etc." but complain that their true view is that both social conditions and evidence play a role.[20] Using Mannheim's classic sociology of knowledge program as a model, Michael Lynch has recently outlined the program as involving two steps:[21]

1. Employing historical comparisons to show that an "immanent theory" cannot entirely explain the contents and historical development of the system of knowledge in which it is situated. This procedure is used to demonstrate that such a theory cannot unequivocally and exhaustively attribute the state of its knowledge at any given time to "the nature of things," "pure logical possibilities," or an "inner dialectic."
2. Specifying the social conditions (the local historical milieus, class interests, and group "mentalities," rhetorical strategies, etc.) that influenced the development and content of the given state of knowledge.

So we see that evidence does enter scientific practice, as sociologists claim, but it does not explain why scientists adopted a given view, because evidence does not restrict the options available to only one choice. Theories are underdetermined by evidence.

There remains a tension in this view, however. Imagine a traditional philosopher complaining: "I never said that social factors did not come into play; I admit that they do. I am only saying that social factors always underdetermine the choice of a scientific theory, and that in order to explain why scientists made the choices that they did, consideration of empirical evidence and methodological standards will have to be brought into play."[22] All that one must do is ask which type of consideration is decisive, and we are brought right back to a debate between rational and social accounts of science. And this demand seems natural, since the sociologists have already claimed that methods and evidence are never decisive (step 1 above). It seems as though the social explanation has to be doing all the work; evidence and reasons are there, but they leave too many options open to be of use in an account of the development of science. I suggest that this debate is strongly miscast and that the local nature of scientific practice has been lost:

1. Choices are at least as underdetermined by social factors as by internal ones.

2. Social factors do not explain anyway.[23]

3. To say that either internal or social factors explain the development of a science amounts to a reduction of scientific practice to a single dimension.

4. Both sides in the debate over whether science is rational or social assume that a "reconstruction" of scientific practice is possible.

5. No difference between "content of a theory" and "social conditions" can be maintained under the current historiographic standards.[24]

We have to move beyond proving general epistemological points with case studies; understanding the development of science is what is at stake. We should not be debating between an internalist history of science and a social history of science, since neither kind of history is possible without demarcating social reasons from purely scientific ones. Instead, we should take account of the heterogeneous and contingent nature of scientific practice. I suggest that we must follow the consequences of point (5) to the end; there is no room for general epistemological arguments in the current historiography. If we could distinguish the purely social from the purely scientific in our accounts, then we could continue a battle over whether the development of science can be accounted for by evidence and methodology alone, or whether social factors must enter. But since we cannot even separate the two, the debate makes no sense.

A look at Andy Pickering's replies to criticisms of his work will be a good way to focus the issues involved, since Pickering advocates a particularist view of scientific practice and he has been embroiled recently in battles over relativism. The site of analysis is what Pickering calls the interactive stabilization of three elements that are involved in the production of experimental results—a material procedure, an instrumental model, and a phenomenal model.[25] Three main features of scientific practice, according to Pickering's account, are that it is:

situated: Scientific practice takes place in a specific, concrete setting that cannot be captured either by internal scientific developments or by social influences.

temporal: There is no fixed element; theories, experimental apparatus, procedures, goals, plans, and "nature" are all open to reinterpretation, and all can change in the development of a scientific practice.

contingent: There is no reliable way to predict the outcome of the practice of using plans, theories, experimental apparatus; of accommodating results, resistances to plans, and so on.

The stabilization of scientific practices is a concrete, local phenomenon, and therefore amenable to particularist analysis. Here I will focus on the contingency of scientific practice and show how Pickering now expresses a position on the objectivity of science that there is no reason for anyone to reject unless she or he holds on to a traditional standard of theory justification and hopes for some kind of foundational element in knowledge. If I understand his particularism correctly, the point of the claim that the play of resistance and accommodation is contingent is that there are no *general rules* for the stabilization of scientific practice, but rather only particular historical explanations.[26] Therefore, the stabilization of scientific practice is not reducible to anything that could allow a prediction of how resistance will be accommodated.

Pickering's rejection of reductionist accounts of scientific practice is connected to the overcoming of realism and constructivism because it implies that the kinds of explanations appropriate to an examination of scientific practice cannot be reduced either to the claim that "the world" plays the key role in determining scientific practice (i.e., discovery of preexisting scientific laws) or to the claim that interests or other social factors play the key role. Thus, in his recent work, Pickering joins Bruno Latour and others in rejecting the relativism and realism debates entirely, which I applaud.[27] The point here deserves emphasis. Traditionalists claim that social factors enter only into explanations of bad science, and that sometimes at least scientists follow purely rational procedures. I agree with the symmetry thesis of the strong program, insofar as it says that both good and bad science are to be explained by local conditions. I am not claiming that the independence of results means that sometimes social factors do not enter; I agree that they are always there.[28] The epistemic disunity of science provides, however, a picture of the social practices of science that is radically different from a relativist view of science.

Pickering claims that the relativism of his view of science has not been refuted by talk of constraints, that is, by the independence of results from the scientific theory. He claims that all that has been refuted is a ridiculous form of subjectivism—the view that anything one believes is true:[29]

Constructivism is thus understood [by its critics] to assert not just a form of cultural relativism, but something much more specific, a self-fulfilling wish or desire-relativism. And through this series of displacements, the critics of constructivism put themselves in a position to say something terribly sensible as if it were news. Scientific practice is difficult, they say sagely, to much nodding of heads, scientists cannot just believe what they like, and we can show it.

Even antirealists' wishes do not always come true. Pickering accepts an everyday notion of objectivity, the idea that what we think does not decide a matter of fact:[30]

To return to everyday usage, to speak of the objectivity of knowledge is to deny that knowledge is a "mere construction," a projection of human fantasy onto the world, and to affirm that knowledge is somehow disciplined by the otherness with which it engages.

We will see that William James adopts "realism" in this sense, and that it is only the metaphysical version of realism that James sees as a problem. Although Pickering accepts this kind of everyday objectivity, he wants to distance himself from any sort of traditional view. Indeed, he is perfectly justified in this sentiment, since, as I will show below, the everyday notion of objectivity, including the independence of scientific practice, does not lead to realism; it is perfectly compatible with several versions of metaphysical anti-realism. However, the independence of results does block the usual epistemological argument that justification of scientific results is circular. Furthermore, since this epistemological argument often unjustifiably motivates metaphysical antirealism, it is not totally irrelevant. I will argue that both the epistemological and the metaphysical arguments for relativism need to be resisted, and that it is quite important to see that the arguments are separate.

Regarding Pickering's complaints about "constraint talk," I think that in large part the debate may be simply a matter of clarifying what is meant by constraints. Pickering seems bothered by the fact that constraints are negative, but the constraints discussed by Galison and others can be positive as well. They not only hold scientists back from certain choices; they also give them choices as possibilities with which to work. A more substantial point involves the complaint that constraints are fixed elements in scientific practice that may seem prior to practice, while Pickering's resistances emerge in practice.[31] Pickering rightly wants to emphasize problems that emerge in real time, rather than prior theoretical and metaphysical presuppositions, which were widely discussed in the work of Feyerabend, Holton, Kuhn, Lakatos, Toulmin, and others when philosophy of science took its historical turn in the 1960's. However, the independence of the results of experimental practice does not come from any fixed presuppositions that act as constraints. The independence of experiment and theory is a claim that there is an epistemic disunity in science, a claim that, *pace* Kuhn, there is no single "disciplinary matrix" that determines scientific practice. The independence of results follows from an anti-essentialist view of science, exactly the kind of view that Pickering is trying to formulate.

Of course, I am not claiming here that scientific results are always accurate, but rather that the ordinary checks on scientific practice cannot be dismissed by a general argument that would always undercut any justification of results by pointing to "other options" or by questioning the procedures and standards by which justification takes place. Such an epistemological argument is global and based on a bad model of scientific practice. Such a model covers up the situated and contingent nature of scientific practice that Pickering and others are trying to highlight. Without such a neo-Kuhnian model of scientific practice, the epistemological argument for relativism does not work. Constraints are part of the social structure of scientific practice, but Galison argues persuasively that the local, contextual nature of these constraints, far from implying relativism, supports its objectivity.[32] Pickering helpfully emphasizes the contingency of scientific practice, providing one more way to reinforce the claim that the results of experiment are not built in.[33]

In response to Kuhnian arguments for relativism (highlighted by Doppelt), I have argued that the epistemic disunity of science allows nonfoundational, fallible evidence for a result to provide good, "objective" grounds for a theory without necessarily begging the question, but I should emphasize that I do not intend anything more.[34] I am arguing against relativism, but I certainly do not want to be taken as a realist, at least not in several of the senses of this overly debated word. The independence of results provides no support for a metaphysical realism in which the world "out there" determines the results of scientific inquiry, since the results are still part of scientific practice. Furthermore, it does not provide an argument against theories of truth that define true beliefs as those that cohere with other beliefs or as those that are agreed upon by a relevant community (epistemic theories). The independence of results from scientific practice can be accounted for with (almost) any theory of truth and hence is separate from these metaphysical issues. It is obvious that we cannot vote to make a fact true, or wish something into existence, which is Pickering's point in response to his critics. What we think about something (a value judgment) or whether we are in a position to settle a question (an epistemological issue) is irrelevant to the claim that there is some state of affairs right now. We must have evidence for our beliefs, and even when we do have evidence we can always be wrong. However, even though everyone agrees that a local consensus theory of truth would be absurd, many theories of truth that define truth according to some epistemic property can account for the fact that what is true is independent of what an individual or group believes. Hence, there is no automatic support for a nonepistemic theory of truth from our accepting a "difference between what is true and what is merely thought to be true." Putnam uses this phrase as a definition of realism at one point, but, as it stands, it is clearly inadequate.[35] As many realists have recognized, antirealists do not say that any thought is true, but rather that only those that meet the standards of their definition, such as coherence with other knowledge, consensus by relevant experts, and so on, are true.[36]

Any reasonable demand for independence can be accounted for by (most) theories that equate truth with some epistemic property such as coherence with other beliefs or consensus by relevant experts; therefore, definitions of truth are not relevant to epistemological issues. Without strong and untenable further claims about the nature of knowledge such as Kuhnian holism about science, or global skepticism about knowledge, the independence of beliefs from particular sets of practices comes naturally. Thus, metaphysical positions are neither necessary to overcome these epistemological arguments, nor are they, by themselves, capable of overcoming them. However, the separation of epistemic and metaphysical arguments cuts both ways. Realists are wrong because the kind of objectivity that follows from the everyday notion that the results of experiments are not (always) built in does not by itself lead to a metaphysical version of realism—to a correspondence or even to a minimal theory of truth. Realists need a separate argument that truth should be interpreted nonepistemically (nonrelativistically). Nevertheless, as I have argued above, the independence of results breaks the circle that the traditional skeptical argument raised, and therefore undercuts the epistemological argument for relativism, even if it does not refute a metaphysical one. The epistemic disunity of science therefore provides a mechanism for explaining the correspondence intuition that truth is independent of our beliefs and practices, and this mechanism does not rely on metaphysical facts to make truth "radically nonepistemic," since epistemic independence can be found within science.[37]

Pickering still holds out for some form of relativism, at least as a rhetorical device to separate himself from traditional views:[38]

So the historicity of knowledge emerging from the mangle is a culturally-situated historicity, and knowledge is, in this sense, relative to culture. But this relativity cannot be summed up in any enduring social principle like interest that specifies the link between present and future. The mesh-ing of contingency and structure in the mangle points, then, to a kind of hyper-relativism.

I think that it is a mistake to claim that scientific practice is relative to culture in any sense.[39] When I defend the thesis that reasons for acceptance can be independent of a specific theory or set of prac-tices (which is why the results are not built in and cannot be predicted in advance), I am not claiming that they are independent of all culture; nothing is. That thesis is intended only to refute the epistemological argument for relativism—the argument that any evidence for a theory will always beg the question. Besides, culture is just as heterogeneous as science, and, as Pickering himself points out, it is not causally effective.[40]

Once the epistemological argument for relativism is rejected, what is the status of the claim that situated knowledge is somehow determined by culture? One way to understand such an argument for cultural relativity is to hold that it must amount to an epistemic theory of truth—either a coherence or consensus theory;[41] but the realist demand for an explanation of the success of science leads to many problems if social explanations replace the old metaphysical ones. Since claims that scientific practices are epistemically inde-pendent of each other are compatible with both realist and (some) antirealist views about truth, constraint talk need not be interpreted as intending anything about realism. I will turn now to philosophi-cal theories about truth and show how a deflationary account of truth can lend support to an antireductionist program for science studies.

Truth and Metaphysics

The epistemological debate described above focused on the ra-tionality of science, but the argument for the social character of scientific knowledge has shifted as the strong program moved through its three phrases. Generally, the argument has moved from epistemological considerations in Edinburgh, represented by Barry Barnes and David Bloor, to inclusion of arguments about skill and tacit knowledge in Bath, represented by Harry Collins and Trevor Pinch, and finally to a reflexive stage, represented by Malcolm Ashmore and Steve Woolgar, who explicitly says that he is extending social-constructive arguments to what were traditionally considered metaphysical issues.[42]

Metaphysical realism is an ontological claim, the view that there is only one actual world, that it has definite features independent of our knowledge of it, and that facts exist in this world prior to any human activity. I do not endorse metaphysical realism. Quantum mechanics, at least on some interpretations, will refute the formulation I have given. Of course, quantum mechanics is not the usual source of relativism, though it may provide a reason to give up metaphysical realism. Metaphysical antirealism is the view that the world is a creation of human activity (or mind, to use the older language) and can be seen in the claim that reality is just a text to be interpreted, in claims that facts are created, and in claims that scientists holding radically different views live in "different worlds."

Until recently, philosophers had done a bad job in formulating positions within which one could break the dilemma of "world-out-there" realist accounts and social-constructive relativist accounts, but some progress has recently been made by focusing on truth. Both realists and antirealists treat truth as a real property with a real definition. The former define truth as something external to all social character and the latter as something internal to a theory (disciplinary matrix, conceptual scheme, form of life, etc.). A minimal or deflationary account of truth displaces both metaphysical realism and metaphysical antirealism by rejecting the demand to explain why sentences are true.[43] Truth should be defined neither as transcendent nor as a social construct; rather, truth should not be defined at all.

Traditionally, truth was defined as correspondence between statements and the world, but what correspondence meant was left open and could be quite far from naive realism.[44] Traditional skeptics always raised the problem that appearance may not match reality, and the problem of access: we are never in a position to compare appearance and reality, because we only have access to appearances.[45] It is with Kant that the debate over truth gets started. Kant claims to solve the problem of skepticism by literally defining reality as the totality of the phenomenal realm. In other words, he makes the world accessible by defining the world as whatever we can reach. Definitions of truth follow an analogous pattern in the writings of the nineteenth-century idealists, who defined true beliefs as those that cohere with other beliefs. There are other versions of such epistemic theories, all of which make what is true dependent on human thought in one way or another. William James defined truth as utility (whatever that means beyond coherence). James has been accused of holding a local consensus theory, but maybe he never really did. A major theme of James's classic article "Pragmatism's Conception of Truth" is that correspondence—agreement with reality—can be interpreted without metaphysics and in a completely everyday sense. But James also seems to make truth epistemic, and he makes truth "mutable" (changing from one epoch to another), a point that especially bothered Peirce.[46]

The epistemic definitions of truth make realists' blood boil and are indeed problematic:

1. They may be self-referentially inconsistent. (I think this can be avoided—you need a universal and a negation to create a problem analogous to the liar paradox.)
2. They do not fit the "grammar of truth." (For example, iteration: for any true sentence P, it is true that it is true that P, etc.; but under the epistemic definitions iteration does not necessarily hold.)[47]
3. They do not fit our intuitions that:
   a. our agreement or acceptance is irrelevant to the truth of a statement.
   b. no amount of coherence guarantees that you have the right solution.[48]
   c. there is only one world; we do not create facts, and so on.

Bertrand Russell debated the central point with James: those who advocate epistemic accounts of truth confuse evidence for truth with the definition of truth.[49] They feel a need to identify justified belief with true belief in order to avoid the problem of access.[50] But this line of argument is totally unnecessary. Fallibilists say that we never have anything but evidence, never truth itself. Unless one wants certainty, which is impossible, good evidence, though fallible, should be enough. Any theory, no matter how well confirmed, could be wrong, but that does not prove that science is relative to anything, or that facts are created. The point is analogous to the theory-ladenness of observation. If we could see the world the way it really is, without any infection from theory, then we could define reality as what we see. As soon as we reject foundationalism, we create a gap between evidence and knowledge. Those who advocate epistemic theories try to close the gap with their definitions of truth in order to get back what they lost when they gave up certainty, but the fact that any of our beliefs may be wrong should not be considered problematic. The fact that any statement could be wrong does not lead us to sit in our tub, but rather to reject the old view that knowledge must be certain.

The minimal theory of truth gives us the resources to avoid a dilemma that clearly parallels the dichotomous choice of rational or social accounts of scientific practice.[51] If we say that correspondence with the world makes a sentence true, we have the problem of access and are left with skepticism. If we say something accessible makes sentences true, we make truth relative to a criterion that can be seen as part of a framework. How are we to respond to this dilemma? Deny that anything makes sentences true by adopting a minimal account of truth and refusing the request for explanation demanded by both metaphysical realists and metaphysical antirealists. Those who adopt a deflationary or minimal account of truth claim that truth is not an essential property at all, so there can be no account of the meaning of the word in terms of properties in the world. Just about the only thing that we can say about truth in general, and there is not very much, is that every statement P can be considered equivalent to the statement that P is true.[52] The minimal account of truth is compatible with ordinary-language usage of truth, facts, and so forth, and with our everyday intuition that what we think does not decide a matter of fact.[53] So this minimal position is already strong enough to block a metaphysical version of relativism, the claim that there may be multiple worlds that depend on our conceptual scheme. One might think that defenders of such a minimal realist position would be committed to a correspondence theory of truth, but it has been argued persuasively that this is not the case.[54]

Defenders of the minimal account claim that the ordinary notion of truth is worth saving but the independent world (the metaphysical correlate) is not. Why do these philosophers want to hold on to a notion of truth that accords with ordinary language? After all, ordinary language can be wrong. In the first place, we have been given no good reason to give it up. The epistemological problems of access can never give us evidence to support the metaphysical claim that there are multiple worlds. As I have argued above, a proper understanding of fallibilism makes epistemic definitions of truth unnecessary, for there really is no problem of access. The problem of access is an artifact of the representational account of knowledge. Furthermore, in order to make the argument from lack of access to multiple worlds, antirealists must employ some principle of verificationism. They must argue that there can be nothing outside what is accessed by our knowledge-producing practice. In Hegel's version, "The rational is the real, and the real is the rational."[55] But all formulations of verificationism have been shown to be problematic, and this strong epistemological principle certainly should not be used without careful discussion and defense. We can always respond that access has nothing to do with the way the world actually is—in other words, we can embrace fallibilism in order to block relativism.

Realists and relativists both assume that if there is only one world, then there must be only one correct description of reality (one correct explanation of a phenomenon, etc.). Realists believe in one world, and apply the assumption and *modus ponens* to arrive at the conclusion that there can be only one correct description of the world. (Usually a causal description is privileged.) Relativists who believe that there is no uniquely correct description of the world use this assumption and *modus tollens* to argue that there is more than one world. Thus, metaphysically inclined relativists adopt an essentialist definition of truth by equating the truth of a sentence with some property of human thought.[56] I claim that this conditional assumption is the link between metaphysics and epistemology that stands behind all essentialist definitions of truth, and that this link must be rejected. Instead, we should remember that scientific description and explanation are always incomplete. If the description of the world is an endless process, then there cannot be a single correct description; but this is no reason to claim that there are multiple worlds or that facts are created. Do not solve your epistemological problems with metaphysics; instead, solve (or dissolve) them by rethinking epistemology.

In arguing that we need neither to drop talk about facts nor to talk about their creation, I am not claiming that there is anything special about ordinary usage; I am far from an ordinary-language philosopher. I have blocked the usual arguments in favor of a relativist usage, but arguing for linguistic inertia given that no argument leads away from ordinary usage is not completely satisfying. The important positive argument in favor of the ordinary usage of the word "true" is that the disunity of science makes facts independent of our knowledge production, as I have argued above. There is no metaphysical independence of results from social practices, no transcendental world out there, but there is still independence in as robust a sense as we need or should want. William James recognized this very clearly in accepting "the epistemological dualism of common sense," and in calling his theory of truth "realistic."[57]

Also important for the topic at hand here, however, is the point that any argument in favor of multiple worlds, whether metaphysical or epistemological, must be a global argument.[58] It must be either a general epistemological argument or an *a priori* metaphysical stance that is imposed on case studies. By contrast, the kinds of thin explanations of the truth of statements given by a minimal theory are particularist. In the trivial explanation given by the disquotation scheme, the minimalist simply repeats whatever particular fact is expressed by a statement: Why is the statement "Snow is white" true? Because snow is white. In ordinary kinds of situations, including scientific ones, the minimalist accepts the ordinary justification of the truth of a given statement (for that context). Of course, in disputes the context can be broadened as black boxes are opened. What is denied is the demand for a general account of why statements are true. Neither the realist account nor the constructivist account is satisfactory. The demand for a general account of why statements are true leads immediately to a dilemma in which we are caught between skepticism and relativism. The nicest feature of the minimal conception of truth in regard to the current debate is that it allows one to block relativism without committing oneself to a realist position that leads to other types of problems. Thus, the problem of access and the inevitable skepticism that result from the correspondence conception of truth are eliminated.

I have argued previously elsewhere that Ludwik Fleck's argument for the construction of facts fails.[59] Briefly, his argument is that the skill needed in the experimental laboratory is independent of theory and that this implies that knowledge is culturally conditioned. Skill is an irrational element of science that is not captured by theory on Fleck's view, and he argues that this irrational element is determined socially by the "thought style."[60] I claim that something quite different follows from Fleck's argument. I accept all of Fleck's epistemological points—observation is theory-laden, scientists do not discover preexisting facts about nature, social pressures influence every aspect of science, and so on—but I deny his conclusion that facts are the creation of a particular "thought style." I have extended this argument to make a much more general point here.

If Fleck's statement that facts are the creation of a particular thought style is thought of metaphysically he is wrong, and Fleck clearly invites a metaphysical reading when he adopts the rhetoric of the creation of facts. In an everyday sense, a certain group of scientists and technicians did create a new phenomenon in the laboratory—the Wassermann reaction. In an everyday sense, they changed the world. But facts are not created by a thought style. Facts are neither internal to culture nor external to it; nor can they be said to be a simple mixture of factors. Facts are the result of the stabilization of a practice, in this case the Wassermann test. The situation here is completely analogous to the definition of truth, and philosophers have taught us that neither an internal nor an external account will serve. The situation is also analogous to social and theoretical accounts of science, where sociologists and historians have taught us that neither evidential nor social accounts of science are sufficient. While analysis of science cannot be reduced to analysis of discourse, nature does not enter through material experimental practice, either. The internal and the external, the material and the conceptual, and the social and the evidential are all inextricably mixed in scientific practice; none can be privileged.

The traditional realist view was that correct scientific belief is determined by discovery of facts about the world, facts that are independent of all scientific practice. The constructivist view is that social conditions are always involved. Of course, in an everyday sense, it is true that scientific practice is always social. It is very important that this explanation not be presented as a new answer to a metaphysical question, however; and this means eliminating talk about "creating facts" in the laboratory, as well as eliminating the offending metaphysical realism. I think that it is clear that if science studies is to overcome the debate over rational and social accounts of scientific practice, we must give up the demand to explain what makes scientific practice stabilize.[61] We must say that some of the contingency in scientific practice cannot be eliminated, that some practices just work, full stop.

The minimal theory of truth that I have described performs a mostly negative function, providing a way to avoid both relativism and realism. Something toward a positive account of the epistemology and metaphysics of scientific knowledge is provided by Bruno Latour. Latour has described one part of getting beyond the determination of scientific practice by either social or purely scientific factors in this way:[62]

We do not have, on the one hand, a history of contingent human events and, on the other, a science of necessary laws, but a common history of societies and of things. Pasteur's microbes are neither timeless entities discovered by Pasteur, nor political domination imposed onto the laboratory by the Second Empire social structure, nor are they a careful mixture of "purely" social elements and "strictly" natural forces. They are a new social link that redefines at once what nature is made of and what society is made of.

Latour replaces both the natural and the social with the single category of actants (a term taken from semiotics), and by doing so, he overcomes the passive/active dichotomy.[63] The traditional view has an active human (individual or social) who acts upon passive nature. On this view, we make constructions with our actions, but we discover a reality that is passive. Latour's actants include entities that would traditionally be considered agents, such as Pasteur, and also entities that would traditionally be considered passive objects, like the microbes.

Latour has made a real advance by avoiding the active/passive dichotomy, since debates over realism and social construction can be thought of in terms of whether knowledge is passively discovered or actively created by scientists. For example, in *Constructing Quarks*, one formulation of Pickering's complaint with "the scientist's account" or retrospective realism is that it makes a scientist into a passive observer or a passive discoverer of reality, instead of an active agent who creates scientific practice.[64] However, by my lights, Latour relies too much on ontology to overcome the debate over rational and social accounts of science. He makes a claim about what there is in the world (actants) to settle what remains essentially an epistemological debate. It is the representational picture of knowledge that underlies the debate over rational and social accounts of science, not the passive/active dichotomy. In the representational account, knowledge that is grounded by something outside practice ("really in the world") is taken as valid; the denial of such a grounding makes knowledge a mere construction. We can settle the epistemological issues independently of metaphysics. The minimal theory of truth fits nicely with Latour's rejection of the demand to say what is the decisive factor in determining scientific practice, but Latour sometimes comes close to advocating a coherence theory of truth.[65] Furthermore, Latour's arguments are most persuasive when seen as methodological proscriptions for science studies, as how to study science without taking either the natural world or society as a given unexplained explainer. It is not clear, however, what justifies a move from methodological proscriptions to metaphysical claims about the world. In any case, Latour has said there is more philosophical work to be done in science studies, and he even says that ontology is the easy part.[66]

The aspect of Latour's work that I would like to highlight here is the way in which the emphasis on temporal changes in ontology can begin to dissolve the dualist representational account of knowledge. First, using recent historiography, ethnomethodology, and semiotics as a guide, Latour starts with the actors' own categories, with what is said and done in the laboratory, or with what is written in a historical text.[67] He does not allow the importation of our current knowledge of what is correct. Second, he allows all the actants an equal role, refusing to privilege the natural world or society in explanations of scientific practice. Finally, Latour shows that the social/natural split is a result of the actants' negotiation. That is, a newly stabilized set of scientific practices emerges with human and nonhuman agents embodied in it, and material and social elements defined by it.

When I said earlier that there is no simple mixture of internal and external factors that can account for a scientific practice, I was accepting Latour's point that the division between nature and society comes only after the fact. Therefore, the stabilization and development of scientific practice are primary. Traditional metaphysics demands a prior and static classification of entities into natural and social (or material and mental), and this classification makes genuine historical change impossible. We can only say that those who held views that are now discredited were wrong, that we know what they did not; in short, we can only be whiggish. Latour denies the view that there is an *a priori*, atemporal distinction between nature and mind/society, so in a sense, he rejects the question of what makes science stabilize, just as the advocate of the minimal theory of truth rejects the question of what makes sentences true in general; but his is a more radical rejection. Advocates of the minimal theory allow a form of dualism when they say that truth is independent of human activity, but they minimize the effect of dualism by denying any effective role for the independent world. They deny that it is the cause of the truth of sentences. Latour goes farther, by not allowing the dualism until after the fact.

In another sense, Latour has an alternative explanation for what makes sentences true in general, since the negotiating of actants always determines what stabilizes. It may seem to some that Latour is advocating a form of idealism by classifying only after the fact, but this charge misses the point. One simply begs the question by demanding that there must be an *a priori* distinction between the social and the natural. Of course, we do regularly make a distinction between the social and the natural; that is why overcoming dualism seems so difficult. The lesson from the local studies of scientific practices is that we only have the distinction after the stabilization of a practice, and then only for a short time. Latour's "answer" is surely a minimal one; it is like saying that all kinds of things (heterogeneous local conditions) make sentences true. Here we have a nominal epistemic answer to the question of what makes sentences true in general, analogous to the nominal realist answer that James accepted. Minimalists can say that correspondence with reality makes sentences true, as long as they refuse to pin down what that means with a theory of reference. I suggest that both realist and epistemic *nominal* answers are acceptable.

Like Latour, David Gooding attempts to overcome dualism and the realism-versus-constructivism debate. Gooding shifts the discussion of scientific practice away from the traditional debates by focusing on detailed accounts of experimentation. Experimental practice is both embodied and cognitive and does not lend support either for traditional realism or for social construction. Relativism only follows with the addition of strong epistemological or metaphysical assumptions.[68] I want to focus here on the new interpretation of the correspondence theory of truth and realism that Gooding presents. Rather than correspondence between theories and a reality that is taken as a given, we have correspondence between theories and experimental practice on Gooding's account,[69] and he shows further that experimental phenomena are neither given natural objects nor mere social constructs: "Faraday did not discover correspondences as ready-made relationships between the concept of some thing (in reality) and the perception which it denotes. The meaning-relation depended on invention and disseminating procedures. Their meaning depended in turn on shared practical skills as well as shared assumptions."[70] Phenomena are the result of an experimental practice, which is the product of many factors, some cognitive, some material, and some social.

I think that Gooding has made a real advance in showing how a nominal or minimal theory of truth can work, although he has not put it that way. After an experimental practice has been stabilized and a way to present it to others has been found, we can say that the theory represents or corresponds to the phenomenon;[71] therefore the correspondence relation is completely unproblematic. All the work goes into creating the stable experimental practice. There is no more mystery left at the linguistic level. The puzzling relation of words to the world is now a puzzle about experimentation, on Gooding's account. The mix of social and cognitive factors in science is located in experimental practice, making it harder to force a dichotomy between realism and constructivism by asking which factors explain stabilization, since both kinds of factors are involved in every step of the process. Furthermore, since the mind and the world are now in one place, the site of experimental practice, their relation can be studied successfully in the concrete and local setting in which we find them. Gooding sums up his view by saying:[72]

I argue that the correspondence of representations to their objects is the end product of a process of making convergences which are subsequently reconstructed as correspondences. Reconstruction has a material and a verbal aspect. The material embodiment of skills into apparatus and techniques . . . establishes the objective distance of phenomena from representations of it, while narrative reconstruction imparts the logical structure found in published reports.

We still need to be careful to avoid being pressed into saying what makes something true, however, or in this case, what makes an experiment work. Gooding says that "certain features of experimentation . . . indicate that something outside the cultural net ('nature') is involved."[73] Clearly, "nature" is intended to block the claim that experiment and observation are always open to interpretation. In experimentation one often encounters something different than what is expected, a point that I take to be similar to Galison's claim that experiment is independent of theory and not infinitely interpretable. However, Gooding also says that "the phenomenal outcomes appeared to be real because they brought a stable resolution to the problem of making sense of all of the other phenomena produced up to that point." I am surprised by this statement, since it depends so strongly on coherence as a criterion. Making sense of the phenomena must surely be an interpretive enterprise, and there will inevitably be alternative stable resolutions. I realize that I am stepping back to a global perspective to make the point, but it seems to me that coherence is irredeemably antirealist in character.

Gooding claims to be a realist, but only "locally."[74] The point is that objectively real phenomena are made, not discovered, through experimental practice. In fact, the embodied and heterogeneously influenced experimentation does not lead to relativism or to traditional realism. The correct account of scientific practice corresponds to science and society, to material practice and thought, and to the way that both changed together. Thus, Gooding advocates an ontological ambiguity concerning phenomena that is similar to the ontological status of Latour's actants.[75] The answer to the question of what explains scientific practice can be found in experimental practice, but heterogeneous factors, both social and cognitive, make up the concrete practice that provides the answer.

Toward a Grand Disunified Theory of Science

The picture that emerges here is one of both holism and disunity applied at different levels of abstraction. Metaphysical realists argue for metaphysical oneness, the most abstract kind of unity imaginable, and social constructivists for multiple worlds, an abstract kind of disunity. The fracture of science into independent domains implies the disunity of science on the middle level of abstraction. The disunity of science is all that we need to obtain the epistemic independence of results that we expect in adopting the everyday use of the word "true." As I argued above, the epistemic independence of different scientific practices is crucial to blocking traditional skeptical arguments. Finally, on the local level, we have holism again in the local stabilization of the different elements of theory, experiment, and instruments into a concrete scientific practice.

The first lesson from the analysis of epistemological and metaphysical debates that I have given above is to give up both kinds of debates over realism. Studies of local practice can lead to good analyses of ways in which scientific practices develop and are stabilized. The debate between philosophers and sociologists of science has always hinged on whether to hold on to some notion of objectivity, on an independent element that determines scientific practice. I have argued that these debates are miscast. A related but different story tells how both realism and instrumentalism in the philosophy of science are equally wrong.[76] As I mentioned in the beginning, some topics are not amenable to particularistic analysis. I will now return to considering the possibilities and the problems facing a genuinely local account of science.[77]

I have been endorsing and defending a deflationary framework that will please neither traditional philosophers nor the sociologists of the strong program but that could help us get beyond the debate over realism and relativism. The central problem that remains is answering questions about the historiography and methodological models that should be used in science studies. We have to have some historiography or ethnographic framework for doing our studies, but general methodological frameworks run counter to particularism and may seem reductionist, especially if science studies opts for a "privileged discipline" that will stand as the methodology of science studies, which many naturalized philosophers of science have sought.[78] Here I will sketch answers to objections raised around two issues: first, the problem of reflexivity or the status of our claims as science-studies inquirers, and second, the problem of how to take a critical stance toward science.

I began with an account of two models for the relation of case studies to general accounts of science. The first approach I called a "ground-up" procedure. No general claims seem to follow from this procedure; at most, what follows is the falsification of the general claims of others. In the second approach one adopts an *a priori* program and develops case studies within it. I called this a "top-down" procedure. The problem with this approach is that unless one can claim special insight into methodological matters, there is no way to justify a purely *a priori* stance. There are two related models of inquiry that may avoid the problem of grounding a forthright stance. The first is to adopt an interest model, which could be Marxist, feminist, or the like.[79] Although the interest model is "top-down," and usually privileges the standpoint of a special community, it could be seen as less than absolute, despite the fact that the stance taken is self-consciously value-laden. Temporarily adopted standpoints, rather than absolute positions, can be seen as a methodological tool kit for science studies, with no prior assumptions where the particular methodologies will apply best in the particularity of science-studies practice. Richard Rorty calls such a position "ethnocentrism." He emphasizes that even though we must use our present position to provide criteria for truth, we do not have to claim that these provide a definition of truth,[80] thus meeting Russell's requirement that I cited above. We can avoid metaphysical antirealism by refusing to adopt an epistemic theory of truth.

A second approach is to use case studies as evidence for what kind of methodology works best without self-consciously adopting a standpoint from the beginning. We can learn how to learn, as the naturalized philosophers of science say.[81] While naturalism exists in several forms, a general formulation will be adequate for the discussion here.[82] Traditionalists hold that methodology is an autonomous, *a priori* discipline, whereas naturalists hold that methods and cognitive aims can be informed by scientific knowledge. Evidence for a methodological model includes the success of case studies carried out under the model, and such evidence for the model does not have to be established before the model's use. Of course, one must have some standpoint, but naturalized philosophy of science can adopt a methodological standpoint pragmatically and temporarily. As C. S. Peirce said long ago, you can only begin inquiry from wherever you are now; there is no privileged starting point.[83] Such models may even be opportunistic, adopting whatever methodological stance seems interesting or appropriate to a given case study. Current science studies may be in such a stage now, with case studies developed in a wide variety of methodological styles.

Reflexive doubts about the status of methodological criteria and of our own stance as science-studies researchers can be answered by an extension of fallibilism. Our criteria could be wrong, just as any of our factual beliefs could be wrong. Fallibilism follows when we give up certainty in knowledge, and fallibilism should continue throughout the traditional methodological hierarchy of facts, methods, and aims. According to a traditional philosophical model of science, methodology determines what counts as a fact and the basic aims of science determine (through philosophical analysis) what counts as proper methodology. This model leads to a kind of relativism in the views of Popper and some of the logical positivists, since they see aims and methods as conventional, but for many others the traditional hierarchy leads to a foundational and autonomous version of epistemology.[84] These traditionalists accept the fact that scientific knowledge is fallible, but when fallibilism is applied to methodology and criteria, they balk. Why? Nothing follows from the fact that our best-justified methods may later turn out to be wrong, since accepting fallibilism gives us no reason to deny that we have knowledge. These traditionalists clearly assume that as one progresses from one level to the next, one must meet stricter epistemic standards, which is a version of the traditional regress argument for self-justifying foundations. Traditionalists are so wedded to the hierarchical model of theories, methods, and aims that they cannot imagine that fallibilism could apply everywhere, and they simply beg the question when they criticize non-foundational positions. The solution is to reject the hierarchy. Methodological criteria are just as fallible as all other beliefs, and methods and practices have changed dramatically through the history of science. But such a contextualization of scientific practice does not require us to throw it into question.

The problem of reflexivity that has been so widely discussed in the sociological literature will not seem so pressing if contextualization is no longer considered delegitimation. There is no reason to think that fallible, historical, and localized practices should be thought of as illegitimate, so our own practices can be considered fallible, historical, and local as well. It is also important to forestall the possible objection that there is a self-referential problem with the formulation of an antiessentialist view of science. Antiessentialists are not claiming that all universal claims are false (or unwarranted, etc.). Such a claim would be blatantly incoherent. Instead, they claim that there is no universal scientific method. This claim is not itself a method, so there is no self-referential problem here at all. The same point follows if we formulate antiessentialism as the view that there are no universal aims of science, as does Fine.[85] There are limits to logic chopping and no reason to think that one can force everyone who is skeptical about some particular aspect of knowledge into an inconsistency.

The restriction of the adopted standpoint to a local community may appear to rule out any generalization, but one cannot assume that this is automatically the case. The issue is whether any generalization that appears in a "top-down" account would be imported from the self-consciously adopted stance. It may be possible to find a general regularity that has nothing to do with the consciously adopted stance, and is therefore epistemically independent of it. In other words, we must look case by case to see if presuppositions beg the question. Regarding normative stances toward science, my reply is similar: we are all in the same position when we try to convince someone; local standards and norms always apply. We do not need to claim a special, privileged position from which to criticize science. Calling arguments universal or transcendental will be no more convincing than simply presenting them without these labels. While I stress that there are no criteria that are not tied to local aims and values, I argue that we can always localize our methodological disputes to the context of a local community. Following the naturalists, I argue that local agreement about criteria can be used as a basis for further discussion of more controversial methodological criteria. Of course we cannot guarantee agreement by this method; but no one is able to guarantee agreement, not even the foundationalist. The only thing that could be grounds for settling a dispute is some agreed-upon standard by which evidence can be judged, and appeals to rationality principles are in fact appeals to standards that are assumed to be acceptable to everyone. It is claimed or hoped that the critic will agree with some general principles and that the dispute can be settled amicably according to these principles. However, there is no need for these principles to be certain, or universal, or external to the debate at hand. A non-foundationalist or, indeed, even a relativist can search for common ground with which to settle a dispute. Whether anyone will be successful in convincing others is a complicated, messy affair.

When there is no agreement, the traditionalist wants to use rationality principles as an argument to convince a third party that the skeptic is not only wrong, but wrong in an especially deviant way—which is again nothing more than an attempt to find agreement about general principles, this time with a third party. There is no reason to think that this strategy will always be successful, since it again requires agreement, this time agreement on rationality principles by the third party. Those who are tempted to argue that universal standards are required think that they know in advance that only certain types of arguments will be convincing, and furthermore that these standards can be projected onto future debates. If we could find the standards of rational debate in a Platonic heaven somewhere, that would be the case, but short of such foundations I do not see how the position can be maintained.

Returning to the problem facing "ground-up" methods, it is often thought that a local, nontheoretical analysis will be inadequate to ground a critical stance toward science, that local knowledge can only be descriptive.[86] These worries are too abstract, and based on a commitment to a foundationalist model of grounding. Local analyses are as large as they are made. If one can find a way to make connections between disparate events, one will have an extended analysis. As Callon and Latour have emphasized, the lesson to learn from the demise of intellectual history is that connections between historical players and ideas must be shown in the local setting. However, there is no theoretical limit to the size of the analytical network that one creates.[87] Extending this idea to the political consequences of analysis of science with the tools of science studies, we see that there is no limit to the critical force of an analysis. The political force that a local analysis will have is a practical, concrete question, not an abstract issue of grounding. To make one's position critical and have an effect, one must muster allies, be read and cited, and so on, and make a practical difference. One cannot say in advance what will work in a given community.

I do not deny that any set of practices is open to criticism, and I endorse science studies as a critical force. I only insist that the criticism not amount to general philosophical arguments. Criticisms have to be much more engaged with the practices that they are criticizing to be compelling and, of course, traditionalists can no longer get away with merely blocking general skeptical arguments in order to support science. Science is both too encompassing and too heterogeneous to allow a straight-up-or-down vote. Even detailed engagements with science on its own terms can be led astray by prior metaphysical commitments to realism or anti-realism and by prior stances for or against science. The local concern of the science-studies community that I have dealt with in this essay is how to end the tiresome battles over realism and relativism. In order to build a new discipline, we must radically reframe old questions that each of the science-studies disciplines took as their own, and find issues that all three disciplines share. I have been arguing that the time to give up the debate over the rationality of science is long overdue, and suggest that if we are to take the local, contingent nature of scientific practice seriously, and I think that we should, then all the science-studies disciplines will have to redefine the issues that they address, and continue to learn new tools of analysis.

KARIN KNORR CETINA

The Care of the Self and Blind Variation: The Disunity of Two Leading Sciences

In this essay, I want to present two stories about kinds of empiricism: about the ways a science, experimental high-energy physics, understands and enacts empirical research, and about how this understanding differs from that of another science, molecular biology. An ethnography of the empirical in different sciences has never been written. One reason for this surely lies in the fact that the meaning of empirical procedure is thought to be common to all experimental sciences, describable in terms of a few injunctions, and spelled out in any textbook introduction to a particular field. By studying scientific laboratories, the new sociology of science overcame the textbook image of science,[1] but it did not break away from the assumption that all sciences conform to similar procedures, exemplify similar attitudes to the empirical world, and form part of one culture. It also focused on the role played by contingencies, interpretation, and negotiation in the creation of scientific knowledge. What got left out of the picture was the construction of the empirical machineries involved in this creation.

What are the differences in the empirical procedures of experimental high-energy physics and molecular biology? The question itself may seem impossible to answer. After all, the experimental natural sciences deal with their matters in a deep sort of way. The facts they produce are intricate in the making; the things they handle are handled in detailed and complicated ways; the chains of processing involved are infinite and divided into many components. The task to see through the thick growth of experimental manipulations in search for the cultural switchboard that sets the directions is overwhelming, and the sociological revenue may be unclear at first. The help I enlisted was that of the comparative method, which I used less as an asset in generalizing results than as a humble supplier of frameworks of seeing: I looked at high-energy physics equipped with a good view of molecular biology, and at molecular biology from the viewpoint of high-energy physics. Through such a comparative optics, an ethnographer can discern not the essential features of a field, but differences between fields, which seem far more tractable anyway, ethnographically speaking, than essential features. The focus of observation was on the rough build of the empirical machineries at work in the two sciences; it was not on the level on which single screwdrivers are fidgeted with when they are pointed at individual screws. To characterize these machineries, I shall use the analogy of "blind variation and selection by success" to describe the referent-oriented epistemics of molecular biology. My general picture of experimental high-energy physics will look different. I shall use the analogy of a closed system that interacts with the world only mediated through interactions with itself to designate what one might call the *liminal* and *recursive* epistemics of high-energy physics. In designating these differences, I do not draw on philosophical labels such as realism, instrumentalism, pragmatism, conventionalism, and the like. This project attempts to provide a richer description of epistemic practice, something like, if I may misuse a term Geertz once made popular, a "thick" theory of knowledge. If anything is suggested with respect to the philosophy of science, it is that there exists no "scientific method" that extends to all fields. The disunity of scientific practices can be witnessed on many levels: on the level of their orientation toward and treatment of signs, of their relation to themselves, of the forms of alignments they institute between subjects and natural objects, of their general approach to capturing and engaging truth effects in inquiry. It is also located in how these practices set up and include the referent—whether they attempt to form, with the referent, a common life-world or leave the work of dealing with the referent to an interposed machine.

The Closed Universe of Particle Physics

There is an analogy that I think appropriately describes the "truth-finding" strategy of particle physics. This is the analogy of the brain as an informationally closed system. The neurophysiology of cognition is based on results developed in the nineteenth century according to which states of arousal in a nerve cell in the brain represent only the *intensity*, but not the *nature*, of the source of the arousal. Maturana and Varela applied these results to the experimental study of perception.[2] They concluded that perception must be seen as a cognitive process that is energetically open but informationally closed. Perception is accomplished by the brain, not the eye, and the brain can only construe what it sees from signals of light intensity that arrive at the retina. In order to form a picture of the nature of the source of these signals, the brain makes reference to its own previous knowledge and uses its own electrochemical reactions. Phrased differently, in perception the brain interacts only with itself and not with an external environment. It reconstructs the external world in terms of internal states, and in order to accomplish this the brain "observes" itself. Consciousness, according to this theory, is a function of a nervous system capable only of recursive self-observation.

I want to argue that like the brain, particle physics operates within a *closed* circuitry. In many ways, it operates in a world of objects separated off from the environment, a world entirely reconstructed from within the boundaries of a complicated multilevel technology of representation. A detector is a kind of ultimate seeing device, a sort of microscope that provides for the first level of these representations. The representations themselves show all the ambiguities that afflict any world composed of signs. Yet particle physics is perfectly capable of deriving truth effects from its representing operations. I want to specify more concretely three aspects of this world before I go on to discuss the strategies particle physics has developed in moving within its boundaries: first, its experience of objects as signs and the associated technology of representation; second, its turn toward the negative, that is, the character of these signs as simulators and deceivers; and last, the issue of the "*meaninglessness*" of *measurement*, which is part of particle physics' technology of representation.

When Objects Are Signs

In particle physics, natural objects (cosmic particles) and quasi-natural objects (debris of particles smashed in particle collisions) are admitted to experiments only rarely, perhaps for a few periods of several months in an experiment that may last anywhere between eight to sixteen or even twenty years. The proposals for UA1 and UA2, the two large collider experiments at CERN (the European Laboratory for Particle Physics in Geneva, Switzerland) were approved in 1978, after several years of preparatory work, and both experiments were dismantled in 1991, although analysis on some of the experiments' data continues. During the upgrade period in the 1980's in which the detectors were rebuilt, the experiments had four "runs" (data-taking periods) between 1987 and 1990 of about four months each. Thus experiments deal with the objects of interest to them only very occasionally, while most of the experimental time schedule is spent on design, installation, testing, and other work outlined below. Second, these objects are in a very precise sense "unreal" or "fantasmatic";[3] they are too small ever to be seen except indirectly through detectors, too fast to be captured and contained in a laboratory space, too dangerous to be handled directly. Furthermore, they usually occur in combinations and mixtures with other components, which mask their presence. Third, most subatomic particles are very short-lived, transient creatures that exist only for a billionth of a second. They are subject to frequent metamorphosis and to decay, which makes their existence always already past, always already history rather than present.

Now these phantasmatic, constantly changing historical occurrences can be established only indirectly, through the signs they leave when they fly through different pieces of equipment. Physicists deal with them through a technology that creates and exploits representations on three levels. The first level of representations results when particles interact with detector materials through, for example, liberations of electrons and the emission of light by electrons. The work on this level is done by the particles themselves; the experimental procedure lies in the design and construction of the apparatus in which the particles register. Physicists, however, don't start with the particles; they start with the detector. A second level of representations involves *representations of the detector*, that is, "off-line" manipulations of the signals extracted from detectors after data have been taken that *reconstruct* the events in the detector and slowly mold these signals into a form that echoes the particles of interest to physicists. The signs produced by detectors are strewn all over different pieces of equipment and are generally meaningless without further elaboration. They must be assembled, interpreted to have certain (energy) meanings, and coordinated to yield consistencies between different representations, that is, "tracks." In a sense they must first be put together and brought into shape *as signs* before their analysis can begin. Physicists' *representational vocabulary*, their reference to energy and track "reconstruction," to electron "identification," and more generally their implementation of a "production" program that performs the major portion of the work of "producing," from signals meaningless in themselves, signs that can be associated with physics events, exemplifies this work. But this work is not the whole story, either. There is a third level of representations: from the reconstruction of events *in* the detector physicists create "variables," which are no longer interpreted in terms of the signs that register in detector materials but are designed and analyzed in terms of physics distributions and models (e.g., expected distributions for certain kinds of particles).

The Antiforces of Research

The representations physicists deal with are nonarbitrary; in accordance with their own use of the word, they are signals. If there is no relationship between a sign and the object it stands for (a tree and the word "tree"), then there is also, in this respect anyhow, no problem. Problems arise when the relationship is thought to "be there" and one wants to use it to point back to the objects. Frequently, and not only in high-energy physics, the passage between signal and object is uncertain, strewn with obstacles, and difficult to control.

The obstacle that collider experiments face with their sign-catching instruments is that the signs of interesting events are muffled and smeared by signs of other occurrences in the detector. These other signals derive from uninteresting *parts* of events, from other *classes* of events, or from the *apparatus* itself—from the signals it emits in addition to signals evoked by real objects. Furthermore, there are limitations of the apparatus that affect the signal. All these phenomena are a threat to interesting events. They may falsify their signature, misrepresent their character, jeopardize their identification. They deceive detectors and analysts about the presence of events, tamper with the shape of their distributions, and substitute false information for the real. They are tricksters, fakers, and impostors, or just plainly deteriorating factors—factors that worsen the results that one could get in a better world. They aggravate the analysis and cause infinite problems to researchers.

There are forces that stand out in this picture. The most insidious force surely is what the physicists call *background*: competing processes and classes of events that interfere with a signal. The physicists in proton-antiproton collider experiments see themselves as "buried in background": "The nature of the problem is to deal not really with the signal so much as the background. You have to deal with the horrible case that you didn't want to see."[4] Their task, as they see it, is to get the proverbial needle out of the haystack. The signs of the events of interest are almost muted by the background. If you think of these signs in terms of footprints, it is as if millions and even billions of different animals had stampeded over a trail from whose imprints one seeks to discern the tracks of a handful of precious animals—those one is really looking for in the experiment. In the search for the particle at CERN in the early 1980's, less than one event was retained out of every ten billion interactions.[5] In the search for the top quark during the upgrade of the collider experiments UA1 and UA2, for example, UA2 expected on the order of forty top-quark events in six million selected electron triggers (electron candidates), which is already a vastly reduced number compared with the number of interactions.

The "Meaninglessness" of Measurement

An internal universe of signs of "external" occurrences in which these signs are buried in other signs and appearances—these were the first two aspects that feed into the analogy of the closed universe. Let me raise a third issue, one that lies at the core of the universe considered. In fact, it might be its most crucial component; it sets high-energy collider physics of the kind described apart from many other sciences. In many fields, measurements, provided they are properly performed and safeguarded by experimenters, count as evidence. They are held to be capable of proving or disproving theories, of suggesting new phenomena, of representing more or less interesting, more or less publishable "results." This holds irrespective of the fact that measurements are theory-laden, prone to raise arguments in crucial cases, and sometimes subject to reinterpretation. What I have in mind is the role of measurements as, one might say, end-of-the-line verdicts—verdicts to which experimental work leads up in intermediary and final stages, from which it takes its clues, at which it pauses and starts afresh. In high-energy collider physics, however, measurements fall short of these qualities. They appear to be curiously immature beings, more defined by their imperfections and shortcomings than by anything they can do. It is as if high-energy physics recognized all the problems with measurements that philosophers and other analysts of scientific procedures occasionally point their fingers at. It is as if, in addition, they had pushed one problem to its limit and drawn a conclusion that other sciences have not taken: that measurements are to be considered as no more than a stage in a cycle of stages, that they are to be pushed back behind the lines of what counts as a result, that they are not to be displayed in public unless accompanied by other elements. Purely experimental data, as physicists say, "mean nothing by themselves." Not only are there few quantities that can be measured relatively directly, but even those that can cannot be taken as they are. They must be further refined by or in some other sense combined with nonmeasured quantities, such as theoretical ratios and Monte Carlo simulations. As one physicist put it, a little indignant at my insinuation that one might "just measure" the mass of the W particle: "You cannot read off a detector how big the mass of a particle is like you can read the time off a watch!"[6]

For example, with respect to the strong-force coupling constant, Alpha S, in effect a measure of the probability for the emission of a force-carrying particle, what is interesting is not the experimental value but "the theoretical ratio in relation to the experimental ratio for a given detector configuration."[7] This, of course, sounds much more complicated than a simple experimental measurement. And it is. First, one must determine the ratio between the number of W + 1 jet events and the W + 0 jet events, which, with the search for the top quark in the experiment studied, one could measure; second, one has to assemble a Monte Carlo program that includes all necessary theoretical calculations *and* simulates the detector, the "fragmentation," that is the breakup of quarks and gluons into jets, the underlying event, and so on. From this, one obtains the same ratio as the experimental one, in theory. The theoretical ratio is a function of, among other things, the coupling constant. It increases when the coupling of relevant particles increases. The experimental ratio, on the other hand, is a constant. The "real" Alpha S derives from intersecting the experimental value with the Monte-Carloed curve of the theoretical ratio.

Measurements in high-energy physics always walk on crutches. They are a sort of *amputated* quantity, a quantity that, without the nonmeasured parts that are missing from it, is not worth much as an experimental result. It is not a final figure that can stand on its own but a position in a structure of relations, in which the other positions must be filled before the whole becomes useful. With respect to the analogy of the closed universe this means that *measurements are placed firmly*—and obviously—*inside the ranks and components of the experiment rather than outside*. They are not cast as external evaluations of internal propositions, not even as outposts through which one can make independent contact with the world, but rather as elements and stages that are held in check and turned into something useful only through the interaction of these elements with other features of the experiment.

The Structure of the Care of the Self

How does a science like high-energy physics nonetheless derive truth effects from the appearances it deals with? The answer is, in a nutshell, that it substitutes for the care of objects *the care of the self*.[8] By this I mean the preoccupation of an experiment with itself, *with observing, controlling, improving, and understanding its own components and processes*. Confronted with a lack of direct access to the objects they are interested in, caught within a universe of appearances, and unwilling to trespass beyond the boundaries of their liminal approach, high-energy collider physicists have chosen to switch, for large stretches, from the analysis of objects to the analysis of self.

Self-Understanding

This switch can be seen, for example, by merely looking at an experiment's expenditure of time. More time in an experiment is spent on designing, manufacturing, and installing its components, and in particular on predicting their performance and understanding every aspect of their working, than on anything to do with data. Time expenditure, however, is only one indicator. Another aspect, more significant perhaps, is the importance credited to self-analysis in practices and discourse at all points of experimental activity. This is codified in the native terminology and prescription of "understanding" each aspect of the experiment, for example in *understanding the behavior of the detector*, which comprises a major portion of the care of the self. The detector is an apparatus that is created and assembled within the experiment. Nonetheless, the behavior of this apparatus—its performance, blemishes, and ailments—is not self-evident to the physicists. Such features must be learned, and the project of understanding the behavior of the detector spells this out (see Fig. 1).

What exactly does one mean by "understanding the behavior" of the detector? First, in the words of physicists, this means "knowing when some physics process of some kind happens in [the detector and accordingly in] what comes out of it." It is "being able to do a perfect mapping" of it, and "trying to unfold what has happened between an input and an output" of results.[9] Understanding the behavior of the detector begins when its first components, for example the silicon crystals in a silicon detector, arrive and undergo test-bench measurements; it continues through steps such as testing, characterization, and installation; and it culminates (in terms of time spent) when the "response" of a detector is determined, and its changes are understood, through calibration. Second, in case of *problems*, which continuously occur, the cause of the problems is found out, and the problem cured or otherwise taken care of. Terms such as "testing," "check," "cross-check," and the performance of "a study," are subcategories of understanding. Thus "understanding" refers to a rather comprehensive approach of *unfolding* what happens in every relevant part of the material, how what happens changes over time, and why these things happen. This approach, and this attitude, is maintained even when understanding is *not* necessary for the successful completion of ongoing work.[10]

Self-Observation

The care of the self has a threefold structure that includes, besides self-understanding, also *self-observation* and *self-description*.

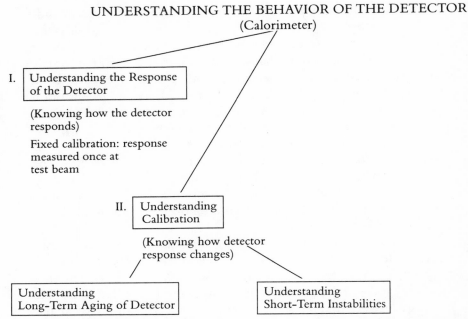

Fig. 1. Some components of understanding the behavior of the detector.

These are not three different avenues to the same goal, but rather sets of practices that supplement each other. They are united through the principle of reentry of the results of self-understanding, self-observation, and self-description into physics calculations (see Fig. 2).

Self-observation is a step back behind the hardware and soft-

THE STRUCTURE OF THE CARE OF THE SELF

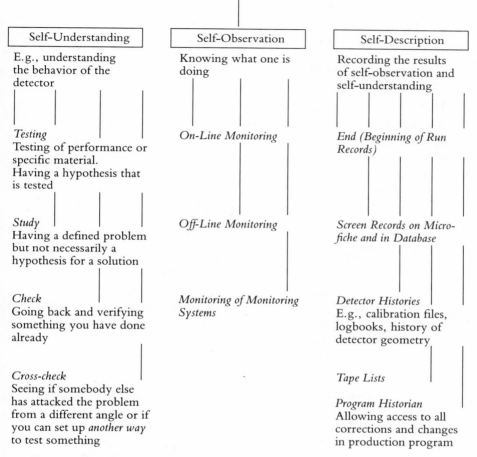

Self-Understanding	Self-Observation	Self-Description
E.g., understanding the behavior of the detector	Knowing what one is doing	Recording the results of self-observation and self-understanding
Testing Testing of performance or specific material. Having a hypothesis that is tested	*On-Line Monitoring*	*End (Beginning of Run Records)*
Study Having a defined problem but not necessarily a hypothesis for a solution	*Off-Line Monitoring*	*Screen Records on Microfiche and in Database*
Check Going back and verifying something you have done already	*Monitoring of Monitoring Systems*	*Detector Histories* E.g., calibration files, logbooks, history of detector geometry
Cross-check Seeing if somebody else has attacked the problem from a different angle or if you can set up *another way* to test something		*Tape Lists* *Program Historian* Allowing access to all corrections and changes in production program

Fig. 2. The tripartite structure of the care of the self in high-energy physics.

ware, and the human beings, to watch what these components are doing with the purpose of checking whether they are performing correctly. Self-observation is a form of *surveillance*, present at many levels of experiment but especially at its later stages, and during runs. Physicists call the most clearly specified and codified part of this self-observation *monitoring*. There is a hierarchy of watching embodied in different monitoring tasks; in this hierarchy, the tasks get more complicated as one proceeds from human observers who watch data going through "the system" (of data taking and read-out), to computers watching and sampling the status of the system, and back to humans watching the results of their own choices and creations implemented in the hardware and software.

Self-Description and Reentry

Self-understanding and self-observation are joined by meticulous efforts at self-description. By this I mean not only the well-known phenomenon of logbook keeping. There are also many forms of computerized recording and track keeping that accumulate to "histories" of experiment. First, there is "bookkeeping" information, compiled in order that people can find their way around in however many data tapes they have. Second, physicists put "an incredible amount of information out for each data run," for example:

the type of record, run number, start time of the run and end time, which tape it was written on, what software triggers and hardware triggers were enabled, number of events in the run, length of tape used, number of words in the experiment, which gates, what type of trigger (whether we are running on beam or whether it was a clock trigger), which triggers were enabled at each level, which processing was enabled, which detectors were active, what the pre-scaling factors were on the level 1 triggers, what all the thresholds are in the level 2, what all those parameters were on level 3, etc.

"So in principle we know exactly what was going on on-line," as one physicist said in the experiment studied, who was heavily involved in analyzing these records. Besides tape records and run records there are "detector histories," which include the logbooks physicists keep on all the tests and monitoring tasks they have performed. Important components in relation to detectors are the calibration files—records of the energy scales through which raw signals are transformed into physics quantities and their changes over time. When "running production" one has to know "all the calibration files for all the experiment at all times." Some are considered as kinds of *photographs* of the status of the experiment at the moment the tape was written. Finally, there is "the historian." This is a program that maintains the main production program and all its older versions, to which it gives access. As "the historian," native references to "history keeping," and similar terms suggest, physicists are well aware of the effort they maintain not only to know what they are doing, but also to store this knowledge and keep it available for future uses.

Records such as the above fulfill many purposes. For example, they allow backtracking in error searches, and in searches for old solutions to recurring and newly appearing problems. Yet the most interesting use of history keeping is perhaps not the possibility of going back in the history of an experiment, but the reentry of the recorded information in the experiment.

Negative Knowledge and the Liminal Approach

This brings me to my next topic, a further move high-energy collider experiments make as a means of assurance of success, one that puts to use and extends the analysis of self. This is a turn toward the study of liminal phenomena, by which I mean phenomena that are neither the phenomenal, empirical objects of positive knowledge nor objects in the formless regions of the unknowable, but something in between. *Limen* means "threshold" in Latin. The term "liminal" has been used in the past to refer to the ambiguous status of individuals during transitional periods of time.[11] I shall use the term to refer to (knowledge of) phenomena and objects that are at the fringe and at the threshold of the objects of interest to high-energy physics. That discipline has enlisted the world of disturbances and distortions, of imperfections, errors, uncertainties, and limits of research into its project. Through the care of the self, it has lifted the zone of the unsavory blemishes of experiment into the spotlight, and applied itself to study its features. It cultivates a kind of negative knowledge. Negative knowledge is not nonknowledge, but knowledge of the limits of knowing, of the mistakes we make in trying to know, of the things that interfere with our knowing, of what we are not interested in and do not really want to know. We have already encountered some forces of this kind in the background, the underlying event, the noise, the smearing of distributions. All these are limitations of the experiment, and many (but by no means all) are linked to a transitional object, the detector, the outpost of experiment that bears the brunt of the incoming particles. High-energy collider physics *defines* the perturbations of positive knowledge in terms of the limitations of *its own* apparatus and approach. But it does not do this just to put the blame on these components, or to complain about them. Rather, it teases the fiends of empirical research out of their liminal existence; it draws distinctions between them, elaborates on them, creates a discourse about them. It puts them under the magnifying glass and presents blown-up versions of them to the public. In a sense, high-energy experimental physics has *forged a coalition* with the evil that bars knowledge, by turning it into a principle of knowing.

In Christian theology, there was once an approach called apophantic theology, which prescribed that the study of God was to be in terms of what He is *not* rather than what He is, since no positive assertions could be made about His essence. High-energy experimental physics has taken a similar route. Through developing liminal knowledge, it narrows down the region of positive, phenomenal knowledge. It specifies its boundaries and pinpoints the uncertainties that surround it. It delimits the properties and possibilities of the objects that dwell in this region through the properties of the objects that interfere with and distort them. Of course if one asks a physicist in this area he or she will say that the goal remains to catch the (positive, phenomenal) particles that are still on the loose, to measure their mass and other (positive, phenomenal) properties, and nothing less. All things else are ways and means to approach this goal. There is no doubt that this is indeed what physicists wish to achieve, and occasionally succeed in achieving, as with the Nobel Prize–winning discovery of the vector bosons at CERN in 1983. My point is by no means to deny such purposes or their fulfillment. However, what is of interest when working one's way into a culture is precisely by what ways and means a group arrives at fulfillment. The upgrading of liminal phenomena, the torch that is shone on them, the time and care devoted to them, is a cultural preference of some interest. For one thing it extends and accentuates what I called high-energy physics' negative and self-referential epistemics. For another, the fields in which the preference is not shared, among them molecular genetics, seem to be the majority. Third, it is quite remarkable how much one can do by mobilizing negative knowledge.

Knowing one's limitations: Efficiencies and acceptance, errors, and limits. There are three areas in which the liminal approach is most visible: those of *errors and uncertainties*, of *corrections*, and of *limit calculations*. Limit calculations are analyses in which the goal is to identify the boundaries of a region within which a certain physical process can be said to be unlikely. Limit analyses offer a way out of negative results: if the top quark for which one searches in one's data is not there, it is at least possible to say that up to a certain mass for which the terrain has been searched the top quark is unlikely to occur. Limit analyses are perhaps the most frequent output of collider experiments. Even in experiments designed to produce precision mass measurements of known particles, such as the Large Electron Project (LEP) experiments at CERN, limits may be the most frequent result: LEP is said to produce a stream of papers where they produce limits on all sorts of things. Added to the calculation of limits are the calculations of other limitations, the analyses that physicists perform of corrections and of errors and uncertainties. Corrections are ways of putting to work all the knowledge the experiment has gained about itself through the care of the self. Corrections refer, mostly, to the calculation of efficiencies and acceptance—figures that indicate whether, if an event is produced in a detector, this event is identified, and how well it is identified. With typical analysis, for example an analysis published on the search for the top quark, each particle that is part of the signature of the top will have a string of several (I counted up to nine) efficiencies attached to its identification. However, self-knowledge too is subject to limitations. To a substantial degree, errors and uncertainties are ways of addressing, on a second level, the blemishes of the above analysis of (efficiency and acceptance) limitations. All sciences, presumably, recognize some measurement errors, but few have such elaborate conceptions of systematic errors and the urge to pursue them into their finest details. Systematic errors point to a systematic problem, for example, using a ruler that is too short in a measurement of the length of some object, but the problem is unknown. As one physicist in the experiment studied, who was heavily involved in the analysis of supersymmetric particles, put it, "the systematic error is just a way to measure our ignorance. . . . [It] is our way to try and estimate what we've done wrong. And if we knew what we did wrong we could correct it rather than to allow for it in the error." There is a lot of ignorance high-energy physics takes stock of in a typical analysis, as the lists of error terms in published and unpublished analyses show.[12] Interestingly, the difference between a first and a more refined analysis with higher statistics (more data) often concerns shifts in the error and correction portion—not in the sense that the list of error terms becomes shorter, but that it becomes longer, and the terms more precise (some errors may turn into corrections). That measurements are subject to long lists of corrections and have long tails of error terms highlights once more their status as figures that count for nothing if they are not surrounded by an (albeit quantitative) account of their circumstances, conditions, expectancies, differences, and so forth. Consider an example of the origin of one such error term, one in which physicists feel they have to take into account the difference between different theories. In early 1991, about forty-five sets of structure functions that describe the density of quarks and gluons (partons) within the proton and are needed for calculating the number of expected events in a proton-antiproton collision were available. They involved different assumptions and counted as different theories about how to extrapolate a few available low-energy data to higher energies. As a physicist in the experiment studied who worked with these structure functions explained: "One structure function might lead you to this value, another to that, etc. If these values would result from measurements you could construct a broad Gaussian out of this with an average and a sigma. . . . But these are not measurement errors, these are different theories, and for the moment we have no way of telling which is right and which is wrong. All of these values are equally probable." What physicists do in this situation is apply (preferably) all these functions to their cross-section measurements. The variation between different structure functions (the spread between the curves) in regard to the contribution of a particular quark or quark combination to the cross-section is then interpreted as the theoretical systematic error or uncertainty associated with the structure function.

To fields that are used to different preferences, this procedure of turning variations between answers to a problem into an error-and-uncertainty estimation is quite stunning. The mere fact that several theories about the same phenomenon are available in an area counts as an error, and the deviations between the predictions of these theories are used as a resource in estimating the size of the error.

Scientificity consists in considering all theories one can get hold of, provided they are not completely outdated by recent measurements. Would sociologists or philosophers care to consider the variability between different theories on a subject as a source for making a calculation of their theoretical error? Different theories in sociology—or in molecular biology—give rise to scientific arguments, and to the formation of different groupings of scientists divided according to their theoretical preferences, but never to error calculations. Would it make sense to these fields to require that the dispersion of these different theories should somehow be ascertained, so that we know, if not what is right, then at least how far we might go wrong? Of course sociologists and biologists do not make primarily quantitative predictions. But this is hardly enough to account for these researchers' preference. There is little concern for exploiting liminal phenomena in these areas, whereas in physics there is.

Molecular Biology and Its Intervening Technology

Experimental high-energy physics is marked by a *loss of the empirical*; recall the nonencounterability of the objects of interest, the diminished role of pure measurement, the construction of the evidential domain as meaningful only when it is *firmly embedded* in theoretical predictions, phenomenological laws, and Monte Carlo simulations. Recall also the care of the self in lieu of setting up reactions with the outside world, and the highly sophisticated exploitation of liminal phenomena and objects. Molecular biology, on the other hand, constitutes itself as a system open toward natural and quasi-natural objects. It shows none of the interest of high-energy physics in self-understanding and none of its virtuosity in separating off and relating to its own components. Instead, it shows a different virtuosity. It upgrades and enhances natural objects in a continuous stream of experimental action.

In experimental high-energy physics, experience appears to provide no more than an occasional touchstone to hurl the system back upon itself, and "success" may well depend on how well—and how intricately—the system interacts with itself. Molecular biology, on the other hand, appears to base progress upon maximizing contact with the empirical world. If in high-energy physics experiments natural and quasi-natural objects are admitted only rarely, in molecular biology they are sought out and encountered on a day-to-day basis. If in high-energy physics experiments it seems no longer a phenomenon itself that is at issue but rather its reflection in the internal megamachinery that envelops and tracks down physical occurrences, in molecular biology the phenomena assert themselves as independent beings and inscribe themselves in scientists' feelings and experience. Experimental high-energy physics can be characterized in terms of a negative, self-referential epistemics built around sign systems. In molecular biology, on the other hand, the epistemic culture is oriented toward positive knowledge built from the manipulation of objects in an analog regime that continuously turns away from sign processes.

Three aspects of this preference stand out in our observations of experimental work. One is the close circuit established between scientists and *objects* through the massive presence of objects in the laboratory and the modes of organization linked to these objects—a mode of organization wherein objects are embedded in processing programs that transform these objects. The second aspect refers to the further enhancement of objects and experience in what one might call an *analog regime*—a regime whose components include the embodied functioning of the scientist, visual scripts, and the narrative culture of the laboratory. The third aspect is the preference for *blind variation*, and natural selection by success, as a strategy deployed in molecular biology when problems arise. Natural and quasi-natural objects are not only present in the lab on a continuous basis; they also provide for the selection of experimental strategies when things don't work out. In the following, I shall very briefly summarize some of these aspects.[13]

A Technology of Intervention

Consider first the maximization of contact with the empirical world through the massive presence of objects in the laboratory and the practices of dealing with these objects. Molecular biology laboratories are archetypal in the way they feature test tubes and pipettes, samples of specimen and chemical reactions, small-scale instruments and craftlike, manual work. Molecular biology does not process signs; it processes substances and organisms in a multitude of steps and substeps. The technology in terms of which work proceeds is not a technology of representation, but of intervention. Nonliving materials are subject to almost any imaginable intrusion and usurpation. They are smashed into fragments, evaporated into gases, dissolved in acids, reduced to extractions, run over columns, mixed with countless other substances, purified, washed, spun round and centrifuged, inhibited and precipitated, exposed to high voltage, heated or frozen, and reconstituted. Cells are grown on a lawn of bacteria and raised in media, incubated and inoculated, counted, transfected, pipetted, submerged in liquid nitrogen and frozen away. Animals are raised and fed in cages, infused with solutions, injected with diverse materials, and cut open to extract parts and tissues; they are weighed, cleaned, controlled, superovulated, vasectomized, and mated; they are anesthetized, operated on, killed, frozen, and cut into sections and slices, and they will have dispensable parts such as tails cut off to test their genetic makeup.

The Analog Regime

Through this technology of intervention (a term taken over by Hacking),[14] natural objects and quasi-natural objects are included in a common life-world in which they thrive, resist, perform their functions, and so on, in direct and often intimate relationships with scientists and technicians. The laboratory is a kind of second-nature biotope—a laboratope, as a student of mine calls it—in which certain things (cells, mice, microorganisms) grow and develop, live through reproductive cycles, and infect and affect each other interspersed with human beings who try to (!) arrange and control some of these processes. The notion of a laboratope stresses the labored nature of this living together: nature is not romantically imitated in the lab; it is split apart, rearranged, and disfigured at the same time as it is laboriously reconfigured. Common life-worlds are built through co-presence, which Schütz saw as an important feature of face-to-face situations, through cotemporality, the possibility of conjoint time, and the possibility of conjoint statuses for human agents and nonhuman entities or objects: in other words, they are built through structural features of the arrangement; we need not assume shared beliefs or some other form of unity. There are some interesting structural *alignments* which I want to point out. For example, molecular biologists function vis-à-vis natural objects often in a sort of analog mode. By this I mean something like the opposite of a digital functioning,[15] of the kind of repetitive, dividable, tractable, and, above all, fully describable mode of operation that can be automated and is sometimes required of factory workers. Analog functioning, on the other hand, is functioning that shuns or refuses description and even cognition. Analog processors are automats too, but because of the dynamic, adaptable, and noncognized nature of the processing they perform, they are less likely to be modeled by artificial-intelligence programs. The concept of analog functioning is of course something of a recasting of the notion of tacit knowledge that Polanyi once described, a notion that incorrectly stresses, I think, the knowledge aspect, rather than the embodied-skills aspect, of experts like molecular biologists.

Molecular biologists meet sensory objects as sensory performers: they register things without consciously marking them off; they act upon things in a conversation of gestures (Mead),[16] not a conversation of words; they emphasize and constantly draw upon "experience" with these objects, without being able to spell out, and without caring to codify, what this experience consists of. There is a native discourse on the role of the analog body in research, a discourse embedded in an abundance of instructions that stress embodied experience—that advise practitioners to perform all kinds of experimental activities in person, by themselves; that warn about misinterpretations of results that occur when one has not been present at experiments; that advise setting aside the time that is needed to "handle"—that is, embody—any method of dealing with objects. There is also a preference for traveling to places where objects are dealt with successfully instead of learning about them from lab protocols, and, if practitioners are asked to solve an experimental problem, to solve it by displacing themselves and attaching themselves to the problem situation. What interests me is the distrust in the mind's being able to figure things out *at a distance*, and in language and communication as supplying, for this purpose, the necessary information. Through this distrust, molecular biologists act in an inclusionary manner toward natural objects; they align themselves with them (or they are aligned by them) by letting their processing capacities be triggered by situations rather than by mental events, and by substituting "behavior" for cognized, premeditated action.

Molecular biologists' analog mode of operation enhances the features and reactions of natural objects. The common life-world is a trick, so to speak, that allows this science to adjust day by day to these features and reactions. There are other forms of enhancement of the phenomenal world: for example, the method of *appresenting* this world and the circumstances that surround it, when it is not present, through visually recalling its features. When molecular biology produces signs, the activity of decoding becomes like opening a window upon the phenomenal reality that supposedly gave rise to the sign: one asks what happened in the lab, which steps were taken, which procedures turned out how, and so on. Appresenting is also notable in molecular biologists' ways of dealing with invisible objects—these are constantly rendered visible through drawing them on paper and blackboards—and in technical discussions, in which scenic descriptions appresenting laboratory objects substitute for detached measurements and technical terms. (For example, participants may not talk in grams to indicate a quantity; they may talk in laboratory dishes.) I am not suggesting, by the way, that appresenting is never used in physics, or that the analog mode of functioning is unnecessary. I am suggesting, however, that these processes, when they occur, are usually detector-related (i.e., equipment-related), and that other epistemic strategies are superimposed upon them.

Blind Variation and Selection by Success

There is another way of enhancing natural objects in molecular biology, which I alluded to in the beginning of this section, and which I want to now sketch out briefly in concluding. Natural objects and processes are also set up in molecular biology work as *selection environments*, to which experimental strategies propose alternatives for selection. This is particularly visible when experimental strategies are unsuccessful, a common occurrence in all laboratory work. This is also the case where the analogy of blind variation comes into the picture. What is the point of the analogy? In evolutionary biology, mutations introduce *variations* in the genetic material, which can be passed on to descendant molecules or organisms. If a given organism always reproduced itself perfectly, its descendants would never change, and evolution would be impossible. Which mutations are beneficial and survive is determined by natural selection—the differential advantage bestowed on those organisms whose qualities, introduced by variation, are more effective in a given environment. Mutations, of course, are "blind": they are random errors not preadapted to the environmental conditions that they encounter.

If there is a general strategy molecular biologists adopt in the face of open problems, it is one of blind variation joined with a reliance on natural selection. They vary the procedure that produced the problem, and let something like its fitness—its success in yielding effective results—decide the fate of the experimental reaction. Variation is blind in a very precise sense. It is *not* based on the kind of detailed investigation and understanding of a problem that is so popular with high-energy physicists. Confronted with a malfunctioning reaction, a problem of interpretation of data, a string of methods that do not seem to work, molecular biologists will not set out, like physicists, to find, through a "study," the reason for the difficulty. Instead, they will try out several variations and rely on the fact that these will result in the end in workable evidence. Note that in physics understanding and convictions are based upon demonstrable data points that detail the crucial aspects of the difficulty. Nothing like demonstrable data points is necessary or sought after in molecular biology.

Blind Variation and the Care of the Self

Blind variation is a strategy of dealing with the resistance of natural objects, equivalent to the master strategy in physics of self-analysis and self-understanding. Let me dwell for a moment on this equivalence. It is important to realize that molecular biology's preference for blind variation and selection by success by no means implies that this method is any less effective than physics' care of the self and negative epistemics. In fact, molecular biology by all standards has been very successful in the last twenty to thirty years, and seems bent on remaining successful in the foreseeable future. Moreover, from the perspective of molecular biology, it is not at all clear that a strategy like that of experimental high-energy physics would work. Molecular biologists will argue that their attempt to understand a life organism of which little is known quickly reaches its limits; and since the machinery used in molecular biology is largely the life machinery of the cell and of organism reproduction, attempts at any self-understanding of the tools and components of experiment are jeopardized by the same limitations as investigations of the subject matter of molecular biology. Furthermore, they will argue that liminal knowledge, so useful in physics to correct for errors and systematic problems, may be less useful with an intervening technology. If an inadequately construed vector (plasmids or viruses that serve to transport and replicate DNA) generates the wrong protein, this cannot be subtracted out of the experiment through remedial calculations—the vector has to be remade until it performs. Biochemical reactions as used in experiments are not formulated mathematically, and hence cannot be calculated with in the ways the reactions in a detector can be computed. What it all boils down to is that for molecular biology to behave like experimental high-energy physics many components of its system would have to change in synchrony with other components. In other words, it would involve a change of the whole epistemic culture. The argument is not that this is impossible. It is just that any central component of a system is often sustained by other components. It is rendered effective by them and works in conjunction with them. Blind variation works with the massive presence of small objects, the intervening technology of molecular biology, and its many ways of placing a premium upon empirical reality while deemphasizing work with representations.

Conclusions

There are several things I might say in concluding. For example, I can highlight the uses high-energy physics makes of reflexivity. Reflexivity has been the rage in anthropology, and science studies, and literary criticism, and other fields in recent years. It is usually discussed, epistemically speaking, as a monster that must somehow be kept at bay, a serious challenge, of which Kuhn says in a recent paper that our inability to answer it is a grave loss to our understanding of scientific knowledge.[17] Yet in high-energy physics we have a field that has long turned reflexivity into a principle of knowing, that brings into focus the possibilities of informationally closed systems in exploiting internal mechanisms and knowledge of the self, and that continuously curls back upon itself while instituting threefold hierarchies of observation elaborated toward the inside, rather than the outside of the system (observation through transitional objects, observation of transitional objects through experiment, and observation of these observations through error calculations). Perhaps it would be time to ask, if we have to have foundations, whether we cannot build a theory of knowledge from circular foundations. Molecular biology, while involving different types of circularities, chooses another road to the referent. It sets up long front lines in which it engages the other side of the referring activity in analog, "body-to-body" exchanges, thus including natural objects in a system in which they are continually enhanced through assuming the status of a selection environment, through appresentation, and through the willingness of scientists to meet them on a sensory, object level. This raises the question of the local ontologies that different sciences institute in meeting the referent, a question I only touched upon by mentioning transitional objects like the detector or the analog functioning of scientists.

There are other issues I could mention—for example, what it means to look from a cultural perspective on both sciences' handling of signs. Signs are prominently present in all sciences, a fact recognized by semiotic and communication-oriented perspectives in science studies. Yet from a praxeological perspective on culture, what matters is not their presence but how they are featured, inserted into different processes, and dealt with in scientific practice. They are treated very differently in epistemic practice. Cultural systems of behavior, as we know, construe the world in which they live differently. If they involve sign processes, as they invariably do, the question is nonetheless on what, figuratively speaking, they place their bets and stake their money—signs or not-signs. They may construct their world in terms of sign processes, or continuously construct it away from such processes. They may choose to combine the care for signs with an elaborate care of the self, or they may show a preference for mechanisms that reduce representations and minimize interaction with the self. Both methods go by the names "empirical" and "experimental." Nonetheless, the disunity of these strategies is apparent: the disunity of the two disciplines involved and of "the method of science." Different sciences of this kind feature different *epistemic cultures*— different ways to approach the world and different ways to derive sources of epistemic profit.

ALISON WYLIE

The Constitution of Archaeological Evidence: Gender Politics and Science

❖

From the time North American archaeologists undertook to establish their discipline as a *scientific* field, a component of anthropology, they have insisted that their goals were not merely to recover "specimens"—the preoccupation of amateurs and antiquarians—but to understand "the whens and the whys and the hows" of the prehistoric cultures that produced these specimens.[1] And as long as archaeologists have embraced this vision of their enterprise, they have debated fundamental questions about the security of archaeological evidence and its limitations as a source of explanatory, anthropological understanding of the cultural past. It is inescapable that archaeological data "do not speak for themselves";[2] they must be richly interpreted to stand as evidence of any aspect of past cultural life. Indeed, these data cannot even be identified as archaeological—as *cultural* material—without considerable "ladening" by background knowledge, variously identified as "middle-range theory,"[3] interpretive assumptions, linking principles, arguments of relevance,[4] a "conceptual scheme."[5] While this seems a straightforward enough condition of practice, many fear that any such ladening compromises archaeological interpretation at its foundation; the empirical "facts" on which reconstruction and explanation are to be based seem themselves an unstable construct, indistinguishable, in this, from the very interpretations they are meant to ground and constrain. The perennial difficulty is, then, that of how to get beyond (mere) fact gathering without lapsing into (arbitrary, unscientific) speculation.

Roughly every twenty years since the beginning of this century concern with these issues has erupted in public debate. In the last two decades this has culminated in a sharply polarized opposition between the defenders of a self-consciously positivist, objectivist program of research that dominated North American archaeology through the 1960's and 1970's, the New Archaeology, and their constructivist critics, who are sometimes explicitly skeptical and relativist. It is in this context that the first recognizably feminist initiatives have emerged in archaeology. From the outset critics of the New Archaeology have advocated feminist approaches as an exemplar of precisely the sort of politically engaged archaeological practice that they hope will displace the pretensions to value-neutrality and objectivity they associate with the New Archaeology. It is striking, however, that although feminist political commitments do clearly animate much of the new research on questions about women and gender in archaeology, few feminist practitioners embrace the constructivist, often ironic view of the research enterprise proposed by these critics. Indeed, they depend on quite conventional appeals to evidential constraints in the arguments they put forward for rethinking explanatory and reconstructive models that often leave women and gender out altogether or depend on manifestly ethnocentric and androcentric presuppositions about gender relations. In fact, even the most explicitly feminist research routinely produces results that diverge sharply from expectations, sometimes calling into question the presuppositions that informed the reframing of questions and the reinterpretation of the data by which they were established. I will argue that there is manifest in this program of research an interplay between evidential constraints and social, political factors that is poorly comprehended by positions articulated at either the objectivist or the antiobjectivist extremes that dominate current archaeological discussion, and that figure in parallel debates in other social sciences and in the sociology and philosophy of science.

In what follows I will propose a model of interpretive practice in archaeology that better captures both its vagaries and its security, based on a consideration of internal archaeological debate and an analysis of some key examples of the new archaeological work on gender. In particular, I will be concerned to show how archaeological data can be "laden" with theory, how they can be interpreted as evidence of antecedent events and conditions that produced them, such that they must be regarded as constructs and yet can still (sometimes) decisively challenge or sustain even quite ambitious models of past cultural systems, their internal dynamics, and their trajectories of development and transformation. Although the results of this process of "constructing evidence" are never final, in the sense of complete or unrevisable, they are by no means radically open-ended.

Archaeological Evidence: The History of Debate

Data Collection versus "Theorizing"

The debates that periodically bring to the surface anxieties about the security of archaeological evidence are generally provoked by impatience that anthropological goals are not being realized, that the collection and analysis of archaeological data have not provided any very rich understanding of past cultures that produced the "specimens" surviving in the archaeological record. On one hand there have always been those who insist that any failure to deliver "ethnological" understanding, as one early commentator describes it,[6] is simply a function of the incompleteness of our empirical knowledge of the archaeological record and not of any "alleged or real deficiency of [the] methods" by which archaeologists analyze and interpret the material they recover from this record.[7] On this view, archaeologists should continue to make the recovery of a secure and complete database their first priority. Indeed, some have insisted that a premature preoccupation with interpretation compromises the enterprise as a whole; as Laufer put the point early in this century, archaeologists should abandon "theoretical discussion" of all kinds on the principle that the "constant brooding" about strategies for interpreting archaeological data leads "to a Hamletic state of mind not wholesome in pushing on active research work."[8]

Twenty years later this line of argument was taken up again, albeit in somewhat less dramatic terms. Archaeology was still, it was claimed, "a youthful discipline whose primary concern must be the accumulation of essential data";[9] it was premature to engage in interpretative speculation—indeed, the business of interpretation and synthesis not only could but should be deferred "to a future time of greater leisure and fullness of data,"[10] a time when archaeologists might have in hand "objective and complete information" about the contents of the record.[11] Even in the 1950's, when North American archaeologists faced something of an embarrassment of riches and problems of data management took on special urgency, many simply extended arguments about the need to accumulate a rich database to the development of expansive typological schemes. Interpretive understanding would follow only when adequate empirical systematization of the data had been achieved. Moreover, many held that patterns of formal variability (as captured by typological units) are inherent in the record, there to be discovered by judicious use of statistical or other forms of empirical analysis.[12] Once again, then, empirical work, now redefined to include descriptive systematization, took precedence over interpretive "theorizing"; indeed, some insisted that every effort must be made to *exclude* from the empirical phases of inquiry (the recovery and analysis of archaeological data) the biasing influence of the problems and interpretive frameworks they might subsequently bring to bear: "So far as archaeologists are concerned, the facts themselves are sacred."[13]

On the other hand, however, there have always been critics of these broadly empiricist standpoints (as they were described internally) who have insisted that archaeology could not be expected to make any progress toward achieving culture-historical or anthropological understanding of the past so long as interpretive "theorizing" is deferred. Unless data collection and systematization are informed by specific problems and interpretive presuppositions, much that might be useful will be lost, typological schemes will be inherently arbitrary and meaningless, and the evidential significance of the data recovered will remain unrecognized. As early as 1913 critiques appeared of "haphazard and uncoordinated" approaches to data recovery that were dominated by a preoccupation with "the specimen"[14] and uninformed by any "reasoned formulation of definite problems."[15] Writing in 1917, Clark Wissler insisted that "the mere finding of things," the accumulation of data, is "impotent to answer the very [culture-historical] questions we are all interested in."[16] With this, the advocates of a "real," genuinely scientific and anthropological archaeology hailed the emergence of what they described as a "New Archaeology" (the title of an article published in 1917).[17] A similar condemnation of persistent tendencies to "obsessive wallowing in detail of and for itself" appeared in the late 1930's.[18] But at this juncture it was underwritten by a principled, explicitly antiempiricist, argument to the effect that "blind" fact gathering is not just inefficient and counterproductive; it is, in fact, impossible. Clyde Kluckhohn, who published one such critique in an early issue of *Philosophy of Science*,[19] insisted that such approaches are predicated on a series of untenable epistemological assumptions: that observational "facts" are separable from (autonomous of) "theory"; that they can be collected as independent of, and as neutral with respect to, theoretical commitments or assumptions about their cultural significance; that their significance as evidence of the past will simply emerge (inductively) as a sufficiently rich body of data is accumulated; and that properly scientific practice eschews theorizing that ventures beyond the systematic description of these facts.[20] This last, Kluckhohn insists, is "a vulgarization of physics and chemistry":[21] there are no theory-independent facts to be accumulated in the social or physical sciences; indeed, "probably no fact has meaning except in the context of a conceptual scheme."[22]

Similar arguments appear through the late 1940's and the 1950's, usually in connection with debates about the status of typological concepts. One critic of the notion that "facts" are a given—a "phenomenological datum"[23]—insisted that all archaeological facts are "abstractions." Both descriptive facts about the record and typological systematizations of them are determined by the "problem at hand" and constituted as "bundles of testable hypotheses about the nature of correspondence of cultural objects to the dynamic culture-historical pattern which bore them."[24] Another especially vociferous proponent of this constructivist view, James Ford, objected that those who believe types can be "discovered" in the archaeological data (statistically, empirically) operate as if "nature [had] provided us with a world filled with packaged facts and truths that may be discovered and digested like Easter eggs hidden on a lawn."[25] Albert Spaulding, the chief advocate of "statistical techniques for the discovery of artifact types,"[26] replied that Ford's position "carr[ies] the logical implication that truth is to be determined by some sort of polling of archaeologists."[27]

Although Spaulding clearly intended this as a *reductio* of Ford's position, a number of others evidently took quite seriously the proposal that the identification, systematization, and interpretation of archaeological data might be essentially conventional. By the late 1950's several had come to the conclusion that all aspects of their practice depend on the application of ethnographic categories to archaeological data; data are identified as archaeological on the basis of general ethnographic analogs, and any typological sorting of these data depends (implicitly or explicitly) on a number of more specific analogical inferences about their functional, social, cultural, or other significance. At least one proponent of this view argued explicitly that archaeological inquiry must therefore be recognized to incorporate an irreducibly "subjective" element;[28] indeed, he insisted that, given the underdetermination of typological systems and of their application to specific bodies of data, any assessment of credibility (in description, systematization, or interpretation of the record) must ultimately turn on "an appraisal . . . of professional competence."[29] Such considerations led, in some cases, to profound skepticism about the enterprise as a whole. One archaeologist working in the United Kingdom concluded that it is "a hopeless task" to attempt any reconstruction of antecedent cultural conditions and activities; this can only be ever a matter of creative speculation given that "between the human activities we should like to know and their visible results there is logically no necessary link."[30]

The New Archaeology and Its Critics

The most recent round of debate has been generated by a concerted effort, on the part of the (American) New Archaeologists of the 1960's and 1970's, to turn this tide of methodological discontent. They objected that "traditional" archaeology had failed to realize any very satisfying anthropological understanding of the cultural past because it persisted in an essentially unscientific, "inductivist" (specimen-focused) mode of practice. To rectify this they came down strongly on the side of problem-oriented, theoretically informed practice, but unlike previous advocates of such an approach, they proposed a self-consciously "positivist" research program the hallmark of which was to be its "deductivism."[31] In reaffirming a commitment to explanatory goals, the New Archaeologists insisted that the central aim of archaeological research must be to understand the cultural systems and, above all, the cultural processes that produced the archaeological record: "The process theorist is not ultimately concerned with 'the Indian behind the artifact' but rather with the system behind both the Indian and the artifact."[32] Most important, this "processual" understanding was to conform to a Hempelian covering-law model of explanation (a deductive-nomological [D-N] model, specifically); archaeology was to establish a body of laws of culture process and system dynamics capable of explaining deductively, by subsumption, particular cultural lifeways and their archaeological traces.

The methodological program of the New Archaeology—its call for problem orientation and for fuller use of archaeological data as evidence of the past—was also characterized in deductivist terms. The New Archaeology was to invert the "inductivist" practice of traditional archaeology and institute a hypotheticodeductive (H-D) testing methodology: "The generation of inferences [inferential conclusions] regarding the past should not be the end-product of the archaeologist's work"; rather, they should "stand at the beginning of research," as test hypotheses to be "verified against independent empirical [archaeological] data" through systematic, "problem-oriented" investigation of the record.[33] Existing data might serve as a basis for formulating explanatory models, but these should be specified closely enough to yield quite specific expectations about additional material that should be present in the archaeological record if these models are accurate, and any investigation of the record should be designed, on an H-D model, to "test" these implications.

It is a significant irony, however, that despite their positivist commitments, the New Archaeologists also routinely invoked Thomas Kuhn to support the insight that archaeological data never stand as evidence for or against a test hypothesis except under interpretation, that "facts do not speak for themselves."[34] Far from seeing this as cause for skepticism, the New Archaeologists considered it grounds for optimism. However fragmentary and ephemeral archaeological data might seem, they were confident that if only they could secure a sufficiently rich body of background knowledge—if only they could create or borrow appropriate "linking principles" by which "meaning" (specifically, evidential import) might be reliably ascribed to archaeological data—they would be in a position to investigate virtually any aspect of the cultural past.[35] Given their Hempelian allegiances, this background knowledge was to take the form of low-level laws capable of covering (on a D-N model) what amount to retrodictions from archaeological data (the surviving "statics" of the archaeological record) to the antecedent conditions and events that would have been necessary and, ideally, sufficient for their production (the "dynamics" of past activities and events capable of producing them). Properly constructed and grounded, these ascriptions of evidential significance would approximate "deductive certainty." The hallmark of the "processual" New Archaeology—its "strongly positive" view of the prospects for inquiry—is, then, the conviction that archaeological understanding is constrained not by limitations inherent in its database but by limitations in the conceptual framework that informs inquiry.

Critics both within the New Archaeology and outside it were quick to point out the incongruity of demanding both aggressive theorizing, informed by appeals to Kuhn, and strict conformity to a deductivist, positivist testing methodology. It proved very difficult to implement the recommended testing program in any interesting cases. Archaeological data proved to be no less underdetermining when used to test explanatory and interpretive hypotheses than when invoked as inductive support for them (i.e., it became patently clear, in practice, that the H-D method is not deductive); only rarely, if ever, could archaeologists rely on biconditional laws to establish a unique linkage between archaeological "statics" and particular antecedent cultural "dynamics." Since the mid-1970's Lewis Binford (perhaps the most influential advocate of this view), and numerous others sympathetic to the general aims of the New Archaeology, have turned their attention to "actualistic" research—to ethnographic and experimental studies of how material culture is produced, used, discarded, and preserved or transformed—in an effort to secure the background knowledge necessary to set archaeological interpretation on a firmer foundation. Binford is adamant that this will vindicate his deductivism, providing a set of "uniformitarian assumptions," a "rosetta stone" for reliable archaeological code breaking.[36] But in the process of defending his position, he has retreated to an increasingly narrow ecodeterminist conception of the cultural subject, restricting archaeological inquiry to precisely those material, technological, and adaptive features of past cultural systems that are most promising of the security he values. Most others (and Binford himself in some contexts)[37] operate with the more modest ambition of establishing some limited interpretive control over the range of variables that enter into the creation of an archaeological record.

Less sympathetic external critics reject the New Archaeology program out of hand. They insist that if the recognition of Kuhnian theory-ladenness proves anything, it proves too much; it establishes that archaeologists have always, of necessity, actively *constructed* (not reconstructed, recaptured, or represented) the past, no matter how deeply committed they may have been to the quest for one true story, for an account of "the past as it was."[38] Several such critics have insisted that any use of archaeological data to evaluate statements about the unobservable past is viciously circular; one of the most influential, Ian Hodder, regards such inference as patently "tautological and self-fulfilling."[39] It depends on an "edifice of auxiliary theories and assumptions" that archaeologists accept on purely conventional grounds.[40] Indeed, Hodder insists that these linking assumptions are essentially "unverifiable,"[41] and therefore yield a body of evidence that is "without any ability to test . . . reconstructions against the past."[42] Taking up this line three years later, Michael Shanks and Christopher Tilley insist that "there is literally nothing independent of theory or propositions to test against. . . . Any test could only result in tautology."[43] Indeed, they invoke Nietzsche and Spivak, declaring that truth is a radically transient, ephemeral ideal, a "mobile army of metaphors."[44] Echoing the conventionalists and skeptics of the 1950's, they conclude that archaeological data can be interpreted in any number of ways, and can thus support myriad reconstructive and explanatory hypotheses; archaeologists necessarily find in the record just, or only, what their conceptual framework prepares them to recognize as evidence.

In some cases such "postprocessualist" (or "antiprocessualist") critics take the further step of endorsing an uncompromising social constructivism. What counts as true or plausible, indeed, what

counts as a "fact" in any relevant sense, is determined by con-
textually specific interests: individual, micropolitical interests, as
well as class interests, broadly construed. Shanks and Tilley characterize archaeology as an inherently imperialistic enterprise by
which the past is colonized and made to serve various recognizably
presentist ends, which are usually unacknowledged and oppressive;[45] positivist and objectivist rhetoric simply serves to mask the
political agendas that archaeologists actually (if unwittingly) serve.
Where Shanks and Tilley advocate a self-consciously political construction of the pasts that are necessary for "active intervention in
the present,"[46] Hodder enjoins archaeologists to avoid "writ[ing]
the past" for others, or for societies in which they are not themselves prepared to live.[47]

Gender Research and Theoretical Ambivalence

At various junctures, the recent critics of "scientific," processual
archaeology have all endorsed feminist initiatives, usually in the
abstract and in prospect, as exactly the sort of politically engaged
research they hope will displace the scientism and objectivism of
the New Archaeology.[48] Although they themselves have done little
to develop a program of feminist research, the first recognizably
feminist initiatives in archaeology did appear in the decade after
they began to challenge the (perceived) hegemony of the New
Archaeology.[49] The first paper to make the case for feminist approaches to archaeology was published by Margaret Conkey and
Janet Spector in 1984,[50] and the first collection of essays devoted to
work in this area appeared six years later, the outgrowth of a small
working conference organized by Conkey and Joan Gero in 1988.[51]
In organizing this conference, Conkey and Gero approached a
number of colleagues working in widely different areas and asked if
they would be willing to explore the implications of taking gender
as a focus for analysis in their various fields; even several years after
the appearance of Conkey and Spector's paper there was little more
in print or in process. Most had never considered such an approach
and had no special interest in feminist initiatives, but they agreed to
see what they could do. In effect, Gero and Conkey commissioned
a series of pilot projects on gender that they hoped might demonstrate the potential of research along the lines proposed by Conkey

and Spector in 1984. Their motivation was explicitly political; they sought to engage some potentially sympathetic and influential colleagues in the investigation of new questions they thought should be asked concerning women and gender, questions they had come to see as important because of their own feminist commitments. Although a number of other contextual factors of a sociopolitical nature came into play—enabling conditions and factors responsible for the subsequent ground swell of interest in work in this area—these deliberate efforts to mobilize support for research on questions about gender and women in prehistory were a crucial catalyst for the considerable body of work that has since begun to appear.[52] In short, there could hardly be a more explicit case of political interest playing a crucial role in shaping the direction of research.

I will argue, however, that despite the explicitly political origins of the diverse programs of research that now address feminist questions about women and gender, these do not unequivocally support, or instantiate, the strong claims of the postprocessualists. Feminist commitments do not displace evidential considerations; if anything, they enhance a commitment to empirical rigor, especially in the critical inspection of the (sexist, androcentric) presuppositions that have framed much otherwise exemplary research in the field. Indeed, the new research on gender frequently reflects a wariness of strong constructivist conclusions, and in this feminist archaeologists are not alone. Feminist practitioners in a number of other contexts have expressed considerable ambivalence about the relativist implications that are often presumed to follow from their own wide-ranging critiques of extant traditions of scientific practice and its claims to objectivity. Even those who recommend a postmodern stance as a resource for feminist research acknowledge the "dilemma" that this creates for feminists or for any who would use postmodern insights "in the interests of emancipation."[53] In this connection, and with special reference to feminist critiques of science, Sandra Harding argues the need to tolerate "ambivalence," to embrace both "successor science" projects, which serve to expose and displace androcentric bias in entrenched bodies of knowledge, and the vision of alternatives embodied in postmodern disruptions of these projects.[54] Many who are less sympathetic express concern that, at the very least, such positions have "both emancipatory and reactionary effects"; indeed, they may be "especially dangerous for the marginalized."[55] The worry is that the deconstructive arguments intended to destabilize Enlightenment "myths" of objectivity and truth are themselves "merely an inversion of Western arrogance";[56] specifically, they are an inversion that serves the interests of those who have always benefited from gender, race, and class privilege: "The postmodern view that truth and knowledge are contingent and multiple may be seen to act as a truth claim itself, a claim that undermines the ontological status of the subject at the very time when women and non-Western peoples have begun to claim themselves as subject."[57]

The tension between postmodern and emancipatory projects is evident in much feminist practice in the social and life sciences. On the one hand, feminist critics of science have exposed such pervasive androcentric bias that they seem inevitably to call into question not just "bad science," but much that passes for "good science," even exemplary science.[58] Where this erodes any conviction that scientific method might be self-cleansing, a guarantor of objectivity, it is often presumed that feminist critics undermine any possibility of claiming, for their own insights, greater credibility in any but a purely political sense. And yet, the feminist researchers responsible for these critiques are by no means prepared to concede that their accounts are simply equal but different alternatives to those they challenge. Where women and gender have been characterized in stereotypically androcentric terms, or simply ignored, in what purport to be humanly inclusive accounts of a society or cultural group (as in hunting-focused accounts of foraging societies),[59] a historical epoch (as in the Renaissance that Joan Kelly-Gadol suggests women did not have),[60] psychological processes,[61] or physiological and cognitive capacities,[62] the result has frequently been pervasive error and misrepresentation, measured by such standard criteria as empirical adequacy and internal coherence. Indeed, the claim made on behalf of research informed by a feminist "angle of vision" is often that it is, quite simply, *better* science in quite conventional terms.[63] In a close analysis of exactly how and where androcentrism arises in biology (evolutionary theory and endocrinology), Helen Longino and Ruth Doell argue that such critiques of science should not put feminists in the position of having to choose categorically for or against science; we should not have to "turn our backs on science as a whole . . . or condemn it as an enterprise."[64] Their reason for cautioning against such simple, polarized critical responses is immediately relevant for understanding feminist practice in archaeology: "The structure of scientific knowledge and the operation of bias are much more complex than either of these responses suggests."[65] Longino has since argued the viability of a sophisticated "contextual empiricism" in general terms,[66] and as an option for feminist practice—for "doing science as a feminist"[67]—that would preserve some (mitigated) claims to objectivity.

Even without the benefit of the methodological/epistemological analysis offered by Longino and Doell, it is perhaps unsurprising that feminist practitioners and critics of science should be reticent to embrace a thoroughgoing constructivism about empirical inquiry (despite providing considerable evidence of the social, political nature of science). At its best, feminist research grows out of a commitment to understand, accurately and in detail, the institutions, attitudes, and practices that oppress women in a diversity of contexts and ways, *so that we can be effective in changing them.* And in this its roots and inspiration lie in the varied experiences of constraint and dispossession that mark women's lives. Uncompromising constructivism or relativism often trivializes these experiences, and deflects attention from questions about how and why they arise, about the structures and conditions that constitute, for any who lack power, intransigent realities that impinge on their lives at every turn; in this respect, such a position embodies what seems patently an ideology of the powerful.[68] Certainly it is a central part of the activist experience of women who attempt to change oppressive conditions of life that we require, first and foremost, a sound understanding of the forces we oppose. In short, a commitment to the emancipatory potential of feminism, and respect for the very real constraints we encounter in practice, persistently force us back from the extremes of both objectivism and relativism that emerge in abstract debate about the status of empirically grounded knowledge claims.[69]

These sorts of concerns, which are ubiquitous in discussions of the apolitical, even reactionary, implications of (some) deconstructive and postmodern positions,[70] are not lost on the proponents of an explicitly political (postprocessual) archaeology. In fact, each of the radical critics of objectivist, processual archaeology can be seen to retreat from an uncompromising constructivism as soon as it becomes clear that such a position threatens to undermine their *own* social and intellectual agendas just as surely as it does those of the positivists they repudiate.[71] Three years after his strongest critiques of processual New Archaeology had appeared, Hodder qualified his arguments from underdetermination with the striking observation that, despite all facts' being constructs, there does exist a "real world," and, what is more, "the real world does constrain what we can say about it."[72] His students Shanks and Tilley likewise declare themselves realists in the same contexts in which they promote postprocessualism, and invoke a "dialectical relationship" between object and subject that ensures that archaeological construction is not "free or creative in a fictional sense."[73] In fact, they are quite clear that they do not intend to endorse an anything-goes pluralism or relativism; they are not prepared to allow that all claims about the past must be considered "equal."[74] Indeed, they see "the archaeological record itself" as a source of constraints that may "challenge what we say as being inadequate in one manner or another."[75] The difficulty, however, is that having made these qualifying statements Shanks and Tilley and Hodder do not return to consider the premises about theory-ladenness and the role of social/political interests in knowledge construction that led them to insist, in the first place, that archaeological data, constituted as evidence, can never provide a test of explanatory or reconstructive claims about the past. They simply juxtapose with these claims a series of counterassertions that seem more reasonable. The result is incoherence.

The challenge that arises in all these contexts of practice, a challenge especially pressing at the intersection between feminist practice and archaeology, is to give an account of how science can be conceptualized so as to recognize, without contradiction, *both* that knowledge is constructed and bears the marks of its makers, *and* that it is constrained, to a greater or lesser degree, by conditions that we confront as external "realities" not entirely of our own making. A fruitful point of departure is a grudging consensus, evident on all sides of the current debate,[76] that although archaeological data must be richly interpreted to stand as evidence, they do (sometimes) have a capacity to challenge and constrain what we claim about the past: they routinely turn out differently than expected; they generate puzzles, pose challenges, force revisions, and

canalize reconstructive and explanatory thinking, sometimes raising doubts about even the most well-entrenched presuppositions.

A concern with just this nexus of problems can be discerned in
the work of those philosophers of science, including feminist philosophers of science, who have undertaken analyses of how observational and experimental results are stabilized such that, in
practice, they may show less instability and arbitrariness of construction than has been insisted upon by some of the stronger
sociological critics.[77] In this philosophers of science reassess what it
means to say that observations, or evidential claims generally, are
"theory-laden." I have in mind, for example, Longino and Doell's
analysis of the role played by background assumptions in travers-
ing the "distance" between data, evidence, and hypotheses,[78] and
Dudley Shapere's analysis of the role played by "prior information" in determining what will count as an "observation" in physics.[79] Shapere insists that although nothing can provide observation
an "absolute guarantee" of efficacy,[80] it is simply not the case that
observational beliefs are all (equally) doubtful or unstable. These
points are developed, in feminist terms and with reference to biological research, in Longino's discussion of objectivity.[81] Much
recent philosophical and historical work on experimental practice
likewise takes, as its point of departure, a concern that positions
formulated in reaction against positivism often simply invert positivist priorities, privileging theory or interests over observation,
with the result that they yield just as "partial" an understanding of
observation as do the accounts they displace.[82] I find, in this literature, the elements of a general model of how archaeological observations are constituted as evidence that is instantiated, I will argue,
in the new archaeological work on gender.

Theory-Ladenness Reconsidered

The key to understanding how archaeological evidence can (sometimes) function as a semiautonomous constraint on claims about
the cultural past is to recognize that the ladenness of archaeological evidence by various kinds of background knowledge is never
monolithic or all-pervasive. When archaeologists make judgments
about the efficacy of claims about the cultural past, they exploit an
enormous diversity of evidence, not just different kinds of *archae-*

ological evidence, but evidence that depends on background assumptions derived from a number of different contexts, that enters interpretation at different points, and that can be mutually constraining when it converges, or fails to converge, on a coherent account of a particular past context.

Most often it is assumed that evidential claims derive their stability and autonomy from the security of the sources on which is based the imputed linkage between the surviving record and the antecedent contexts, conditions, events, or processes that produced it. Security is a complicated matter, however. On one hand, what counts is security in the sense described by Shapere in connection with his analysis of what physicists mean when they say they "observe" events at the center of the sun. It is the degree to which the assumptions linking these solar events (ultimately) to instrument readings on earth can be regarded as "free from doubt,"[83] given the results of inquiry in the contexts from which they derive. On the other hand, however, what is at issue is something like the additional considerations captured by Longino and Doell's spatial metaphor of inferential distance,[84] and by Peter Kosso's account of the directness, immediacy, and amount of interpretation required to bridge the gap between the accessible "data" and the events or conditions under investigation of which these data are effects or traces.[85] In archaeological cases there can be no question of literally "interacting in an informationally correlated way" with the cultural past; in this sense any direct measure of "immediacy" is beside the point.[86] Nevertheless, the length and complexity of the causal chain by which archaeological remains are produced—considerations of the number of interactions and of different kinds of factors involved—are clearly a relevant analog to the "directness" and degree of "nesting," or amount of interpretation, that Kosso finds crucial to the credibility (*qua* potential objectivity) of physically mediated observation. Archaeologists discuss these directly in terms of the various "transforms" (cultural, natural, depositional, and so on) responsible for the production of an archaeological record.[87] More often, however, the sort of security that concerns archaeologists is not so much the length or interpretive complexity of the linkage between a surviving record and the subject past, as the *nature* of such linkage. The ideal of security in this third sense is realized when background knowledge supports (or provides) a biconditional linking principle to the effect that a surviving archaeological trace could have been produced by only one kind of antecedent condition, event, or behavior; this is, in essence, the "deductivist" ideal endorsed by the New Archaeologists. Longino and Doell treat these latter two considerations of security together when they characterize the concept of "distance" between data, evidence, and hypotheses as a "logical notion of being more or less directly consequential."[88]

There are, then, (at least) three sorts of security at issue in archaeological assessments of evidential claims: security as a function of the entrenchment or freedom from doubt of the background knowledge about the linkages between archaeological data and the antecedents that produced them; security that arises because of the overall length and complexity of these linkages (the factors identified by Kosso in connection with "directness," and degree of "nesting"); and security due to the nature of these linkages—specifically, the degree to which they are unique or deterministic.

While methodological discussions among New Archaeologists have concentrated on considerations of security (especially security in the third sense), occasional references are made, notably by Binford, to the importance of making "ascriptions of meaning" to archaeological data on the basis of principles that presuppose background knowledge about "processes that are *in no sense dependent* for their characteristics or patterns of interaction upon interactions [that constitute the subject of the reconstructive hypothesis under evaluation]"—in Binford's case, interactions between "agricultural manifestations or political growth."[89] In this Binford appeals to exactly the sense of independence that Ian Hacking[90] and Kosso[91] find crucial in determining whether an observation can stand as evidence for or against a given test hypothesis—namely, independence between the constituents and the conclusions of an inference that runs on what amounts to a vertical axis from some element of a database to claims about its linkage to (and significance as evidence of) some aspect of the cultural past. Longino and Doell's concept of "distance" also covers considerations of independence in this sense: "The less a description of fact is a direct consequence of the hypothesis for which it is taken to be evidence, the more *distant* that hypothesis is from its evidence" (emphasis added).[92] When independence is realized in this sense, it ensures that, while observations are clearly "loaded with theory" (as Hacking puts it),[93] the theory with which they are loaded has no connection with the subject under investigation, on current understanding of the relevant subject domains; in archaeology as much as in physics this means that evidence may (sometimes) have considerable autonomy from the hypotheses it is used to support or evaluate.[94]

There is, finally, another sort of independence that plays an important role in stabilizing evidential claims in archaeology that is rarely discussed by archaeologists, and only briefly in philosophical contexts.[95] This is the independence of linking hypotheses *from one another* that arises when a number of such hypotheses are used to establish evidential claims about different aspects of a particular past event or set of conditions. It is analogous to the independence Hacking finds exploited by the makers and users of microscopes,[96] where different physical processes (i.e., different interaction chains, different bodies of ladening "theory") are used to detect the same microscopic bodies, or structural features of bodies, underwriting a localized "miracle" argument.[97] Certainly triangulation on a single aspect of an archaeological subject is sometimes possible and important; the routine use of multiple lines of evidence to date an artifact, feature, or stratum testify to this. But more often diverse (independent) resources are used to constitute evidence of quite distinct aspects of a past context. On the assumption that these are interconnected, the requirement of congruence among lines of evidence (i.e., that they yield a coherent model of the past context as a whole) sets up a system of mutual constraints on a horizontal dimension that can be as important in determining the credibility of any given bit of evidence as independence on the vertical dimension. The significance of considerations of independence in this sense is that they turn on a presumption that constituent elements of the subject of inquiry are interdependent—the subject is unified—at the same time as they exploit disunity among the linking principles used to gain access to them (i.e., disunity among the domains from which these linking principles derive and among the disciplines that establish them).

Archaeological Analyses of Gender

It is a significant irony that the role of these various evidential constraints is nowhere clearer than in the new feminist work on gender, which is so often identified as precisely the sort of explicitly "political" research that the relativist critics of the New Archaeology advocate. I will consider here a number of examples from contributions to the ground-breaking 1988 conference on gender research in archaeology, now published in *Engendering Archaeology*.[98]

Although, as I have indicated, most contributors to this conference remarked that they began with serious reservations about the approach urged on them by Gero and Conkey—they did not see how questions about gender could bear on research in their fields or subfields, given that they had never arisen before—even the most skeptical found that attention to such questions brought to light striking instances of gender bias in existing archaeological research and opened up a range of constructive possibilities for inquiry that had been completely overlooked.[99] One especially compelling critical analysis, due to Pat Watson and Mary Kennedy,[100] exposes pervasive androcentrism in explanations of the emergence of agriculture in the eastern United States. Whatever the specific mechanisms or processes postulated, the main contenders all assume that women could not have been actively responsible for the development of cultigens even though they also assume that women were responsible for gathering plants (as well as small game) under earlier foraging adaptations, and for the cultivation of domesticates when a horticultural regime was established. One model turns on the blatantly *ad hoc* proposal that shamans, who are consistently identified as male, were the instigators of this culture-transforming development; it was the knowledge they developed of plants for ritual purposes that led to production of the cultigens on which Eastern Woodlands horticultural practices were based. In effect, women passively "followed plants around" when foraging, and then passively tended them when (re)introduced as cultigens by men. The dominant alternative postulates a process of "coevolution" by which horticulture emerged as an adaptive response to a transformation of the plant resources that occurred without the benefit of any deliberate human intervention; at most, human patterns of refuse disposal in "domestilocalities" unintentionally introduced artificial selection pressures that generated the varieties of indigenous plants that became cultigens. On this account the plants effectively "domesticate themselves," and women are, once again, represented as passively adapting to imposed change.[101]

Watson and Kennedy make much of the artificiality of both models. Why assume that shamans were men, or that dabbling for ritual purposes would be more likely to produce the knowledge and transformations of the resource base necessary for horticulture than the systematic exploitation of these resources as a primary means of subsistence? Why deny human agency altogether and represent the emergence of horticulture as an "automatic process"[102] when it seems that the most plausible ascription of agency (if any is to be made) must be to women?[103] Indeed, Watson and Kennedy observe that they are "leery of explanations that remove women from the one realm that is traditionally granted them, as soon as innovation or invention enters the picture."[104] The common and implicit basis for both theories is, they argue, a set of underlying assumptions, uncritically appropriated from popular culture and traditional anthropology, to the effect that women could not have been responsible for any major culture-transforming exercise of human agency.

In a constructive vein, a second contributor, who works on pre-Hispanic sites in the central Andes, Christine Hastorf,[105] drew on several lines of evidence to establish that gendered divisions of labor and participation in the public, political life of the highland communities in question were profoundly altered through the period when the Inka extended their control in the region; the household structure and gender roles encountered in historical periods cannot be treated as a stable, "traditional" feature of Andean life that predates state formation.[106] In a comparison of the density and distribution of palaeobotanical remains recovered from household compounds dating to the periods before and after the advent of Inka control, Hastorf found evidence of both an intensification of maize production and processing, and an increase in the degree to which female-associated processing activities were restricted to specific locations within the sites over time. In addition, she reports some quite striking results from a comparison between the sexes of skeletal remains recovered from these sites and the results of a stable-isotope analysis of bone composition for evidence of variability in dietary intake. Although the lifetime dietary profiles of males and females are undifferentiated through the period preceding the advent of Inka control in the region, Hastorf finds evidence of divergence, which is consistent with the results of the palaeobotanical analyses, in the period when evidence of an Inka presence appears. Specifically, males show higher rates of consumption of foods that have the isotope values Hastorf identifies with maize than do females. To interpret this result Hastorf turns to eth-nohistorical records that document Inka practices of treating men as the heads of households and communities, drawing them into ritualized negotiations based on the consumption of maize beer (*chicha*) and requiring them to serve out a labor tax away from their villages compensated with maize and *chicha*. She concludes that, through this transitional period, the newly imposed political struc-tures of the Inka empire had forced a realignment of gender roles on local communities and households. Women "became the focus of [internal social and economic] tensions as they produced more beer while at the same time they were more restricted in their participa-tion in the society."[107]

Parallel results are reported by Elizabeth Brumfiel in an analysis of changes in production patterns in the Valley of Mexico through the period when the Aztec state was establishing a tribute system in the region.[108] She argues, through analysis of the density and dis-tribution of spindle whorls, that fabric production, largely the responsibility of women (on ethnohistorical and documentary evi-dence), increased dramatically in outlying areas but, surprisingly enough, decreased in the vicinity of the urban centers as the prac-tice of extracting tribute payments in cloth developed. On further analysis, she found evidence of an inverse pattern of distribution and density in artifacts associated with the production of labor-intensive and transportable cooked food based on tortillas; the changing proportion of griddles to pots suggests that the prepara-tion of griddle-cooked foods increased near the urban centers and decreased in outlying areas, where less demanding (and preferred) pot-cooked foods continued to predominate. She postulates, on this basis, that cloth may have been exacted directly as tribute in the hinterland, while populations living closer to the city center inten-sified their production of transportable food so that they could participate in the markets and "extradomestic institutions" that were then emerging in the Valley of Mexico,[109] and that required a mobile labor force. In either case, Brumfiel points out, the primary burden of meeting the tribute demands for cloth imposed by Aztec rule was shouldered by women and met by strategic realignments of their household labor. Where the Aztec state depended on tribute to maintain its political and economic hegemony, its emergence, like that of the Inka state studied by Hastorf, must be understood to have transformed, and to have been dependent on a transformation of, the way predominantly female domestic labor was organized and deployed.

Finally, several contributors considered assemblages of "artistic" material, some of them rich in images of women, and explored the implications of broadening the range of conceptions of gender relations that inform their interpretation. Russell Handsman undertook a critical rethinking of the ideology of gender difference, specifically, the "male gaze,"[110] and the presumption of a timeless, natural, and hierarchical opposition between men and women, that infused a British exhibition, "The Art of Lepenski Vir," challenging the notion that gender can be treated in essentialist terms in this or in any context. He suggests several interpretive options that might be pursued in constructing "relational histories of inequality, power, ideology and control, and resistance and counter-discourse" where gender dynamics are concerned.[111] In the process he points to a wide range of evidence—features of the images themselves and associations with architectural and artifactual material that might provide them context—that constitutes "clear signs"[112] of complexities, contradictions, "plurality and conflict,"[113] undermining the simple story of natural opposition and complementarity told by the exhibit. In a similar vein, Conkey has developed an analysis of interpretations of Palaeolithic "art," especially images of females or purported female body parts, in which she shows how "the presentist gender paradigm has infused most reconstructions of Upper Palaeolithic 'artistic' life," yielding accounts in which "sexist twentieth-century notions of gender and sexuality are read into the cultural traces of 'our ancestors'" with remarkable disingenuity.[114] She concludes that whatever the importance of these images and objects, it is most unlikely that they were instances of either commodified pornography or "high art," as produced in contemporary contexts.

Evidential Constraints in Practice

In all these cases, the results—both critical and constructive—turn on the appraisal of evidential constraints. And in all cases, the evidence appraised plays a role that is to varying degrees autonomous and corrective of the expectations and presuppositions that "laden" it, that bring it into view, or give it specific evidential import. This capacity to constrain is, indeed, a function of considerations operating on a number of dimensions, as Kosso argues.[115] Nevertheless, these multiple factors generate cases that fall along a rough continuum defined chiefly by degrees of independence in the first sense (the independence of linking hypotheses from claims or presuppositions about the subject past they help establish) and by the nature of the linkage invoked (i.e., the degree to which it is uniquely determining, establishing security in the third sense).

At one end of the continuum, the end that draws the attention of antiobjectivist critics, ascriptions of evidential significance are entirely determined by theoretical commitments, a set of precepts about the nature of the cultural subject, that are also embodied in the broader interpretive and explanatory claims that this evidence will be used to test or support. In part this is what Watson and Kennedy object to in explanations for the emergence of horticulture in the Eastern Woodlands. Sexist assumptions about the nature and capabilities of women underlie standard models of the horticultural transition (consistently reading women out), and they infuse interpretations of the archaeological data used to evaluate these models, insuring that these data will be seen as evidence for models that project onto the past a "natural" sexual division of labor in which women are consistently passive and associated with plants.

But even in these worst cases it is often possible, as Watson and Kennedy demonstrate, to establish grounds for questioning the assumptions that frame both the favored hypotheses and the constitution of data as the evidence from which they derive support. Two strategies for critique are evident in their analysis. The first exploits nonarchaeological resources, both conceptual and empirical, in an independent assessment of the framing assumptions. In this connection, Watson and Kennedy draw attention to a straightforward contradiction inherent in current theorizing about the emergence of horticulture in the Eastern Woodlands: women are persistently identified as the tenders of plants, whether wild or under cultivation, and yet are systematically denied any role in the transition from foraging to horticulture, whatever the cost in terms of theoretical elegance, plausibility, or explanatory power. To indicate just how high this cost may be, they draw on background (botanical) knowledge about the range and environmental requirements of the plant varieties that became domesticates to establish that they routinely appear in prehistoric contexts that were far from optimal.[116] They argue, in short, that it is most implausible that they could have arisen under conditions of neglect, as suggested in the coevolution model. Where the shaman hypothesis is concerned, their analysis is informed by an appreciation that, over the past three decades, feminist anthropologists have documented enormous variability in the roles played by women, in the degrees to which they are active rather than passive, mobile rather than bound to a "home base," and politically powerful rather than stereotypically dispossessed and victimized. This seriously undermines any presupposition that women are inherently less capable of innovation, self-determination, and strategic manipulation of resources than their male counterparts, and renders suspect any interpretation that depends on such an assumption, regardless of its archaeological implications. In this way they challenge the credibility of the interpretive principles used to bring archaeological evidence to bear on questions about the transition to horticulture, and of the more fundamental framework assumptions that underlie the explanatory models of this transition that they challenge.

In addition, however, even when the problem is "self-accounting" (as Kosso describes it), the archaeological data under interpretation can sometimes function as a locus of evidential constraint; this makes possible a second strategy for critique. The predisposition to interpret archaeological data in terms of sexist assumptions about the nature and capabilities of women does not necessarily insure (indeed, as Watson and Kennedy point out, it has not insured) that the record will obligingly provide evidence that activities identified as male, on these assumptions, mediated the transition from a foraging to a horticultural way of life, however strong the expectation that they must have. Indeed, where the coevolution model is concerned, most of the activities responsible for the creation of the "domestilocalities" in which cultigens emerged were women's activities, if the archaeological record of such sites is interpreted in light of the traditional assumptions about gender relations that Watson and Kennedy find presupposed by this account.[117] If the interpretive assumptions in question constituted a more closely specified theory, the outlines of a "bootstrapping" inference, of the sort proposed by Clark Glymour, might emerge in cases like these, complete with internal-to-theory independence between linking and test hypotheses.[118] In practice, then, even cases in which vicious circularity is a danger can be much more closely constrained by both external and internal evidence than strong constructivists acknowledge.

Straightforward circularity is generally not the central problem in archaeological interpretation, however. Given the state of knowledge in the relevant fields, and the complexity of most archaeological subjects, it is almost unimaginable that a single encompassing theory could provide both the linking principles necessary to interpret a given body of data as evidence of the cultural past, and a suite of hypotheses capable of explaining the events and conditions this first level of interpretation brings into view. Usually the basis for ascribing evidential significance to archaeological data is some form of analogical inference that draws on diverse sources, most of which are understood in terms of highly localized theory. Here the worry is not overdetermination by an all-encompassing conceptual framework, but underdetermination due to a lack of generalizable knowledge about the conditions under which observed linkages between (archaeological) "statics" and (cultural, behavioral) "dynamics" may be projected onto past (or otherwise unobserved) contexts. The inferential "distance" that must be crossed, in all Longino and Doell's senses,[119] remains considerable, and there are relatively sparse resources for helping to bridge it. As has been argued by those who fear that archaeologists will never make full use of their data unless they enrich the "theory" they bring to these data, there is a pressing need to strengthen the grounds for supposing that a particular element of the surviving record is linked to a particular antecedent, as in better-known (i.e., observed or simulated) contexts, or to eliminate alternatives if alternative linkages are known to be possible. In particular, there is a need to establish security in the third sense by demonstrating that something approximating a unique and determining linkage holds between the surviving record and its postulated antecedents.

Typically analogical inference is constructed and evaluated in terms of two sets of evidential constraints that, when effective, serve to establish security in just this sense.[120] These include both

constraints that determine what can be claimed about the analog, based on background knowledge of the source contexts from
which they are drawn, and constraints that derive from the archaeological record that determine the applicability of the analog to a
specific subject context. In associating women with the use of
spindle whorls in weaving, and with the use of griddles in food
preparation, Brumfiel relies on a "direct historic analogy," essentially an argument that there are such extensive and stable
similarities between the association of these artifacts with weaving/
cooking and women in historically related ethnographic and ethnohistoric contexts that it is reasonable to assume that these associations held for prehistoric contexts as well. Similarly, archaeologists dealing with evidence of horticultural regimes routinely
postulate a division of labor in which women are assumed to have
had primary responsibility for agricultural activities,[121] but they
base this not on an appeal to the completeness of mapping between
source and subject (which Brumfiel's case illustrates), but on the
persistence of this association of women with horticulture across
historically and ethnographically documented contexts, however
different they may be in other respects.

In these cases, completeness of mapping and reliable correlation
figure as evidence that a common "determining structure"[122] links a
distinct type of artifactual material to specific functions, gender
associations, or activity structures securely enough in present contexts to support an ascription of the same functions and associations
to the archaeological subject. These interpretive claims can be
as decisively undermined by a change in background knowledge
about the sources of these analogs (e.g., demonstrating that the
material:behavioral linkage projected onto the past is not stable in
source contexts) as by what archaeologists find in the record of the
contexts onto which they are projected (e.g., demonstrating that a
particular association or function could not have obtained in the
context in question).[123] Alternatively, where the linking principles
based on background knowledge of source (or "actualistic") contexts is uncontested, and their credibility is independent of any
hypotheses that archaeologists might want to evaluate against the
evidence these principles help to establish, they can very effectively
stabilize debate.

The power of the challenge posed by Brumfiel to extant models of the economic base of the Aztec empire depends on precisely this sort of stabilizing analogy. Her association of spindle whorls, and of pots and griddles, with women is unproblematic for any she might engage in debate and wholly independent of both the hypotheses she challenges and those she promotes. Given this provisional foundation, she is able to bring into view new features of the structure of otherwise well-understood assemblages—formerly undocumented patterns of distribution and association among components of these assemblages—that cannot be accounted for on standard models even when constituted and interpreted in terms that are shared by proponents of these models. She thus challenges not (just) their conceptual integrity or prior plausibility (as in Watson and Kennedy's case) but their empirical and explanatory adequacy as models of the political economy of states in Mesoamerica and the Andes. Perhaps most important, she identifies, as the source of their difficulty, implausible assumptions about the stability of gender structures, and argues for an alternative model predicated on the thesis that these are not only dynamic—genuinely historical and cultural, not "natural"—but are also crucial codeterminants of political and economic state-formation processes that had been treated as a public, male preserve. Her account is compelling because it effectively fills some of the gaps and solves some of the puzzles that arise for extant theories given their dependence on these assumptions.

The limiting case on this continuum of theory-ladening inferences, the ideal of security in the ascription of evidential significance to data described earlier, arises when archaeologists draw on completely independent, nonethnographic sources that specify unique causal antecedents for components of the surviving record. Hastorf's analysis of bone composition comes the closest to this ideal among the cases considered here. If the background knowledge deployed in stable-isotope analysis is reliable (and this is always open to critical reassessment), it can establish, in chemical terms, what dietary intake would have been necessary to produce the reported composition of the bone marrow recovered from archaeological contexts. Where this can be linked, through palaeobotanical analysis, to the consumption of specific plant and animal resources, and through skeletal analysis to a pattern of sex-linked differences in consumption, the isotope analysis can underwrite the inference of dietary profiles that is very substantially independent of, and can provide a serious challenge to (or a genuine test of), any interpretive or explanatory presuppositions about subsistence patterns and/or gender-structured social practices affecting the distribution of food that archaeologists might be interested in testing. The independence and security of linking arguments based on background knowledge of this physical, chemical, bioecological sort is exploited in many other areas: in morphological analyses of skeletal remains that provide evidence of pathologies and physical stress; in radiocarbon, archaeomagnetic, and related methods of dating; and in reconstructions of prehistoric technology and palaeoecology, to name a few such examples. As this limiting ideal of (vertical) independence between linking principles and test hypotheses or framework assumptions is approximated, archaeologists secure a body of evidence that establishes provisionally stable parameters for all other interpretation and a stable basis for piecemeal comparison between contending claims about the cultural past. Achieving this sort of security was a central ambition of the New Archaeology, and the primary motivation for Binford's emphasis on the need to develop "middle range theory."[124]

It is important to note, however, that the evidence provided by these sorts of linking principles has limited significance, taken on its own. Hastorf must rely on a number of collateral lines of evidence to establish that the anomalous shift in diet evident in male skeletons was due to increased consumption of maize beer, and to link this to the advent of Inka-imposed systems of political control in the region and to a restructuring of gender relations at the level of the household. This reliance on multiple lines of evidence is an important and general feature of archaeological reasoning that cross-cuts the considerations of independence and of security I have described. In fact, archaeologists rarely ascribe evidential significance to items taken in isolation. Context is crucial and is defined in a number of ways; if it is relevantly cultural, rather than geological or ecological, it may be defined in terms of associations among artifacts or features that are recovered together in undisturbed deposits, that have close spatial or temporal proximity, or that show technological, formal, or stylistic affinity even if widely dispersed.

When elements of the archaeological record can be assumed to bear on a particular past context in one of these senses and, most important, when these elements are ascribed significance on the basis of diverse linking principles (i.e., principles derived from independent bodies of background knowledge), then a network of horizontal constraints may come into play *between* distinct (vertical) lines of interpretive inference that can vastly increase their individual and collective credibility. That is to say, each vertical linkage between data, evidence, and hypothesis may be compelling individually—each may be secure in the relevant senses, and independence between linking principles and test hypotheses may insure against nepotism—but if the linking principles determining evidential significance are *independent of one another* in the second sense (i.e., if no one entails the others as a proper subset of itself, or is confirmed by the same evidence), it becomes possible to use essentially different means to triangulate on the same postulated set of conditions or events. Furthermore, if these diverse evidential strands all converge on a given hypothesis about the past, they can reasonably be said to provide this hypothesis decisive, if never irreversible, support simply because it is so implausible that such convergence should be the result of compensatory error in all the lines of inference establishing its evidential support.[125] Most often the problem in archaeology is not that of adjudicating between a number of equally plausible, well-supported, explanatory alternatives, but that of finding one account, one reconstructive or explanatory hypothesis, that is consistent with all lines of evidence constructed, as they are, using diverse resources.

While Hastorf is most explicit in exploiting the constraints imposed by a requirement of convergence across (horizontally) independent lines of evidence, it is clear that Brumfiel relies on them as well. When she identifies an anomalous distribution of artifacts related to cloth production (over time and space), and then reassesses the evidence related to different sorts of food processing, the new (unexpected) convergence she documents provides her own account especially strong support precisely because such convergence cannot be counted on. But perhaps more significant than cases where convergence enhances credibility are those in which error in interpretive inference is exposed because of a failure to converge, where dissonance emerges among independently constituted lines of interpretation. Even when, taken separately, each line of evidence relevant to a particular account of the past enjoys strong collateral support (i.e., from the sources that secure the linking principles on which they depend), undetected error in these interpretive inferences may become evident when one line of evidence persistently runs counter to the others, when dissonance emerges among lines of interpretation. The failure to converge on a coherent account makes it clear that error lies somewhere in the system of background or auxiliary assumptions and linking principles, however well entrenched they may be.

In cases of extreme dissonance, which are approximated by the interpretations of "artistic" images and traditions considered by Handsman and by Conkey,[126] a persistent failure to converge may call into question the efficacy of *any* interpretive constitution of the data as evidence in a particular area.[127] Both Handsman and Conkey conclude that many well-entrenched interpretive options must be abandoned, given the lack of convergence between ascriptions of significance to the art itself and evidence bearing on the larger cultural contexts in which it occurs. Indeed, all indications are that these cultures may have been (indeed, probably were) so vastly different from any with which we are familiar that the images comprising their "artistic" record cannot be assumed to have any transculturally stable meaning continuous with that which it might have in any of our contexts; they cannot be taken as evidence of many or, indeed, any of the range of activities, beliefs, or sensibilities we associate with "art." This may suggest that there is no determinate fact of the matter where the symbolic import of gender imagery is concerned, or, as Conkey suggests, it may require us to acknowledge that we simply are not, and may never be, in a position to determine what the fact of the matter is in such cases. But even in these worst, most enigmatic, cases the data often do effectively resist the imposition of favored interpretations and, in this, undermine a number of formerly plausible claims about the past. Thus, dissonance among lines of interpretation may make clear what we *cannot* claim in connection with a particular past; in fact, such negative results may force a reconsideration of quite fundamental assumptions about the nature of the subject domain and about the limits or prospects for success in its investigation. It is, paradoxically, the fragmentary nature of the archaeological record that is its strength in setting up evidential constraints of these sorts, even in establishing the limits of inquiry.

Conclusion

In the long history of debate animating North American archaeology, practitioners have repeatedly come up against the problem that their data constitute a "record" of the past only when background knowledge provides interpretive principles capable of linking their data inferentially to the events and conditions presumed responsible for their production. Those who have periodically rebelled against practices of "blind" data collection and analysis are (loosely) united, over several generations, by a central concern that no enrichment of the archaeological database will solve the problem of making this data stand as evidence; there is no critical mass of archaeological material that will, in itself, bring anthropological goals within reach. If archaeologists are to get at past forms of life, events, and trajectories of cultural development, much less at the processes that lie behind them, there is no avoiding the difficult task of "theorizing," where this refers, broadly, to the inferential practices required to traverse the distance between data, evidence, and hypothesis (to use Longino and Doell's terms).[128] Internal critiques of narrowly "empiricist" practice—critiques that extend these insights to their constructivist, antifoundationalist limits—have periodically given rise to various forms of pessimism, skepticism (local or global), and now, to an uncompromising relativism according to which archaeology (as a whole) must be understood to be an emperor who has only just-so stories for clothes. But whenever the strongest of such claims have been ventured—alternately as a challenge to the reigning establishment or as resistance to the claims of young upstarts—they have been tempered by an awareness, inescapable in practice, that however mute and malleable the data may be, they do have a capacity to resist theoretical appropriation. They simply do not turn out as expected and cannot always or easily be made to deliver the evidence required by a favored conception of the past, however coercive the processes of description, interpretation, and analysis.

In the latest round of debate one of the most exciting developments has been the appearance of explicitly feminist initiatives that hold the promise of mediating what has become an increasingly polarized dispute between processualists and postprocessualists, objectivists and relativists. While these make clear the centrality of values, interests, and sociopolitical standpoint to archaeological practice, they also throw into relief a range of empirical and conceptual resources whose capacity to resist, constrain, and canalize interpretation can serve feminists (and others) who are committed to the critical evaluation of entrenched presuppositions about the cultural past. These features of "context" foster research that is explicitly political and should be aligned with positions, like post-processualism within archaeology, that repudiate an untenable objectivism. But they are not altogether assimilable to—indeed, they embody a serious, politically and epistemically principled, critique of—the more extreme constructivism associated with these positions. The strategies for building, evaluating, and strengthening the interpretive linkages between data, evidence, and hypotheses that I find deployed by feminist practitioners are common in archaeological practice; they are strategies for constituting data as evidence that insures that these data have a capacity to surprise, to subvert settled assumptions, and sometimes to provide a provisionally stable evidential basis for evaluating competing claims about the past, sometimes yielding rationally decisive, if never final, conclusions. They turn on the exploitation of two kinds of independence and three kinds of security, with secondary evidential constraints operating on both source and subject sides of the equation. When they establish sufficiently stable evidential grounds, it is sometimes plausible to say we have "discovered" a fact about the world, or have shown a formerly plausible claim to be "just false"; the critical analysis by Watson and Kennedy, and the constructive proposals of Hastorf and of Brumfiel are cases in point. In other cases the outcome of inquiry is more equivocal. As Handsman and as Conkey both illustrate, sustained investigation may lead us to question basic assumptions about the accessibility, or even the existence, of certain "facts" about a given subject domain. In short, some objects of knowledge and epistemic situations do sustain a moderate objectivist and realist stand, while others do not; they are textlike in their interpretive openness, and it may never be appropriate to claim evidential security for descriptive or explanatory claims about them.

The main conclusion to be drawn from these observations is, then, that we should resist the pressure to adopt a general epistemic stance as appropriate to all episodes in the development of science, or all evidential claims. Any question about the status of evidence and the relationship between evidential and sociopolitical interests in the construction of knowledge—whether we should be relativists or objectivists—must be settled locally, in light of what we come to know about the nature of specific subject matters and about the resources we have for their investigation.

PART III

Power

JORDI CAT, NANCY CARTWRIGHT, AND HASOK CHANG

Otto Neurath: Politics and the Unity of Science

Otto Neurath was a key figure in the Vienna Circle and the spear-head of the Unity of Science movement. Even a cursory glance at Neurath's work, however, reveals that his philosophical views were quite contrary to the popular impressions of logical positivist philosophy. What people usually associate with the Vienna Circle and the Unity of Science movement is, if anything, closer to Carnap's early view: basic observational statements form a secure, common foundation; all scientific knowledge can be constructed from these by purely logical means. Throughout his career Neurath attacked as "pseudorationalist" this reductionist picture of a single structure built upwards from the "bricks of experience."[1] Even in the midst of his work for the Unity of Science movement he argued:[2]

The development of physicalist sociology does not mean the transfer of laws of physics to living things and their groups, as some have thought possible. Comprehensive sociological laws can be found, as well as laws for definite narrower social areas, without the need to go back to the microstructure, and thereby to build up these sociological laws from physical ones.

For the sociologist the results of modern physics are irrelevant; may atoms be as they may, human actions hardly depend upon microstructure. The instability of human actions may co-exist with exact atomic struc-tures, or human action may be exactly predictable while there is only a dispersion-prediction in the realm of microstructure.

These remarks should be sufficient to alarm anyone who thought Neurath's idea was to unify all of science by reducing it to physics. What kind of "unity of science" can there be if physics is "irrelevant" to sociology?

We wish to elucidate Neurath's philosophical position on the unity of science and to understand it in the context of his involvement in socialist politics. We want to tell two stories. The first, in the next two sections, is a narrowly philosophical one in which we discuss how it was possible for Neurath to make his antireductionism compatible with his advocacy of unity. The other story, in the two sections thereafter, is more historical and contextualist. We trace the evolution of Neurath's thought, following his political involvements in the tumultuous period between the two World Wars; in particular, we focus on his participation in the Bavarian Revolution of 1918–19, and the politics of "Red Vienna," 1919–34. If the first story argues that Neurath did not contradict himself in pursuing unity, the second story should add insights about why and how he felt compelled to push for unity. In the end, we hope to show that Neurath's philosophy and politics were of a piece. But first we turn to describing his philosophical views in more detail.

Neurath's Antireductionism

Many people assume that the unity of science supposes both reductionism and foundationalism. In this picture science is, or ought to be, a single structure, with a common, secure foundation. "Higher-level" statements reduce to and receive their validity from "lower-level" statements, ultimately reaching a self-evident foundation. For science there is only one foundation; hence there is unity in all science. When we speak of "reductionism" in this paper, we mean this kind of foundational reductionism. There are three important versions of this reductionism, which are interrelated: cognitive, disciplinary, and ontological.

Cognitively, direct sense-experience supplies the observational statements, or "protocol sentences," that provide the common basis against which all theoretical statements are tested. Statements that cannot be tested through sense-experience fall outside the realm of science; this is how science receives both its unity and its demarcation from nonscience. Disciplinarily, there is the faith that the laws of physics ultimately explain everything. Other sciences have their own laws, to be sure; but they are only shorthand renderings of unwieldy physical formulas describing complex physical systems, be they human brains, clouds, or clocks. All science is one because different disciplines are just different branches of physics. Ontologically, the reductionist thesis is easy to understand and widely believed. Everything that exists is made up of very small units of matter, of which there are not too many types. If we know the properties of all the elementary particles and the rules about how they interact with each other, then we can know everything there is to know. All knowledge reduces to one common basis, because all things reduce to common components.

Although in principle these three versions of reductionism could be independent from each other, modern scientism has often combined them all into one picture: elementary-particle physics forms the basis upon which all science can be founded, because it studies the ultimate constituent parts of the universe; it is also the most secure of all scientific disciplines, because its theories are thoroughly tested by experiments. It is not certain where this popular view originated; what is certain is that Neurath was opposed to it, quite thoroughly.

Let us examine Neurath's stance on the three versions of reductionism one by one. First of all, he did not believe in the absolute epistemic security of any kinds of statements. Neurath believed that no statement was ever compared "directly" to experience for its justification. Rather:[3]

Statements are compared with statements, not with "experiences," not with a "world" nor with anything else. All these meaningless *duplications* belong to a more or less refined metaphysics and are therefore to be rejected. Each new statement is confronted with the totality of existing statements that have already been harmonised with each other. *A statement is called correct if it can be incorporated* in this totality.

In fact, he opposed any attempt to refer back to experience, saying it led to "a doctrine of 'personal experiences' which then declines into idealistic metaphysics."[4] Science is a community activity. It requires as its basic data reports about publicly available phenomena occurring in shared space and time, not reports of private experience. For Neurath, the acceptance of a scientific statement was never absolute; there was no secure foundation of knowledge, but only the totality of what we take to be knowledge at a given moment and in a given situation. Even a protocol sentence (a sentence recording someone's basic observation report) might be rejected if it contradicted an overwhelming body of previously accepted statements. This antireductionist viewpoint was expressed in his metaphor of rebuilding a boat at sea, which was later revived and immortalized by Quine:[5]

There is no way to establish fully secured, neat protocol statements as starting points of the sciences. There is no *tabula rasa*. We are like sailors who have to rebuild their ship on the open sea, without ever being able to dismantle it in dry-dock and reconstruct it from the best components.

Neurath's rejection of disciplinary reductionism was equally clear; we have already quoted him twice on this score. He criticized those who thought science could or should be done by getting a complete description of everything in terms of the equations of physics:[6]

Many proceed from an "ideal forecast," from the Laplacean mind that knows all initial conditions and all formulas and thus can predict everything. Such a fiction is already metaphysics. For evidently here there are assumptions that are in principle not subject to any empirical test. In reality we are dealing now with more, now with less predictable partial connections, and in some cases we can say nothing about individuals but can say something about groups of individuals.

For Neurath, Laplace's demon is unrealistic to the point of being meaningless, and such a fantastic imagination could never serve as a guide for doing science. It is futile to hope that we would be able to reduce other sciences to physics, because we cannot obtain physical descriptions of the systems whose behavior we want to predict: "If we wish to predict what peoples, states or organizations will do we cannot go back to the ultimate physical elements. We must be satisfied to consider certain rough facts of a complex character."[7] One point needs to be made clear here: Neurath was a staunch advocate of what he called physicalism (*Physikalismus*), but that did not mean a reduction to physics. Rather, what he advocated was the use of a physicalist language, or a physical-thing language. That, in turn, meant speaking only of things located in space and time, but there was no requirement for those spatiotemporal things to be the objects usually treated by physics. He tried to make the distinction clear:[8]

It might be advisable to speak of "physicalist" when we give a spatio-temporal description in the sense of contemporary physics, for example, a behaviorist description. The term "physical" would then be reserved for the "statements of physics in the narrower sense," those of mechanics, electrodynamics, etc.

Perhaps Neurath was not so explicit about his rejection of ontological reductionism as he was with the other versions we have discussed. Unwilling to admit any nonmaterialistic elements into his ontology, he did accept that anything could be smashed into tiny bits of matter: "Our strict sociological account contains nothing which could not in principle be described in a molecular way."[9] However, in Neurath's view, this in-principle possibility of micro-physical descriptions held no importance for science. First of all, he thought microphysical descriptions were not realistically possible for most entities studied by science, as we have shown in the preceding discussion of his argument against disciplinary reduc-tionism. But there is a more radical reason why Neurath thought ontological reductionism was futile. Not only is it impossible to transfer the laws of physics to other realms of science, but the laws of elementary-particle physics themselves are not just about ele-mentary particles. The statements that we call the laws of physics implicitly refer to all kinds of spatiotemporal entities—not only weights, elementary particles, and so on, but all the other things involved in the practice of physics, including measuring instru-ments and physicists:[10]

Mach made the important remark: one should at least mention all mo-ments which cannot be removed during an experiment. This advice taken seriously implies that we have to formulate all laws of mechanics, biol-ogy, sociology as laws of a respective "cosmic aggregation," as I propose to call a rather indefinite mixture of sun, moon, earth, plants, animals, men, streets, houses, telescopes, watches, etc.

In other words, the subject of *any* kind of science is a cosmic aggregation; no science ever deals with just tiny bits of matter and their mutual interactions. We might say, even if all other sciences could be reduced to elementary-particle physics, it still would not mean that the subject of all science was reduced to elementary particles.

Unity as Coordinated Aggregation

In the preceding section we have mostly examined what Neurath's view on the unity of science was not. To see what unity did mean to Neurath, we should turn to the question, Why should the different sciences be united? Why are they not, and ought not to be, just separate activities existing side by side with no relation to each other? Neurath thought that the main function of science was to predict and control. In order to make predictions about a given concrete situation, we most often need to bring many scientific disciplines to bear upon it. This is especially true when we leave the domain of controlled experiments and look to events in natural settings that we wish to control:[11]

Certainly different kinds of laws can be distinguished from each other: for example, chemical, biological, or sociological laws; however, it can *not be said of a prediction of a concrete individual process that it depend [sic] on one definite kind of law only*. For example, whether a forest will burn down at a certain location on earth depends as much on the weather as on whether human intervention takes place or not. This intervention, however, can only be predicted if one knows the laws of human behaviour. *That is, under certain circumstances, it must be possible to connect all kinds of laws with each other*. Therefore all laws, whether chemical, climatological, or socio-logical, must be conceived as *parts of a system*, namely of *unified science*.

A concrete phenomenon is not likely to belong solely to the do-main of one science. Concrete events, then, provide the necessity for the different scientific disciplines and theories to work together; "therefore unified science is the stock of all connectable and indeed logically compatible laws."[12]

For scientific laws to be so interconnectable, it is essential that they should be formulated in the same language: "To establish uni-fied science . . . a *unified language* with a *unified syntax* is needed."[13] The necessity for a unified language was widely recognized among logical positivists; however, exactly what this unified language should be was the subject of lively debate. As we have already seen,

Neurath's choice was the physicalist language, which describes spatiotemporal entities. In opposition to the attempt to derive the physicalist language from a solipsistic "phenomenal language," Neurath insisted on speaking in terms of physical things throughout, rather than sense-data or experiences.[14] Near the end of his life he wrote, looking back on his position in this dispute with Schlick and Carnap (usually referred to as the "protocol-sentence debate"):[15]

My suggestion seemed to have the advantage that the "when, where and how" attitude could be maintained from the bottom to the top. This I call the "physicalist" approach, which has nothing to do with "mechanism" or anything like that; it only pretends that we can use the everyday language which we use when we talk of cows and calves throughout our empiricist discussions. This was for me the main element of "unity."

One advantage of the physicalist language, vis-à-vis the phenomenal one, is that descriptions in it do not have to refer to how the knowledge under discussion was obtained, or by whom. As Neurath put it, "unified science expresses everything in the unified language that is common to the blind and the sighted, the deaf and those who hear, it is 'intersensual' and 'intersubjective.' It connects the statements of a man talking to himself today with his statements of yesterday; the statements he makes with his ears closed, with those he makes with his ear open."[16]

Neurath's conception of unity explains the form that the chief product of the Unity of Science movement took: the *International Encyclopedia of Unified Science*. He emphasized over and over again that the *Encyclopedia* was intended not as a single and absolute system of science, but as a collection of various types of scientific work:[17]

If we reject the rationalistic anticipation of *the* system of sciences, if we reject the notion of a philosophical system which is to legislate for the sciences, what is the maximum coordination of the sciences which remains possible? The only answer that can be given for the time being is: *An Encyclopedia of the Sciences*.

An important purpose of such a collection was to make it easier for scientists to collaborate with each other in building more and firmer connections between the different sciences. For that purpose, it was crucial that the *Encyclopedia* draw attention to "gaps, difficulties and points of discussion, thereby avoiding the false impression that one wanted to replace a speculative system by 'the system of science.' "[18] Such an overt exhibition of disunity would lead to an effort to overcome it, as much as possible:[19]

The fruitfulness of the unity of science movement is indicated by the fact that more and more students of special fields participate in the discussion of the relation of their discipline to other domains. In this way bridges between different fields are constructed at points where only isolated domains were found before.

Neurath struggled, to the end of his life, to find ways of overcoming the disunity he saw in science. This was a struggle because he could not bring himself to believe in foundational reductionism, which offered an easy way to conceptualize and dictate unity. But why was the unification of science so important to Neurath? That is the question that forms the subject of our second story, and we believe an important key to the answer lies in Neurath's ideological commitments and political activities.

Unity of Science for Unity of Action: Neurath in the Bavarian Revolution

Neurath was a sociologist and economist by training, and he was always interested in social policy and political action. His political interests became enormously strengthened during World War I, and in its wake. In the centrally planned structure of war economy Neurath saw a greater rationality than in the haphazard and short-sighted conduct of capitalism. It was this experience that converted him to socialism; central planning, rather than nationalization, remained the centerpiece of his socialism throughout.

By the end of 1918 Germany and Austria-Hungary were plunging into a general chaos, with the exhaustion from a prolonged total war and the specter of defeat. On 7 November 1918, the Independent Socialists of Bavaria, led by Kurt Eisner, took power in a bloodless coup, and subsequently set up a coalition socialist government, declaring the founding of the Bavarian Republic.[20] This preceded the declaration of the German Republic in Berlin, and in fact was one of the key events that catalyzed the collapse of the German empire. Eisner promised a "radical transformation of the constitution and entire life of Bavaria."[21] However, his faction did not possess the extensive organization necessary for such major changes, and the more conservative Social Democratic Party (SPD, the "Majority Socialists"), which had a somewhat stronger organization, did not cooperate. Moreover, the Bavarian economy was so impoverished from the war that it seemed the question of socialization had to be placed behind the question of simple survival. Eisner himself stated in the program of the new Bavarian Republic: "There can be no socialization when there is scarcely anything to be socialized."[22]

In short, the new socialist government of Bavaria had no real economic plan. Otto Neurath did. On 23 January 1919, Neurath went to Munich to confer with Eisner. Two days later, he lectured before the Munich Workers' and Soldiers' Council. Neurath stressed what was the key to his thought: central information and central planning. There had already been a great deal of propaganda about Marxist socialization of the ownership of land and industry, he explained, but not enough about a different Marxist concern, namely the need for central planning.[23] Apparently Neurath convinced the workers, but not Eisner, who continued consulting the distinguished liberal economist Lujo Brentano. Eisner had appointed Brentano as the chairman of the Bavarian Socialization Convention, which met for the first time on 22 January. Brentano had the opinion, which he expressed in a letter to Eisner, that even a strong regime would be unable to cope with the fundamental irrationalities of the Bavarian economy; he thought the superiority of private enterprise was undeniable.[24] Neurath lamented: "How is it possible that Eisner, who desired socialism with every fibre of his heart, put this vital office into the hands of a liberal who obviously did not want socialism?"[25] In turn, Brentano called Neurath the "utopienschwärmenden ['utopia-enraptured'] Dr. Neurath," accusing him of wanting to institute an economic organization, like that of ancient Egypt, in which "everyone lives directly or indirectly from the king."[26]

Neurath's plan had originally been drafted in Saxony. There the political economist Hermann Kranold was asked by the Social Democratic government to draw up a socialization plan, and Kranold invited Neurath and Wolfgang Schumann to work with him on it. The product of their collaboration was the Kranold-Neurath-Schumann Program, which formed the basis for Neurath's proposals in Bavaria.[27] Before this effort Neurath had been willing to consider partial socialization schemes. But together Kranold, Neurath, and Schumann became convinced that only total, or full, socialization (*Vollsozialisierung*) could succeed. It takes a single, central plan to secure the stability of the economy against crises of overproduction and underproduction, and unemployment; it also takes a central plan to coordinate the distribution of raw materials, and the production of the right kinds of goods in the right amounts to insure general welfare.

For Neurath, central planning was the primary concern, and nationalization was not essential: "Nationalizing basic production . . . and the ripe industries does not help the starving." Neurath argued that it was all the same to the worker "whether he burns state coal or private coal in his stove; what matters to him is looking after the proper distribution of the coal so that coal is not used for luxury purposes."[28] Besides stressing that the nationalization of industry can help only when coupled with planned utilization, he also emphasized the need to break the secrecy with which industry operated, and insisted on the importance of publicly accessible statistics. He also noted the calculation that a worker needed to work for only ten to thirteen years in order to secure the necessities of an entire life.[29]

Not only did Neurath publish this plan, but he worked energetically to communicate its ideas to the masses. Wolfgang Schumann recalls:[30]

Neurath began to lecture at mass meetings. He was extremely fascinating and made an enormous impression, especially in the mining areas of Southern Saxony where his lecturing tours were like triumphal processions. How different it was to talk to party bosses or members of the government; with them there was nothing but the most annoying resistance. Neurath had begun to publish leaflets in Munich, and there his ideas met with more response.

The Kranold-Neurath-Schumann Program fell by the wayside in Saxony as political coalitions shifted, but Bavarian politics took some unexpected turns. On 21 February Eisner was assassinated, and in the ensuing turmoil the convocation of the diet (*Landtag*), which had been scheduled for the same day, was postponed. The newly created Central Revolutionary Council (*Zentralrat*) held power for a brief period, but the *Landtag* did meet on 18 March and appointed a cabinet under the Majority Socialist Johannes Hoffman, who had been minister of education and culture under Eisner. The Independent Socialist Josef Simon, who was appointed minister of commerce, trade, and industry, called Neurath from Saxony to direct socialization plans for Bavaria. On 25 March, Neurath delivered a long lecture on full socialization to the Committee on Socialization, outlining his scheme in detail and urging the creation of a Central Bureau of Economics (*Zentralwirtschaftsamt*). The committee accepted Neurath's proposal, adopted a resolution to create the bureau, and appointed him to direct it. The bureau's immediate task was to conduct a study of the housing, clothing, and provisioning of the Bavarian people.[31]

Hoffman expressed his antipathy to this development and threatened to resign if Neurath stepped beyond the bounds of his competence, making it clear that any socialization law would need the approval of the cabinet and the *Landtag*. This did not deter Neurath's activities. He made moves toward the socialization of the press, arguing that the existing press did not give people sufficient access to information. He continued negotiations initiated earlier with the conservative Farmers' Party over partial socialization of farming. He met several times with faculties to discuss the socialization of their technical colleges and the introduction of courses that would be of help in social planning.[32] He also contributed to plans to deal with immediate housing and food shortages, and played a complicated role in attempts to control the flow of currency.[33]

Neurath's efforts notwithstanding, the radicals found the Hoffman regime as a whole much too conservative and resistant to socialization. On 3 April the leaders of the revolutionary councils had a meeting in Augsburg, called by *Zentralrat* chairman Ernst Niekisch. To Niekisch's own surprise, a large majority there resolved to establish a Bavarian Soviet Republic. In response to this development some one hundred political heavyweights, including Neurath, met on 5 April in the Ministry for Military Affairs to discuss the situation, and most agreed to support the declaration of the Soviet Republic. The Communist leader Eugen Leviné opposed this move, saying the time was not ripe, but he was out-numbered. For several days following the declaration there was no structured government, and various orders, proclamations, and regulations were issued in confusion and mutual contradiction. Now the Bavarian Communist Party moved to replace this "pseudo" Soviet Republic with a "real" one, under the leadership of Leviné, and of Max Levien, who had initially organized and led the Communists in Bavaria. Hoffman, who was out of Bavaria when the Soviet Republic was declared, was determined to regain control. First he imposed a blockade, and then requested the intervention of the *Freikorps*, the ultraconservative private paramilitary force that was employed so often to put down left-wing revolts all over Germany in this period. The *Freikorps* moved in on May Day 1919; after a brutal assault that left at least six hundred dead, the Soviet Republic was destroyed.

Neurath was arrested. During the First Soviet Republic, he had pursued his original program energetically; during the second, he attended primarily to the organization of the various on-site economic councils. Neurath's motivation was again his commitment to full socialization and his drive for cooperation. Fearing that the individual councils would overreach themselves, compete with each other, and destroy the possibility for a unified organization, he hurried to organize them into a central scheme.[34] Neurath had mixed feelings about the Soviet Republic; he was especially critical of the Second Soviet Republic, later stating that the Communist leaders "applied Russian experiences to the German situation" and moreover "wanted to shed blood between the 'proletariat' and 'bourgeoisie.'"[35] Even during the First Soviet Republic, he had considered socialization as a "social service, independent of political constitution"; accordingly, even as he worked energetically he declared his "neutrality" to the *Zentralrat*.[36] At the end of June Neurath was tried for assisting high treason before a mixed court (civilian and military) and sentenced to a year and a half in a fortress. It was a light sentence, in comparison with many others'. Leviné was executed. Rudolf Egelhofer, who had organized the red army, was shot after an "on-site" trial. Gustav Landauer, commissar for education and enlightenment, was murdered in prison. Others received from five to fifteen years in prison.[37] Neurath never served his sentence, since his friend and supporter Otto

Bauer, then Austrian minister of foreign affairs, negotiated his release to Austria. Neurath was forbidden to return to Germany, which certainly ended his chances to take up a position that he had already been offered at Heidelberg. He never had an academic career after that, except for his stint in the Workers' University in Vienna.

To see how Neurath's view on the unity of science connects with the political history just provided, let us look back at Neurath's thought in 1910. He had been reading August Comte and shared in the general interest in questions about the relationships among the sciences—especially questions about a hierarchy of the sciences. In his review of Wilhelm Wundt's *Logik*, Neurath remarked: "The history of science shows us that the thought of a universal science comes up again and again."[38] But it is never carried off. Much of the article is a review of the failed efforts, even within such disciplines as mathematics, where we would expect the task to be easiest. Neurath remarks that we say science today proceeds "by the division of labor [*Arbeitsteilung*]," but what we actually have would be more accurately described as "the segregation of labor [*Arbeitstrennung*]." For "the concept of division only makes sense in the context of a whole, when the separate efforts work together towards some joint effect."[39] That is precisely what we do not have.

Here already the idea appears that will later be the key to unity—the idea of working together toward a joint effect. But that is not what Neurath *stresses* in 1910. His focus is on the hunt for the threads that tie the different sciences together. Science should find these threads and try to expose and strengthen them. If this can be achieved, then "the world will once again stand before us as a whole."[40] We see here a somewhat wistful yearning for a unification that Neurath expects never to obtain. But notice that it is unification into a single theory that he is imagining, and it is a single coherent picture of reality that he wants. In the course of the Bavarian Revolution, his concern shifts to using science as a tool for action. It is no longer the construction of a single picture of reality that he wants from unified science, but rather the assembly of a usable tool kit. At the peak of his efforts in the Unity of Science movement, his vision is not that of a single unified theory, but the patchwork of the *Encyclopedia*, consisting of various elements of scientific knowledge gathered together in one place to facilitate joint application as much as possible. What he now wants by way of unity is a "kind of instrumentarium, a stock of instruments for science in general whose effective uses will be shown at the same time."[41]

Neurath closed his involvement in the Bavarian Revolution with a sentence that represents a leitmotif of his entire intellectual work: "A unified program could have co-ordinated and unified action."[42] This statement provides the key to understanding Neurath's mature view on the unity of science. Different laws from different disciplines cannot be united in a single science, but at the point of action the disunity must be overcome. To illustrate this point, he used an example that we have quoted already: "Whether a forest will burn down at a certain location on earth depends as much on the weather as on whether human intervention takes place or not. This intervention, however, can only be predicted if one knows the laws of human behaviour." So we have a pressing demand: "It must be possible to connect all kinds of laws with each other. Therefore all laws, whether chemical, climatological, or sociological, must be conceived as parts of a system, namely of unified science."[43] We have already noted how, in Neurath's developed view, the necessity for prediction is what demands that the laws from different sciences be coordinated with each other. He said: "All laws of unified science must be capable of being linked with each other if they are to fulfil the task of predicting."[44] Hence unification is important for every single prediction, and every single action.

The importance of unification only increases when we consider the social coordination of actions. First of all, we need to coordinate many different actions, occurring in such disparate realms as mining, farming, industrial production, housing, banking, politicking, and the distribution of goods and services. Neurath also emphasized the necessity for unity arising from his desire for social planning: "Moreover, common planned action is possible only if the participants make common predictions. A common greater error made by a group often yields better results than mutually antagonistic small errors of isolated individualists. Common action presses us toward a unified science."[45]

With these insights about the purpose of unity for Neurath, we can make better sense of his emphasis on physicalism, or the use of physicalist language. It was meant not only to expel metaphysics, but to standardize our tools for constructing predictions and solutions. One of the chief tasks Neurath wanted scientists from different disciplines to undertake was the mutual normalization of their separate languages to insure that they used the same terms when they were talking about the same things. For instance, he said:[46]

The program of physicalism shows us the possibility of building up a uniform scientific language with a uniform terminology; this is in perfect harmony with the circumstance that we have to connect statements and terms of different disciplines for the deduction of individual predictions, and that we have to connect the statement of the theories with the individual predictions.

The tools for problem solving are the different bits of knowledge and methodology we develop in the different scientific disciplines. At the point of action, when we want to put out a forest fire or construct a totally planned economy, we need to be able to deploy our tools together. We do not want to find, as it were, that we have English bolts and metric wrenches.

We have observed a shift in Neurath's philosophy of the unity of science, from a passive longing for a unified picture of the world to an active advocacy for a kit of standardized tools with which one can transform the world. The shift had intimate links with his work on full socialization. On the one hand, as we have stressed already, his desire for socialization demanded that the sciences be unified. On the other hand, his confidence in the development of unified science enhanced his optimism about socialism. Neurath thought that socialism was possible because a stage of scientific and technological knowledge had been reached that would allow us to construct a rational economic plan that could work. For instance, consider a remark of Neurath's in 1920: "Socialism would gain if the political leaders could draw on socio-technical engineers, who . . . construct the economic organization that would best realize the socialist economic plan."[47] Even after the failure of the Bavarian Revolution, Neurath claimed: "The factual relations of development are, in my opinion, ripe for socialism; it is only the organizational moment that comes into question."[48]

The unification of the sciences and the socialization of the economy formed one task for Neurath. He taught that the role of theory was decisive in the shaping of life, insisting that "laws are not statements; they are directions for obtaining predictions."[49] Needless to say, these predictions are to serve as guides to action. In the Vienna Circle Manifesto of 1929 he urged: "We have to fashion intellectual tools for everyday life, for the daily life of the scholar but also for the daily life of all those who in some way join in working at the conscious re-shaping of life. The vitality that shows in the efforts for a rational transformation of the social and economic order, permeates the movement for a scientific world-conception too."[50]

In the framework of socialism, science is not just about "reality" but, more important, about "realizations." It was crucial for Neurath that science deal with possible arrangements of things, not just with those that happen to be realized. If empiricism should mean just sticking to what has already happened, then it would be conservatism: "Those who stay exclusively with the present will very soon only be able to understand the past."[51] Mechanics gives the machine builder information about machines that have never yet been constructed; the variety of sciences to be gathered together in the *Encyclopedia*—psychology, sociology, political economy, hydrodynamics, and so forth—will give us information about socialist economic orders that have never yet been realized. There is no possibility of constructing a single, abstract, unified science. But the sciences can be unified at each point of commonly planned action, and there they should be unified, to build a socialist economy that works. It is in this sense that unified science was, in Neurath's words, "the great task of consciously cultivating the future and the possible."[52]

Unified Science in a Unified Culture: Neurath in Red Vienna

In the summer of 1919 Neurath returned to Vienna, saved from prison by Otto Bauer. Bauer was at that time the Austrian minister of foreign affairs, and a leader of the Austrian Social Democratic Party. Like Neurath, Bauer had designed a socialization plan that had met with an early failure due to similar historical circumstances. In this section we will describe the political groundwork of

Austrian socialism from 1919 to 1934, a period known as "Red Vienna." During this period Neurath expanded and began to publish his views leading up to the Unity of Science movement; the Austrian Social Democratic regime embarked on a sociocultural program to turn the city of Vienna into a model of the future "socialized humanity." We want to show how Neurath's ideals of social change fit into the dominant political milieu of Red Vienna, and how his philosophical ideas continued to develop in tandem with his political activities.

In the municipal elections of May 1919 the Social Democratic Party gained an absolute control of the Viennese government; furthermore Austria, a republic since November 1918, became a federation, and Vienna one of its independent states. These circumstances made Red Vienna possible. Among the European socialist parties, the Austrian Social Democratic Party was unique not only in the amount of power it held, but also in its policies, based on Max Adler's dictum that the future of democracy did not lie in politics but in pedagogy. This emphasis on education (*Bildung*) came to define the widely admired program of *Bildungspolitik*, designed to reform the whole of working-class life, through policies affecting everything from housing to higher education.

An overview of the unique achievement of *Bildungspolitik* is given by A. Rabinbach in his study of Austrian socialism:[53]

The designers of the new Vienna understood the concept of *Bildung* in the broadest sense, from the sweeping educational reforms of Otto Glockel to the creation of a "new socialist individual." . . . By the end of 1933 the city had built more than 61,175 new apartments, mostly in the form of large housing blocks, the *Wiener Hoffe*, with parks, swimming areas, schools, kindergartens, gymnasia, health facilities, and community centers. . . . Moreover, the municipality had introduced an extensive program of adult education that supplemented the party libraries, bookstores, and *Bildungs-commissions* that were set up in all Vienna districts. But most important, the construction of a socialist Vienna was oriented towards the realization of the ideal of a "socialized humanity," not at the level of economic expropriation but at the level of cultural appropriation.

The construction of a new order of life was undertaken with an optimism that is reflected in a remark by Neurath, recalled by one of his students at that time: "What will we do, when all the world's energy will come from such a small center?"[54]

The emphasis on education in general and scientism in particular had long been elements of Marxism. Since Ferdinand Lassalle proclaimed the "alliance between science and the worker," the task of the "enlightenment of the worker" had occupied a prominent place among the activities of the socialist parties. But only the Austrian labor movement carried the slogan "Education is power" ("Bildung ist Macht"). What led the Austrian Social Democrats to adopt *Bildungspolitik* as the central political strategy in the construction of Red Vienna? We can identify four different factors.

The failure of Bauer's socialization plan. In January 1919 Bauer presented a plan for the socialization of the economy of the new Austrian Republic. In March, the central government created the National Socialization Commission, with Bauer presiding. The plan called for a gradual socialization without nationalization, controlling the administration of key industries such as wood, coal, iron and steel, and army supplies. The activities of the commission were hastened by the developments in Hungary and Bavaria, which produced the fear that the Communists might attempt a Bolshevik revolution with the support of increasingly radical workers.[55] Socialization applied effectively to the factories manufacturing army supplies, which were critical for maintaining control over the army in the case of a Communist uprising. However, the pace of socialization had to be slowed down because of the desperate conditions of the Austrian economy. Then, by the summer, the collapse of both the Bavarian and Hungarian revolutions put a virtual end to attempts at socialization in Austria.[56] Deprived of the possibility of implementing radical economic reforms, the Social Democratic Party focused on social and cultural policies.

The constitution of the party. The Austrian Social Democratic Party, founded in 1889, had grown out of a number of cultural societies, the *Bildungsvereine*, that appeared in Austria during the period of political liberalization in the 1860's. These societies stressed the virtue of self-improvement and the education of the workers. Education, leading to the acquisition of a rational conception of the world, was promoted as a mechanism enacting social equality, as well as a tool in the struggle against the obscurantist absolutism of the Church and the aristocracy. This acculturizing role of the party was preserved and explicitly acknowledged in its constitutional platform in Hainfeld:[57]

The Austrian Social Democratic Party, working for the whole people without distinction of nation, race, or sex, strives to free people from intellectual atrophy . . . to organize the proletariat politically, and to fill them with a consciousness of their position and their task, to make them ready physically and mentally for battle and to maintain this readiness.

Education for unification. An underlying motive in Austro-Marxism, particularly in the doctrines of Max Adler, was to bring together the Marxist tradition and the German culture, especially classical German philosophy.[58] For Bauer, the promotion of the traditional German culture served a clear political function: it provided the only authentic bond among the diverse nationalists that composed Austria-Hungary and Germany. In his major work of 1907 on the question of nationalities, Bauer argued that if the empire could not become a nation-state, it could become a community of education (*Bildungsgemeinschaft*): "People tried to discover the nation in our present class society . . . while the growth of the new *Erziehungsgemeinschaft*[59] has not yet been able to unite these small groups into a national whole."[60] As Rabinbach points out, "Bauer's attitude was completely consistent with [Victor] Adler's view that the party's role was to 'keep the proletariat alive, to enlighten it, and to bring it forward, to educate it'—a civilizing mission as much indebted to the traditions of classical German philosophy as to Marxism."[61] The same year, Bauer began to publish in the newly founded journal *Der Kampf* a number of articles touching on different aspects of the education of the proletariat.[62]

Max Adler's revisionism. Austro-Marxist revisionism opposed orthodox—"vulgar"—Marxism, which taught that economic factors were the sole determinants of psychological and intellectual development, thus asserting an absolute determinism unalterable by personal efforts. The intellectual leader of the Austrian Social Democratic Party was Max Adler, who, with his neo-Kantian idealism, advanced an antireductionist interpretation of Marx and Engels that gave an active role to the mind and to ideological and cultural factors: "the superstructure is just as real as the base because both are parts of a real building," and "economy and ideology are certainly different, but at the same time they are parts of a single social-cultural system of human life."[63] Adler stressed the political importance of consolidating the cultural life of the proletariat: class movement "can achieve its own class interests after the abolition of enslavement of labor only if general cultural interests triumph."[64] Notably, Adler, like Neurath, regarded science as a crucial component of the new culture, as the means of dealing with situations and taking action to change them. He wrote in 1895: "The proletariat, in contradistinction to [other political groups], act through clear, scientific knowledge of the existing situation. . . . The procedure of the socialist party must be determined by scientific knowledge and from such a basis Social Democracy—for it is a scientific socialism—stands over all parties, as is always the case with a science."[65]

Neurath's works clearly reflect the social and cultural atmosphere in Red Vienna. It was during this period that his interest in unifying science broadened its political scope from the aim of the socialization of the economy to the task of unification shaping the whole of a new life: "How to organize human life socially—that is the great question which people are asking today with ever greater insistence."[66] This is perhaps not surprising, as Neurath played an active and decisive role in *Bildungspolitik*.

From 1920 to 1925, Neurath was deeply involved in the housing movements in and around Vienna. To the art critic Franz Roh, he wrote in 1924: "Just now I am dictating letters about the redesign of workers' housing through propaganda. . . . 25,000 apartments were just built. . . . Can you send me information about graphics, color lithography, pictures, etc. concerning such worker housing?"[67] Mass accommodation not only met the immediate needs of the workers; through the realization of the idea of unity of building, it was also part of the planning of the collective life of the "new men" (*neue Menschen*).[68] Neurath stressed that the achievement of true socialism required that people transform the whole of their practical and intellectual lives, and he saw the housing project as an important part of this endeavor: "Socialist yearning is not enough, nor talent alone, nor mere Marxist thinking. The coming man demands a living connection with the social transformation."[69] With restless optimism he wrote: "Traditions go, a new day rises: new equipment, but also new homes with new comfort together with new men."[70]

In the area of education in the narrower sense, Neurath proved no less active and concerned. Not only did he teach at the newly founded Workers' University, but he also undertook, with the support of the city government, the foundation of the Museum for Housing and Town Planning, and the world-famous Social and Economic Museum of Vienna, which he directed until the tragic end of the Social Democratic regime. A central goal of the museum was to promote a unified education through a new visual language, which he dubbed ISOTYPE (International System Of TYpographical Picture Education). Neurath located the museum project in the larger context of municipal planning:[71]

So, out of the actual needs of the learners arose the Vienna method of visual education. The city of Vienna has a progressive municipal government; it has built sixty thousand tenements (to rent at two dollars a month for three rooms), hundreds of kindergartens, playgrounds, health centers, and many other things. But how, it may be asked, is it possible in any city with a democratic government to achieve so much of benefit to the masses unless the people understand what it is all about, at least in its larger outlines, and unless these enormous expenditures out of tax revenues are approved on the basis of a constant accounting to the people? Hence, general social education became a necessity for this city. It is out of this need that the Social and Economic Museum of Vienna was born.

One might note that Neurath's earlier insistence on central planning regarding the economy persisted through the museum project as well: "Only through a unified, planned, central control of all museums and educational institutions it is possible to lead the public from one museum to another with the greatest benefit to its education."[72]

Neurath's social plans of this period provide new foci for the political role of unified science, in two ways. First, science serves as a weapon against metaphysics. This idea was widely shared among the members of the Vienna Circle, yet there was a difference. Whereas for most others the motivation for fighting metaphysics was chiefly epistemological, for Neurath it was strongly political. Neurath's targets were the metaphysical doctrines of political and clerical absolutism:[73]

The cultivation of scientific, unmetaphysical thought, its application above all to social occurrences, is quite Marxist. Religious men and nationalists appeal to some feeling, they fight for entities that lie beyond mankind. To them the state is something "higher," something "holy," whereas for Marxists everything lies in the same earthly plane. The community of the state is nothing but a kind of large association, whose statutes do not possess special holiness. What may appear in the way of feelings and ideas is regarded as a piece of this order of life and is not put above it.

Though this identification of scientism with Marxism was common among Marxist intellectuals, it was not a plank in the Vienna Circle's platform. Regardless, Neurath left no doubt about it: "Many who came from the bourgeoisie are worried whether the proletariat will have some feeling for science; but what does history teach us? It is precisely the proletariat that is the bearer of science without metaphysics."[74]

Second, Neurath thought unified science would serve to enhance unity among people. Recall that the fundamental element in Neurath's unification of science was the unified language of physicalism. With a unified language, science promotes the communication and solidarity necessary for cooperation and unified action. It is worth noticing the internationalist character—common to many socialist programs—that Neurath wanted for his visual-education project:[75]

Variety within a community has to be based on some structure of a society, and therefore we have to look at the social elements which make contacts possible between human beings from group to group throughout the world and a communication of understanding from nation to nation. A certain amount of trust is needed for successful cooperation, and also the possibility of making contact through language. If there is not one common language, then languages which are partly translatable into one another are wanted.

Neurath's idea that unified science can unite people received its most succinct expression in his slogan "Metaphysical terms divide; scientific terms unite."[76] We emphasize the novelty involved in this theme. In Neurath's earlier view, unified science was politically important simply because it provided the tools that allowed and aided the execution of common actions. Now unified science, through its unified language of physicalism, would also contribute to the building of general trust among people, which is required for successful cooperation. This idea mirrors the Red Viennese doctrine that the whole of life needs to be transformed for successful socialization.

Summary and Conclusion

We began with the puzzling observation that Otto Neurath was thoroughly antireductionist while actively advocating the unity of science. We have told two stories in order to resolve that puzzle. In the first, we showed how it was possible for Neurath to hold such a seemingly self-contradictory position, by elucidating the conception of unity he held. Neurath's view of the unity of science started with a clear recognition of the existing disunity. By unity he emphatically did not mean a single system of science, which he deemed impossible; instead, what he sought was the linking of different bits of science at various points of prediction and action. To make this possible he urged an "orchestration" of the sciences that was partly to be realized in the form of the *Encyclopedia*.

In the second story, we attempted to show why Neurath was compelled to work toward unity. We did so by placing his philosophical views in the context of his life as a whole, in which politics, science, and philosophy were intimately linked. We followed Neurath as he went through the Bavarian Revolution and Red Vienna, actively participating in the socialist politics of the time. We observed how his commitments to socialism and to the unity of science grew together. In Neurath's view, socialization required that the sciences be united at the point of action, and socialization was possible because that kind of unification could be achieved. In Neurath's later life, science ceased to be a mere tool for action; rather, the unity of science formed the general basis of a worldview and a way of life that included socialism in the narrow sense but was not restricted to it. Therefore, it would not be quite correct to say Neurath wanted the unity of science simply as a means of achieving socialism. The concluding sentences of the Vienna Circle Manifesto give a more accurate impression: "We witness the spirit of the scientific world-conception penetrating in growing measure the forms of personal and public life, in education, upbringing, architecture, and the shaping of economic and social life according to rational principles. *The scientific world-conception serves life, and life receives it.*"[77]

TIMOTHY LENOIR AND CHERYL LYNN ROSS

The Naturalized History Museum

The claim of this essay is that a historically conceived semiotics is required to understand how science functions as a disunified enterprise. The disunity of science features prominently in recent discussions in science studies. Proposals of a heterogeneous and more fragmented picture in which experiment and traditions of instrumentation have lives of their own independent of the guiding hand of "high theory" have focused on the sites of knowledge production—the laboratory and the agonistic field of scientific controversy—and they have emphasized the negotiated character of science in the making along with the instrument-and-practice-laden character of modern technoscience.[1] Whereas earlier work tended to deemphasize the labor involved in creating instrumentation and in stabilizing and replicating experiment, newer accounts have insisted that the objects of scientific investigation are constructed and stabilized through instruments.[2] A closely related genre of recent studies has emphasized understanding the evidentiary context, the socially negotiated conventions and criteria for coming to local agreement about the outcome of experiments, their replication, and standards of competency, trust, and evaluation.[3] These lines of research have foregrounded the division of labor and the differential distribution and dispersion of skill essential to scientific work. Studies arguing for an enculturation model of scientific work have stressed the economies of skill, attitudes, and values that must accompany the formal elements of mathematics, physical theory, and engineering principles in scientific work.[4] In such studies, theorists no less than experimenters are depicted as practical reasoners.

One of the consequences of such research is that the smooth integration of the different aspects of science taken for granted in theory-dominated accounts has itself become an object of investigation. Stress upon the heterogeneous, disunified structure of science, the practice-and-instrument-laden character of scientific work, and enculturation has left science studies struggling to account for the ways in which the work of theorists, experimenters, and technicians is locally coordinated. No less pressing is the need for an attack on the manner in which local contexts are multiplied in order to account for the striking capacity of science to capture supposedly universal features of the world.[5] Although considerations about language have always been part of science studies in one form or another, some researchers have suggested that semiotics, another tradition stemming from linguistics, can usefully address such problems.[6]

The semiotic turn taken by recent science studies offers a promising route for addressing these issues—when it takes a particular direction.[7] The kind of semiotics that we have in mind, and attempt to illustrate below, is practiced by scholars with an interest not in essences and deep structures but in tangled and layered political and economic histories and the way they become naturalized by signs. Such studies stress both the material and the contested nature of signs, and they avoid arid formalism by insisting upon historical accidents and contingencies. The main strategy of this research is to trace the construction of meaning through the configuration of chains of signifiers linked metonymically and metaphorically, and fused together through complex narratives constructed by contentious constituencies and adapted to particular social struggles. Inscribed into the social imaginary through technologies of writing, photography, film, museum exhibits, teaching materials, and guidebooks, these are struggles to define both society and nature.

This essay focuses upon the historical construction of signs and their simultaneous political and cognitive roles. We intend to demonstrate that the power of a sign, a representation, or an interconnected set of representations to support scientific work is not merely a function of its own internal logic but also of its capacity to forge rhetorical links to representations in other domains by drawing upon metaphor as well as repertoires of tropes and narrative structures.[8] In what follows, we explore the manner in which "nature" comes to stand as the author and legitimator of socially constructed practice through such fashionings of politically and ideologically freighted images of natural order.

In the nineteenth century, professional curators and staffs of scientific specialists financed by state and municipal governments edged out of the field the gentlemanly collector of rarities who displayed treasures in a *Wunderkabinet*; public museums of natural history replaced private cabinets of curiosity. These new museums embodied a number of purposes. Certainly the urge to preserve and exhibit rarities—the motives associated with the wonder cabinets—was not altogether unfamiliar to the early natural-history curator. But, more important, these museums were intended to educate their visitors. Although written in 1958, the following discussion by Lothar Witteborg, Chief of Exhibitions at the American Museum of Natural History, epitomizes the norms followed by the directors of most natural-history museums throughout the nineteenth century:[9]

Specimens, reconstructions, and processes should be exhibited because they have the authentic power to open the visitors' eyes to the movement and meaning of the stream of life. The natural history museum should take elements from nature and from life itself along with the theories, concepts, and philosophies achieved through scientific research, and combine them all into a meaningful presentation which tells a story. Within this basic philosophy it is the job of the museum designer and exhibit specialist to arrange the material into an aesthetically pleasing exhibit.

In Witteborg's view, which follows the nineteenth-century tradition, the museum is a window onto nature, a microcosm standing in for the macrocosmic stream of life. To illustrate and to reflect nature: Witteborg assumes that within the walls of the museum elements "from life itself" are pieced together by science into a meaningful representation. But what is being represented in an exhibit? *Whose* nature is being depicted? Witteborg further urges that a good exhibit should tell a story, and an aesthetically pleasing one, at that. But what role should narrative, rhetoric, and aesthetics play in the depiction of nature with scientific tools? Most interesting is the author's claim that the museum exhibit has the "authentic power" to open a viewer's eyes to nature. What gives the representation its authenticity? The following examines two sets of events in museum history, connected with the establishment of the British

Museum of Natural History in South Kensington and the American Museum of Natural History in New York, in order to suggest some answers to these questions.[10]

Museums and Tourism: Producing Authenticity

An 1833 entry in Ralph Waldo Emerson's journal records one nineteenth-century viewer's reaction to a natural-history exhibit. Still known as a "cabinet," the Jardin des Plantes in Paris elicited this from an American tourist in Europe:[11]

I carried my ticket . . . to the Cabinet of Natural History in the Garden of Plants. How much finer things are in composition than alone. 'Tis wise in man to make Cabinets. When I was come into the Ornithological Chambers, I wished I had come only there. The fancy-coloured vests of these elegant beings make me as pensive as the hues & forms of a cabinet of shells, formerly. It is a beautiful collection & makes the visitor as calm & genial as a bridegroom. The limits of the possible are enlarged, & the real is stranger than the imaginary. . . . Ah said I this is philanthropy, wisdom, taste—to form a Cabinet of natural history. Many students were there with grammar & note book & a class of boys with their tutor from some school. Here we are impressed with the inexhaustible riches of nature. The Universe is a more amazing puzzle than ever as you glance along this bewildering series of animated forms—the hazy butterflies, the carved shells, the birds, beasts, fishes, insects, snakes—& the upheaving principle of life everywhere. . . . I am moved by strange sympathies, I say continually "I will be a naturalist."

As the presence of tutor and students attests, the exhibit is evidently held to be educational, but this quality takes on a very specific form. For Emerson, the museum exhibit is superior to the direct experience of nature. By its selection, juxtaposition, and ordering of elements ("how much finer things are in composition than alone"), the museum exhibit, though a fragment, evokes the experience of nature's meaning and variety more completely than would the thing itself. The exhibits provide not nature itself but rather icons for meditation and study. Nature, as Emerson often declared, is a text, a sign to be read most conveniently within the frame of the exhibit hall. Emerson hints at the spiritual truth he locates in the exhibit: the artificial plenitude of the cabinet witnesses to a prelapsarian fullness and variety in nature—"the limits of the possible are enlarged"—and he stands as a new Adam, with all the world before him, contemplating a restored Garden of Eden.

Emerson's thoughts on the collection betray something more than these moral and spiritual truths, however. The bridegroom of Emerson's diary who calmly gazes at the scene before him is genial in anticipation of more intrusive pleasures to come. Adam stands poised, ready to take possession of his Garden. The museum provides a vantage point, a prospect from which he can *command* nature.[12] The production of desire for a certain approach to nature is the ultimate effect of the collection Emerson views: "I am moved by strange sympathies, I say continually 'I will be a naturalist.'"

Emerson comes to Europe, then, as a tourist and a consumer of natural history. The two phenomena—tourism and natural history—are not unconnected. Indeed, the growth of foreign travel is contemporaneous with the development of public museums of natural history. Even today, museums such as the British or the American Museum of Natural History are major tourist attractions. At a deeper level, too, museums and tourism are connected: they both signify the "authentic."

If we consider further, the production of authenticity through signs is as true of tourist sites as it is of museums: authenticity in touristic experience is not simply there for the taking. In discussing the "semiotics of tourism," Jonathan Culler argues that the reflective tourist learns there is no unmediated, original experience of another culture.[13] Though any spot in London could read "London" by virtue of its mere location in that city, not all such sites are notable as tourist attractions. What the tourist in search of the true London experience encounters is a site turned into a sight: a place *marked* as authentic—as "worth a visit" or "worth a detour"—by previous travelers. The "real thing," the authentic, must be marked as real and sightworthy; some sites are, in fact, made more representative than others by the very labor of marking them. Culler points out that reproductions and representations—"markers" in the form of plaques, souvenirs, postcards, guidebooks, and videotapes—*create* the original. In short, without the markers there is simply no "there" there. In the case of tourism, the existence of markers is what makes the thing marked a recognizably original and therefore real thing.

Tourism, in effect, makes a place into a museum, its markers framing the sights that deserve notice as if placed in an exhibit hall. And just as in an exhibit hall, the aim is to have the viewer look past the marker and see only the thing marked. For the touristic experience to be authentic, it should be perceived to be unmarked: it should be found "off the beaten track" or remain "unspoiled." But even if the tourist explores such a place, its authenticity does not result from the absence of markers; the markers are simply of a less obvious, more sophisticated sort. The "dilemma of authenticity," as Culler observes, is that authentic sights require markers to be recognized as authentic, but our notion of the authentic is the unmarked or unmediated: "We want our souvenirs to be labeled 'authentic native crafts produced by certified natives using guaranteed original materials and archaic techniques' (rather than, say, 'Made in Taiwan'), but such markers are put there for tourists, to certify touristic objects."[14] Ideally, then, markers should remain silent, not calling attention to themselves as such.

An example from the early work of semiotician Roland Barthes can help explain the silencing characteristic of markers or, in his terms, signifiers. In a barbershop, Barthes is handed a copy of *Paris-Match* featuring a cover photo of a black soldier saluting an unpictured tricolor. Barthes explicates this sign, in the context of its association with a popular magazine and its placement in the everyday space of a barbershop (rather than, say, at a political rally), as indicating the greatness of the French empire: all her sons serve faithfully under her flag without any racial discrimination.[15]

In the construction of this signifier, wide and complex histories, including the biographies of both the soldier and the empire he serves, have been silenced and turned into natural states. Even more: once this transformation "from history to nature" has been accomplished,[16] the signified becomes a reference that *establishes* French imperialism; the sign of the photograph "naturally" evokes the concept. The signifier thus gives foundation to the signified; it helps to produce and reproduce it through naturalization. In the case of natural-history museums, such signifying practices amount to the very production of nature: museums produce nature with their storage rooms, laboratories, and staffs of taxidermists, artists, and curators. And they produce it in the light of specific interests. To analyze and deconstruct the semiotics of this kind of museum is to account for the naturalization of the history of nature-production—a history involving, among other things, politics and economics.[17]

Richard Owen and the Invention of the Dinosaurs

The controversies connected with the construction of natural-history displays for the 1854 Crystal Palace exhibition at Sydenham and the design of the British Museum of Natural History provide a glimpse into the processes at work in the naturalizing of natural history. The exhibits of interest here were reconstructions of fierce-looking dinosaurs, built by Benjamin Hawkins under the direction of Richard Owen. As Adrian Desmond has shown, the dinosaurs as an order were literally invented in the years 1838–41 by Owen, the "Cuvier of British Comparative Anatomists" and director of the London Zoological Museum at the time.[18] The story revolves around the notion of evolution—but not Darwinian evolution. Owen's defensive invention of the dinosaur took place during the late 1830's and 1840's, when the threat was not Darwin's theory but Lamarck's notion of evolution—much more dangerous in its association with radical social and religious reform.

The period had many radical sects; the social and political uses made of science by one leading educational reformer, Robert Owen (no relation to Richard), illustrate the kind of threat Richard Owen's dinosaurs were armed to combat. Crucial to Robert Owen's rationalist, progressive reforms was a belief in the perfectibility of humankind and the self-organizing power of matter according to natural laws, joined to a faith in the environment as a determinant of form and character. Owenites believed that through the appropriate social and material environment, humanity's spiritual qualities could be molded as a prelude to political change.

Lamarck's evolutionary theory provided a useful and necessary supplement to this environmentalist position. In his *Philosophie zoologique* of 1809, Lamarck had argued for an ascending scale of animals increasing in complexity without gaps or breaks between each level. Lamarck maintained that organic matter and individual animals have innate drives to complexity, so that each generation will produce a successor generation infinitesimally more complex than its progenitor. The increase in complexity is made possible by the inheritance of acquired characters. According to Lamarck, changes in the environment—in climate, nutrition, or social organization—induce new habits in the organism. The nervous fluid stimulates the development of new structures to accommodate these new habits.

Relying on Lamarck's theories, newly available in an 1831 English edition, Owenites projected a cooperative society in which egalitarianism, female emancipation, secularization, and educational institutions called "halls of science" would be central. Lamarckian biology, with its emphasis on progressive transmutation, supported the Owenite socialists in their accounts of human progress from barbarism to civilization.[19]

The social consequences of Owenite views were apparent to the Anglican authorities. Opponents of Lamarckism in the 1830's attacked the central premise that nature had inherent powers of organization and argued that social and natural change emanated from above. For the Anglican hierarchy, judicial and natural law were divinely sanctioned. Richard Owen's "Report on British Fossil Reptiles," a speech delivered in two parts to the British Association for the Advancement of Science in 1839 and in 1841, in which he coined the name "dinosaur," was informed by the concerns of the Anglican dons who had taken control of the BAAS in the late 1830's.[20] The BAAS supported Owen's work with ample funding in 1838 and awarded additional funds upon its completion in 1841.

Before unveiling his dinosaurs, Owen demolished the Lamarckian transformationist position represented by Etienne Geoffroy Saint-Hilaire and the materialist Robert Grant, who had claimed the gigantic saurians—ichthyosaurs, plesiosaurs, and teleosaurs—had been transformed in a series of evolutionary steps from Crocodilians. Owen used the stratigraphical record to show that the ichthyosaur retained its character unchanged throughout the immense succession of Mesozoic strata, and that it emerged in and disappeared from the strata suddenly. Moreover, throughout these expanses of time the species of teleosaurs, ichthyosaurs, and plesiosaurs had remained completely distinct. Even worse for the Lamarckian transmutationist camp, teleosaurs, supposedly the highest forms in the reptilian group, had disappeared in the Oolitic (our Jurassic) era, before the lower ichthyosaurs and plesiosaurs, which both persisted into the Cretaceous era. The Lamarckian interpretation did not hold, Owen argued, for "if the present species

of animals had resulted from progressive development and transmutation of former species, each class ought now to present its
typical characters under their highest recognized conditions of
organization."[21]

Owen's final onslaught came in unveiling his analysis of his new
creatures, recreated from fossil evidence: the iguanodon and the
megalosaurus. Among the anatomical peculiarities of these species
was their possession of five fused vertebrae welded to the pelvic
girdle, ribs, and extremities with hollow long bones, which "more
or less resemble those of the heavy pachydermal Mammals, and
attest to the terrestrial habits of the species."[22] By 1868, T. H.
Huxley, looking for "missing links" between classes of animals
in defense of Darwin's theory of macroevolution, would focus on
the three-toed character of the fossil hind foot of the iguanodon,
which, he would argue, was bipedal, an evolutionary forerunner
of the birds. But Owen wanted to make the dinosaurs proto-
mammals: "prophetic types" of mammals placed by the Creator on
the earth as precursors, the relation between the two analogous to
the typological relation between the Old and the New Testaments.
Indeed Owen's dinosaurs looked something like rhinoceroses and
elephants. He made this last step by arguing even more tenuously,
speculating on physiological and ecological grounds that since dinosaurs have the same thoracic structure as crocodiles, their degenerate relatives, they may be assumed to have had four-chambered
hearts like the Crocodilians, "and enjoyed the function of such a
highly organized center of circulation in a degree more nearly
approaching that which now characterises the warm-blooded Ver-
tebrata."[23]

Owen was lionized for his achievement by the Anglican scientific clergy, receiving two hundred of the three hundred pounds
available from the Civil List for 1842—in addition to the support
he received from the BAAS. Roderick Murchison praised Owen
for having completed "Cuvier's temple of nature" with English
building materials, while Lord Francis Egerton marveled that
Owen's fossil animals were "pregnant with the proofs of wisdom
and omnipotence in their common Creator."[24] Owen's dinosaurs
had, at least temporarily, vanquished the Lamarckian evolutionists
from the scientific field. Furthermore, at the Crystal Palace exhibition of 1854, these creatures stalked, long-legged, into the fray to

reveal God's order in the popular mind as well. Heralded by ominous warnings in the press about invading monsters, the life-size restorations at the Sydenham Crystal Palace gardens, installed on islands in an artificial lake, formed a hugely popular attraction (Figs. 1, 2). Opened by Queen Victoria to an audience of forty thousand, the exhibit was drawing trainloads of spectators within a few months. These enormous markers pointed to a mythic nature; though their originals were absent, invisible, nonexistent—in fact never existent—the markers naturalized a historically constituted order.

Owen's guidebook to the exhibits, available to visitors for threepence, served as an additional term in this signifying system. In a semiotic regression, the guidebook—a marker directing tourists to the sightworthiness of the Crystal Palace attraction—helped to authenticate the dinosaur exhibits, which themselves marked an absent "nature." The exhibits were massive, distant, frightening in the controlled way that a roller coaster at an amusement park is frightening: offering no real danger, just a tantalizing thrill. The printed book that guided tourists through the exhibit functioned as a marker for these markers; where the exhibits' surprising surfaces prevented the distant viewer from penetrating their mysteries, the guidebook—ostensibly only a retrospective description of an antecedent reality—in fact offered authenticating depth, featuring diagrams of the skeletal structures of the creatures and positioning within these illustrations the fossil origins of the reconstructions (Fig. 3).

These illustrations remind us of a crucial aspect of the reconstructions: befitting the original Crystal Palace exhibition's emphasis on products of industry, they were entirely manufactured, not, in a strict sense, *reconstructions*, but constructions of Owen, Hawkins, and others. One thirty-five-foot iguanodon was fashioned from four thick iron columns, 600 bricks, 650 drain tiles, 900 plain tiles, thirty-eight casks of cement, ninety-eight casks of broken stone, and one hundred feet of iron hooping.[25] *There was no fossil trace on view anywhere in the exhibit.* The guidebook illustration suggests that the fossil was imbedded in the interior of the construction, but in fact the massive object was hollow inside, suitable for dinner parties (Fig. 4).[26] How in the face of this evidentiary void was the authenticity of the reconstructions to be established? The

Fig. 1. Richard Owen's Iguanodon, exhibited in the Sydenham Crystal Palace gardens. From Adrian J. Desmond, *The Hot-Blooded Dinosaurs: A Revolution in Palaeontology* (New York: Dial Press/James Wade, 1976), p. 23, fig. 6.

Fig. 2. Richard Owen's Megalosaurus. From Desmond, p. 24, fig. 7.

problem helps explain the guidebook's obsession with origins. In offering signs of the extinct past, Owen continually gestures toward source and origin, the supposed signified behind the signifier. The pamphlet opens: "Before entering upon a description of the restorations of the Extinct Animals, placed upon the Geological Islands in the great Lake, a brief account may be premised of the principles and procedures adopted in carrying out this attempt to present a view of part of the animal creation of former periods in the earth's history."[27] Before the beginning of the tour, then, are inserted the specialized theory and practices that generate and authorize the exhibits. The authority of science is summoned to authenticate the constructions, and in the naturalizing process science itself is reciprocally endorsed. Moreover, the guidebook follows a revealing pattern in introducing each of the exhibits: first comes a detailed account of the origin of the fossil upon which the constructed dinosaur is based, complete with the name of its discoverer and the place it was found, as well as any later accidents that befell it, followed by its present location in a museum or other collection. This written account invariably includes another gesture toward origins: it gives the etymology, in ancient Greek, of the recently coined name of the extinct species.[28] Within this collection of refer-

Fig. 3. Illustration of a Megalosaurus from Richard Owen's guidebook to the Crystal Palace gardens. From Owen, *Geology and Inhabitants of the Ancient World* (London: Crystal Palace Library, 1854), p. 20, fig. 7.

ences to origins of different kinds—what one might call the rhetoric of authentication—Owen's quotations of *Paradise Lost*, the major English poem of primordial origins, find a place. Gillian Beer has argued that Milton's poem, one of Darwin's favorite books and common cultural currency among the Victorian educated classes, provided Darwin with an imaginative resource to conceive, stabilize, and naturalize his evolutionary arguments within a reassuring symbolic context of continuity and tradition.[29] Judging from the poem's appearance in Owen's pamphlet, such deployment of Milton's text and its associations was not unprecedented.

Constructing the New Museum

The Crystal Palace dinosaur exhibit was not the last of Owen's efforts to offer the public such prospects of restored nature. A few years later, he participated directly in the design of the British Museum of Natural History in South Kensington, a project he proposed to his friend Prime Minister William Gladstone. During most of his curatorial career, Owen's collection was housed in the inadequate, cramped space of the British Museum. Public debate on the creation of a new museum, purposely built for natural-history displays, extended over several years; Owen entered the discussion with a pamphlet entitled *On the Extent and Aims of a*

National Museum of Natural History, written in 1862 to promote the project. In this proposal, Owen declares that in an ideal world, perfect knowledge of nature would be acquired by direct, unmediated experience,[30]

even such as was the Paradise in which Adam, as sung by our great poet,

> Beheld each bird and beast
> Approaching two and two; these cowering low
> With blandishment, each bird stooped on his wing.
> He named them as they passed and understood
> Their nature; with such knowledge God endued
> His sudden apprehension.

But Milton's account of Adam's understanding of nature applies only before the Fall. Owen's postlapsarian world becomes "a grand Natural Museum" solely "to the loving eyes of the geological, botanical, and zoological observer";[31] as with Emerson's desiring consumer of natural-history exhibits, the gap in understanding must be filled by desire, produced by a scientific approach to nature. Owen continues:[32]

Fig. 4. Contemporary illustration of a dinner party inside the Iguanodon. From Desmond, p. 22, fig. 5.

Under other and harder conditions we strive to regain that knowledge, needing, and urgently seeking for, every collateral aid in the struggle to acquire that most precious commodity—the truth as it is in Nature, and as manifested by the works of God.

The chief collateral aid Owen advocates is a national museum of natural history. The material world, restored by scientific understanding, would reveal the wisdom and purposes of God. In Owen's view, therefore, a natural-history museum should display and make evident the divine rationality of Creation. Very often, the objects for which Owen wants adequate display room are huge: whales, elephants, rhinoceroses, dinosaurs. As he argues for increased space for his collection—45,000 square meters, or five acres, would just about suffice—his interest as the likely director of this grand institution may seem obvious; he protests, however, that "the larger the Museum, the greater the cares of the Curator."[33] Instead, he continually emphasizes the need to appeal to a wide, popular audience through the sheer size of the exhibits: to shift initial affective responses of awe and wonder to appreciation; to replace the naive sensibility with a cultivated one that would resemble Adam's original understanding of all Creation. In this effort, he ventures, the dinosaur exhibits are irreplaceable:[34]

It is the common experience of officers of National Museums that no specimens of Natural History so much excite the interest and wonder of the public, so sensibly gratify their curiosity, are the subjects of such prolonged and profound contemplation, as these reconstructed skeletons of large extinct animals.

Fortunately, such important exhibits are also economical:[35]

A fossil bone and a coloured plaster-cast of it are not distinguishable at first sight, scarcely by sight at all. The artificial junction of a series of casts of the bones of an unique fossil skeleton, produces a result equivalent, for all the purposes of public exhibition, to the articulated skeleton itself. Thus every capital in Europe, the Public Museum of each civilised community, may show to the people the proportion of the creatures of former worlds that science has so restored.

Fossil dinosaur exhibits, in short, are a curator's dream. Unlike such rarities of nature as gemstones, they can be mechanically reproduced from a single fossil and still retain the trace of the original that seems to authenticate them. The potentially unlimited possibility for multiplication of these signs allows Owen to imagine, with an entrepreneurial air, reciprocal interchange of extinct animals such as his own reconstructions, resulting in "co-adjusted frameworks" among international museum partners. Given the wealth of his own collection, supplied by generous grants from Parliament,[36] Britain's position as a potential trading partner could be a strong selling point in a bid for an enormous museum.

The goal of such a museum, Owen insists, is not to gratify vulgar tastes for size and strangeness but to inculcate "an appreciation of the perfect fitness of the thing to its function" through setting forth "the extent and variety of the Creative Power, with the sole rational aim of imparting and diffusing that knowledge which begets the right spirit in which all Nature should be viewed."[37] Later in his proposal, Owen expands upon this theme: "The most elaborate and beautiful of created things—those manifesting life—have much to teach, much that comes home to the business of man, and to the highest element of his moral nature."[38] Owen believes that "contemplating the extent, variety, beauty, and perfection of Creation Power" in a museum that completely epitomizes nature will have a "humanizing and ameliorating effect" upon "the people of a busy and populous nation."[39]

This result can best be achieved through an ongoing educational effort on the part of the museum curators in designing exhibits, lecturing to the public, and the like. Central to Owen's goal is the spatial organization and architecture of the museum itself: he proposes a central rotunda of glass and iron that would[40]

serve for the reception of an Elementary Collection illustrating the characters of the Provinces, Classes, Orders, and Genera of the Animal Kingdom, and also for the Collections of the Natural History of the British Isles. An Exhibition-room of a circular form is that which admits of the most effective and economic supervision; and the series of specimens there proposed to be displayed are of a nature that would be most profitably shown to, and studied by, the wage-classes after the hours of work.

The classes upon whom the museum's panopticon-like ameliorating effect might be most valuable are here explicitly named. But Owen hopes that the proposed museum will serve others as well:[41]

The local collector of birds, bird-eggs, shells, insects, fossils, &c.,—the intelligent wageman, tradesman, or professional man, whose tastes may lead him to devote his modicum of leisure to the pursuit of a particular branch of Natural History,—expects or hopes to find, and ought to find, the help and information for which he visits the galleries of a Public Museum.

A national museum, then, would serve as a meeting point for a range of social classes, from low to high bourgeois. In focusing on the natural history of Britain, it would promote national awareness and pride. The representation of Great Britain to itself—with an insistence on its power and wealth—emerges as a resounding note in Owen's pamphlet:[42]

The greatest commercial and colonizing empire of the world can take her own befitting course for ennobling herself with that material symbol of advance in the march of civilisation which a Public Museum of Natural History embodies, and for effecting which her resources and command of the world give her peculiar advantages and facilities.

"Resources and command of the world," commerce and colonization: the register of discourse in which natural history was embedded from the beginning of the nineteenth century is here explicitly named. For the British, as for the French and Dutch, the period between 1780 and 1830 saw massive expansion both of empire and of museum collections. Contemporaries were quite clear about the importance of colonial and military conquests for their collections. The size of a nation's museum was an index of its colonial power. The political might and extent of the nation empowered the eye that ranged over the framed view of nature.

Owen was eventually awarded his museum, though the realized edifice differed somewhat from his plan. In the envisioned design for the building, intended as a sanctuary "revealing the advances of the science,"[43] Owen and architect Francis Fowkes used Romanesque elements to suggest a huge domed cathedral (Fig. 5),[44] a place to worship God through nature. When various constraints forced the abandonment of Owen's beloved rotunda, in its place Alfred Waterhouse, who took over as architect after Fowkes died in 1868, designed a rectangular cathedral nave along the lines of German Romanesque churches, with triforia above and arcades below, opening into a series of side chapels designed to enshrine the Elementary Collection. Overhead Waterhouse replaced the Romanesque vault with a roof of glass and iron, unmistakably recalling the architecture of a Victorian railway terminal (Fig. 6). To reach the

Fig. 5. Francis Fowkes's plan for the British Museum of Natural History. From Mark Girouard, *Alfred Waterhouse and the Natural History Museum* (New Haven: Yale University Press, 1981), p. 30. Reprinted with permission from the Natural History Museum, London.

upper galleries from either side, Waterhouse inserted a spectacular bridge (Fig. 7), reminiscent of one of the great railway trestles of the British Industrial Revolution. The building thus synthesized the major European public architectural modes, both sacred and secular, to celebrate the march of national progress.

The outside of the building contributed to the educational function of the collections it housed. Covered entirely in ornamented terra-cotta, the facade's rich decoration featured, in place of Gothic gargoyles, models of monkeys, dodoes, and reconstructed animals such as griffinesque saber-toothed tigers and sphinxlike pterodactyls (Fig. 8). Avoiding any suggestion of Lamarckian transmutation, all were arranged in carefully separated niches rising up the walls of the museum, starting with marine life at the bottom and reaching to the parapet crowning the main entrance, over which were originally intended to stand statues of Adam and Eve. Perhaps as a corrective to Owenite feminism, Eve was eventually dropped from the design. Adam remained alone atop Creation.

Fig. 6. The museum's glass and iron vault. From Girouard, p. 30. Reprinted with permission from the Natural History Museum, London.

Fig. 7. Bridge to the upper galleries of the museum. From Girouard, p. 38. Reprinted with permission from the Natural History Museum, London.

Cro-Magnon and the Modern Threat

Our purpose in examining the relationship between important points in the history of evolutionary theories and the museums in which these theories were constructed has been to interrogate the claim we started with, Witteborg's notion of the natural-history museum as a site of authentic representation of nature. For our final example, we shall turn to Witteborg's own institution, the American Museum of Natural History in New York City. The exhibits in

Fig. 8. Terra-cotta ornamentation covering the facade of the museum. From Girouard, p. 19. Reprinted with permission from the Natural History Museum, London.

this museum were constructed by palaeontologists, taxidermists, and artists who accepted evolution—though not Lamarckian or Darwinian evolution. The people who built this museum had a different set of motivations for their work.

Donna Haraway and Ronald Rainger have provided detailed studies of some aspects of the construction of the American Museum and the interests it was intended to serve.[45] The founders of the museum, which was incorporated in 1868, included the father of Theodore Roosevelt, Henry Fairfield Osborn, J. P. Morgan, William K. Vanderbilt, Madison Grant, and John D. Rockefeller III, all philanthropists and supporters of science in the Progressive era, conservationists, and proponents of "the rational management of capitalist society."[46] The function of the museum they endowed was, not surprisingly, to educate. Its educational message was literally inscribed in the museum's monument to Theodore Roosevelt, the president of the most powerful nation of the industrial age, a man who had found himself as a youth and constantly renewed himself as an adult through combat with nature. Roosevelt's conservationism issued in part from his concern over the potential for decadence in modern urban civilization. His museum monument, though in the middle of the largest city in America, edged Central Park, a modern Garden of Eden constructed to restore health to degenerate city dwellers and to infuse manly vigor into pallid metropolitan youth.[47]

Visitors to the museum were guided, by person or pamphlet, through a series of exhibits leading up an evolutionary chain, culminating in the Hall of the Age of Man. The Hall of the Age of Man was designed by Henry Fairfield Osborn, the president of the museum from 1908 to 1933. His guidebook to the hall emphasizes that most of the exhibits, like Owen's dinosaurs, are reconstructions. Again scientific knowledge, which allows the interpretation of such mysterious fragments, is resoundingly endorsed while social interests are silently promoted. The exhibit was governed by Osborn's view that anthropoid evolution had nothing to do with human evolution. The lines linking different specimens reflected Osborn's interpretation that descent from a common ancestor had yielded separate lineages for *Pithecanthropus*, Piltdown, Neanderthal, Cro-Magnon, and modern man. Highlighted among the hall's display cases were busts of the different Palaeolithic races of humankind (Fig. 9). Osborn believed these various races had been unmistakably human, showing no affinity to the apes—though they had been quite distinct from each other. On the basis of Osborn's views, J. Howard McGregor, Osborn's former student and colleague at Columbia University, reconstructed Neanderthal as a heavy-browed, dull-witted-looking creature that stood in marked contrast to the intellectual, high-browed restoration of Cro-Magnon.

These restorations were accompanied by three murals, painted by Charles R. Knight under Osborn's direction, representing distinct stages of human cultural evolution. The first was entitled "The Neanderthal Flint Workers" (Fig. 10). The painting portrayed the members of a Neanderthal family in what Osborn defined as a typical setting. In addition to a mother and child, the painting showed three generations of males engaged in crude flint-making activities. The family is posed in front of a cave; in line with Osborn's demands there is no evidence of human-built shelters or other structures that might suggest an advanced stage of cultural evolution. Following Osborn's instructions, Knight presented the Neanderthal family "to show the characters of the race which differed widely from any existing or modern human type."[48] The painting emphasized what became the classical view of the Neanderthal: a slouching, stocky, unintelligent race with brutish physical features.

In contrast, the mural illustrating Cro-Magnon showed a group of upright, delicately featured people engaged in comparatively refined activities (Fig. 11). In an amusing bit of self-referentiality, Knight depicted the central Cro-Magnon as an artist, drawing on nature for inspiration as well as food, using natural materials for personal adornment (he wears an animal-tooth necklace) as well as for artistic expression (his palette is an animal's shoulder blade). Evidence of intellectual advancement abounds: some figures hold lamps placed in hollowed stones (which, dramatically, provide the only source of light in the painting) while another grinds pigment for the cave murals. The Cro-Magnon group is clearly culturally superior to the Neanderthals.

The last mural of the series, entitled "Neolithic Stag Hunters"

Fig. 9. Display case in the Hall of the Age of Man at the American Museum of Natural History, exhibiting Cro-Magnon reconstructed busts, skeletal fragments, and descriptive information. Neg. no. 310882, courtesy Department of Library Services, American Museum of Natural History.

Fig. 10. Charles Knight's mural in the Hall of the Age of Man, representing Neanderthal flint workers. Neg. no. 39441A, courtesy Department of Library Services, American Museum of Natural History.

(Fig. 12), depicted a northern European tribe, with exceptionally muscular bodies, after a successful hunt. The leader exults over a fallen deer, while his blond son holds a wolf-dog on a leash. In the guidebook, Osborn wrote of the Neolithic race:[49]

This race was courageous, warlike, hearty, but of a lower intelligence and less artistic skill than the Cro-Magnons; in a rigorous northern climate, it was chiefly concerned with the struggle for existence, in which the qualities of endurance, tribal loyalty, and the rudiments of family life were being cultivated.

The mural features robust natural man in combat with the elements of nature. In books such as *Men of the Old Stone Age*, Osborn praised the intellect, talent, and strength of earlier races, in particular the Cro-Magnon and the Nordic race of Neolithic times, which resulted from their closeness to nature. In the foreword of his 1924 annual report, entitled "The American Museum and Education," Osborn explicitly locates unparalleled educational value in such direct experience of nature:[50]

The cave boy certainly had advantages which our boys have ceased to enjoy; he was surrounded on all sides by vibrant nature, full of inspiring and wonderful phenomena, which filled him with reverence and awe if not with superstition. . . . In the growth of our large cities, in the press, and in the minds of teachers who depend upon the press, civilization has reared a Frankenstein which shuts out the direct vision and inspiration of nature and banishes the struggle for existence.

In fact, contact with nature that in turn inspires its representation, present in the details of the Cro-Magnon mural, was a major motif throughout the entire museum. The museum itself, in Osborn's vision, was to serve as the missing link in the educational experience of modern city youth: "The great function of the American Museum is . . . to restore the vision and inspiration of Nature, as well as the compelling force of the struggle for existence in education."[51]

Taken together, in their insistence on the inspiring qualities of nature and the improving stresses of the struggle for existence, Knight's Cro-Magnon and Neolithic murals represent the ideal effect Osborn hoped his museum would have on its schoolboy visitors. But even though they embodied human strength and nobility in direct confrontation with nature, the Neolithic hunters nevertheless represented a decline from the more intelligent, highly cultured Cro-Magnon artists, whom Osborn elsewhere terms "the Paleolithic Greeks."[52] A lesson too could be learned from this representation of an early stage of human civilization: decline from a high state was always possible. In various dioramas and informa-

Fig. 11. Charles Knight's mural illustrating Cro-Magnon artists. Neg. no. 322602, courtesy Department of Library Services, American Museum of Natural History.

Fig. 12. Charles Knight's mural of Neolithic stag hunters. Neg. no. 37952, courtesy Department of Library Services, American Museum of Natural History.

tional exhibits in the Hall of Mammals, the museum space just before the Hall of the Age of Man, the message reverberated: Osborn showed the extinction of impressive animals such as the titanothere and the Irish elk as a result of overspecialization.[53] Osborn's published writing provides a gloss on these ominous signs of the specter of decline. He frequently urges the analogy between prehistory and history, as in the concluding passage of his *Men of the Old Stone Age:*[54]

The rise and fall of cultures and of industries, which is at this very day the outstanding feature of the history of western Europe, was fully typified in the very ancient contests with stone weapons which were waged along the borders of the Somme, the Marne, the Seine, and the Danube. No doubt, each invasion, each conquest, each substitution of an industry or a culture had within it the impelling contest of the spirit and will of man, the intelligence directing various industrial and warlike implements, the superiority either of force or of mind.

The ringing oppositions—rise and fall, spirit and will, force and mind—belong to a different order of prose than a palaeontologist's dispassionate discussions of fossil evidence. This rhetorical turn appears in Osborn's work when the past is being naturalized in the service of present interests. It is characteristic of his prefaces, written in 1916 and 1917, to *The Passing of the Great Race*, a study by his friend Madison Grant, chair of the New York Zoological Society as well as founder and trustee of the American Museum. Osborn summons the crisis rhetoric of the World War to endorse Grant's research:[55]

European history has been written in terms of nationality and of language, but never before in terms of race; yet race has played a far larger part than either language or nationality in moulding the destinies of men. . . . The Anglo-Saxon branch of the Nordic race is again showing itself to be that upon which the nation must chiefly depend for leadership, for courage . . . for self-sacrifice and devotion to an ideal. . . . In no other human stock which has come to this country is there displayed the unanimity of heart, mind and action which is now being displayed by the descendants of the blue-eyed, fair-haired peoples of the north of Europe. . . . We shall save democracy only when democracy discovers its own aristocracy as in the days when our Republic was founded.

The theme of Grant's work, to which Osborn here lends his powerful support, is that the "insidious victories arising from the crossing of two diverse races" have had more permanent results than all the "spectacular conquests and invasions of history."[56] America's own inherently superior Nordic peoples must protect themselves against the dilution of their hereditary qualities by a melting-pot ideology. Grant offers the Cro-Magnons as an object lesson of "the replacement of a very superior race by an inferior one" and warns:[57]

There is great danger of a similar replacement of a higher by a lower type here in America unless the native American [sic] uses his superior intelligence to protect himself and his children from competition with intrusive peoples drained from the lowest races of eastern Europe and western Asia.

Osborn shared Grant's view that the Cro-Magnons were a notably pure and superior race who conquered rather than interbred with the declining and inferior Neanderthals.[58] The markers of his Hall of the Age of Man portrayed the need for preserving natural and racial purity, jeopardized by war and immigration. Immediate action was required to halt the rapid acceleration toward decline, racial suicide, and extinction.

What kind of action might counter this degeneration? On one level, the museum itself provided an educational site to promote Osborn's program. But, ironically, Osborn believed in fundamental limitations in this process:[59]

In the United States we are slowly waking to the consciousness that education and environment do not fundamentally alter racial values. We are engaged in a serious struggle to maintain our historic republican institutions through barring the entrance of those who are unfit to share the duties and responsibilities of our well-founded government. The true spirit of American democracy that *all men are born with equal rights and duties* has been confused with the political sophistry that *all men are born with equal character and ability to govern themselves and others,* and with the educational sophistry that education and environment will offset the handicap of heredity.

With these words, Osborn opened the Second International Eugenics Congress, over which he presided, in 1921. Specially financed by J. P. Morgan, the opening of the Hall of the Age of Man coincided with that congress; the exhibit was specially adapted to emphasize eugenical issues, and on the floor below the Hall of the Age of Man the temporary exhibits of the congress itself were displayed. With this conjunction, the museum became the site of a powerful constellation of interests bent on exercising social control in order to preserve the national human resources that had made America great: its priceless gene pool.

Conclusion

Since their appearance in the nineteenth century, natural-history museums have provided icons for meditating on nature as well as laboratories and factories for producing nature. They have functioned as cathedrals for worshiping divine purposes and as sites for shaping society. In examining these moments in the history of museum making, we have intended to question the notion of a natural-history museum as a site for the "authentic" representation of nature. We have advanced a different argument: museums provide semiotic markers of nature whose authenticity is guaranteed by having the historical processes that produce them naturalized. Our examples of Owen and Osborn could be enlarged with a treatment of, among others, Cuvier and his Paris museum. Emerson, the Romantic idealist who enjoyed both the Jardin des Plantes on an early tour of Europe and Owen's public lectures on physiology during a later return visit, was fond of the observation that "from whatever side we look at Nature we seem to be exploring the figure of a disguised man."[60] We agree completely.

JOSEPH ROUSE

Beyond Epistemic Sovereignty

Recent work in science studies has often made explicit connections between its conceptions of knowledge and of political practice. For an especially clear example, consider how Shapin and Schaffer describe their work in *Leviathan and the Air-Pump*: "Solutions to the problem of knowledge are solutions to the problem of social order. That is why the materials in this book are contributions to political history as well as to the history of science and philosophy."[1] Yet much of the tradition in twentieth-century philosophy of science has also manifested parallels between accounts of the legitimation of scientific knowledge and conceptions of political order. Karl Popper modeled his notion of the open society upon the critical rationality that he took to demarcate scientific inquiry from other practices.[2] Paul Feyerabend emphasized the indebtedness of his objections to methodological constraints upon inquiry to John Stuart Mill's defense of liberty.[3]

Some of the most prominent defenders of scientific realism have been affiliated with Marxism in ways that have informed their philosophy of science.[4] Thomas Kuhn's account of normal science, although it does not explore such connections, merits close comparison with communitarian conceptions of political justification.[5] Even the logical positivism of the Vienna Circle had strong affiliations between its conception of a unified science and the rejection of nationalist, religious, or historicist conservatism in politics.[6] As Jordi Cat, Hasok Chang, and Nancy Cartwright argued, this affiliation was especially strong in the work of Otto Neurath: "The unification of the sciences and the socialization of the economy formed one task to be performed for Neurath. . . . There is no possibility of constructing a single, abstract, unified science. But the sciences can be unified at each point of commonly planned action, and there they should be unified, to build a socialist economy that works."[7]

The prevalence of these connections between philosophical reflection upon science and upon politics provides a useful context for my project in this essay. I will propose an alternative conception of scientific knowledge, one that is modeled upon a specific critical response to the liberal and Marxist traditions in political philosophy. Michel Foucault on several occasions framed his discussions of power in the 1970's as an attempt to escape the preoccupation of political thought with questions of sovereignty.[8] I will suggest that much of the tradition in philosophy (and sociology) of science has been similarly preoccupied with a parallel problematic, which I call "epistemic sovereignty." This parallel is important, I shall argue, because attempting to understand scientific knowledge without framing it by a conception of epistemic sovereignty offers an attractive and informative approach to the philosophy of science. What is distinctive about this approach is that it offers a *dynamic* understanding of scientific knowledge. This approach may be especially attractive in providing a clear way out of the dilemma that poses epistemological relativism as the only alternative to a general defense of the rationality of science.[9]

This essay is in three parts. The first introduces the notion of epistemic sovereignty as a parallel to Foucault's conception of the problematic of sovereignty as the framework for liberal and Marxist theorizing about power. The second presents a dynamic conception of scientific knowledge, a counterpart to Foucault's proposed analytics of power. The final part shows how such a dynamic conception of scientific knowledge addresses questions of epistemic justification, and why this approach does not entail an objectionable epistemological relativism.

Epistemic Sovereignty

At several points, Foucault situated his own reflections on power within the tradition of political theory in terms of a challenge to the tradition's orientation toward the problem of sovereign power—for example: "At bottom, despite the differences in epochs and objectives, the representation of power has remained under the spell of monarchy. In political thought and analysis, we still have not cut off the head of the king."[10] Foucault notes that in order to understand sovereignty as a political problematic, one must look to the origins of the European monarchies. Although modern political theory posits the questioning of legitimating a sovereign political power as the "original" political problem, Foucault reminds us that the actual role of the modern state, initially located in the person of the sovereign, presupposed a complex prior network of power relations. It is worth quoting his text at some length, for I shall have occasion to refer back to it:[11]

The great institutions of power that developed in the Middle Ages— monarchy, the state with its apparatus—rose up on the basis of a multiplicity of prior powers, and to a certain extent in opposition to them: dense, entangled, conflicting powers, powers tied to the direct or indirect dominion over the land, to the possession of arms, to serfdom, to bonds of suzerainty and vassalage. If these institutions were able to implant themselves, if, by profiting from a whole series of tactical alliances, they were able to gain acceptance, this was because they presented themselves as agencies of regulation, arbitration, and demarcation, as a way of introducing order in the midst of these powers, of establishing a principle that would temper them and distribute them according to boundaries and a fixed hierarchy. Faced with a myriad of clashing forces, these great forms of power functioned as a principle of right that transcended all the heterogeneous claims, manifesting the triple distinction of forming a unitary regime, of identifying its will with the law, and of acting through mechanisms of interdiction and sanction.

The sovereign was a unifying agent, standing above the various conflicting powers as their impartial referee, and guarantor and protector of legitimacy in the form of law, to be enforced against those subordinate powers that overstepped its bounds. Of course, as Foucault suggests, this was a promise that no actual monarch could fulfill, since the monarchy was itself a player in the power struggles it supposedly stood above as neutral arbiter. Hence, the subsequent critique of the monarchy in political theory deployed this conception of the sovereign's role against its nominal occupant:[12]

Criticism of the eighteenth-century monarchic institution in France was not directed against the juridico-monarchic sphere as such, but was made on behalf of a pure and rigorous juridical system to which all the mechanisms of power could conform, with no excesses or irregularities, as opposed to a monarchy which, notwithstanding its own assertions, continuously overstepped its legal framework and set itself above the laws.

Such criticism thus displayed a crucial ambiguity in the conception of sovereignty. It represented both the institutions and practices through which sovereign power is actually exercised, and the analytical standpoint from which the exercise of sovereignty can be assessed. In the end, however, the predominant sense of sovereignty comes to be not the specific position of the sovereign within actual political conflicts, but the theoretical conception with respect to which all participation in those conflicts is to be assessed. This reflects the historical legitimation of the sovereign role itself as standing above particular political conflicts: the illegitimate exercise of the institutions of sovereign power is not sovereignty but usurpation.

Foucault did not argue that sovereignty ceased to be at issue through the political transformations that he claimed to discover in the eighteenth and nineteenth centuries. Rather, he argued that the political forms and practices of sovereign power remained in place, but they were gradually invested by and ultimately sustained on the basis of power relations that functioned on a different scale. A crucial feature of sovereign power is that while there are no limits to its proclaimed scope (all actions, persons, and goods are in principle subject to the sovereign), in practice its capacity to exercise power is discontinuous and solely constraining. The sovereign power can prohibit actions, kill or imprison persons, and tax or confiscate goods, but its productive abilities are quite limited. Increasingly, he argued, the sovereign apparatus (e.g., courts, prisons, military forces) came to be dependent upon what he called the "capillary" power relations through which various "goods"—knowledge, health, wealth, military force, and the like—were actually constructed or enhanced: "The ancient right to *take* life or *let* live was replaced by a power to *foster* life or *disallow* it to the point of death."[13] Thus, in political theory, he argued that theories of sovereignty overlooked the many ways in which power was deployed outside the framework of the state apparatus or class domination; he also claimed that the theory of sovereignty failed in its own terms, since it could not adequately grasp the ways in which sovereign power itself came to be constituted and exercised through tactics on a different scale that were not at the sovereign's disposal.

Yet virtually all the dominant political theories worked within the framework of sovereignty, differing only in where sovereignty was to be located: in natural law, in the people and their representatives, in the real human interests revealed in the situation of a radically oppressed class, or in particular historical communities and their traditions. As Foucault noted, even more radical critics of law and sovereign power did not escape the problematic of sovereignty (today, we might think of Critical Legal Studies in this role):[14]

A much more radical criticism was concerned to show not only that real power escaped the rules of jurisprudence, but that the legal system itself was merely a way of exerting violence, of appropriating that violence for the benefit of the few, and of exploiting the dissymmetries and injustices of domination under cover of general law. But this critique of law is still carried out on the assumption that, ideally and by nature, power must be exercised in accordance with a fundamental lawfulness.

We can now take our first step toward connecting this conception of sovereignty as a political issue to philosophical interpretations of scientific knowledge. Although Foucault himself insisted that knowledge is always intertwined and mutually reinforcing with relations of power, his discussions of power/knowledge never explicitly included an analytics of knowledge comparable to the general reflections upon power to be found in *Discipline and Punish* and parts 4 and 5 of *History of Sexuality, Volume 1*, and elaborated upon in various interviews and lectures. Yet I believe it is possible to construct a parallel discussion of sovereignty as an epistemological problematic.

All the central issues of political sovereignty are reproduced in epistemology: the constitution of a unitary order, based upon legitimacy through law, established from an impartial standpoint above particular conflicts, and enforced through discontinuous interventions that aim to suppress illegitimacy. The problematic of epistemic sovereignty is fundamentally located in the standard contrast between knowledge and belief or assertion. Knowledge arises from a confusing multiplicity of conflicting assertions that circulate through a wide range of communicative interactions. Knowledge is a unified (or consistently unifiable) network of statements that can be extracted from the welter of confused and conflicting contenders and legitimated in accord with rules of rational method, the epistemic surrogate for law. Here is where the notion of epistemic sovereignty is theoretically important. Sovereignty need not be located in any actual sovereign knower, any more than political sovereignty requires a monarch. But just as the sovereign power must be one that *could* consistently be embodied in a single will, sovereign knowledge must be consistently representable propositionally.

Like the political sovereign as arbiter among competing powers, epistemic sovereignty is projected as an impartial referee among conflicting claims. The establishment and especially the deployment of the rational methods of evaluation that distinguish sovereign knowledge from subordinate assertions must in principle be impartial among particular substantive statements.[15] Assertions are rationally justifiable only so long as they can be *independently* shown to accord with the law.

The binary categories invoked by the sovereign power (e.g., legal/illegal) also have clear epistemic counterparts. From the standpoint of rational legitimation, statements are true or false, warranted or unwarranted, rationally permissible or forbidden. And although the question of the rational legitimation of statements is in principle appropriate at all times and places, its application is episodic. The cumbersome procedures of impartial rational evaluation are not deployed to produce assertions, but only to assess them in retrospect, and then only in very limited circumstances. There has always been a close theoretical parallel between the court of reason and the court of law, and in both institutions, the vast majority of possible cases are either never arraigned or else plea-bargained. Few criminal cases are tried, and perhaps even fewer statements and their justifications are rationally reconstructed before the tribunal of reason.

Epistemology as a discipline constituted by the problematic of sovereignty has been organized around four basic issues. First is the question of where epistemic sovereignty is to be located. For the dominant liberal tradition, any rational person may represent the sovereign. The law, in the form of rational method, can be applied by anyone who can place himself or herself in the impartial rational standpoint. Marxism offers an alternative placement of epistemic sovereignty: there is an epistemically privileged class, whose standpoint alone enables the rational critique of ideology. Post-Kuhnian philosophy and sociology of science have revived communitarian conceptions of sovereignty, which situate the rational evaluation of belief within a shared form of life. Even the various forms of relativism are tenaciously located within the problematic of sovereignty, which enables them to recognize or confer equal epistemic "rights" upon individual or cultural worldviews.

There is a second, distinct question concerning where epistemically sovereign judgments are appropriate: Can statements be rationally assessed one by one, or must method intervene only at the level of the theory or research program? A third question concerns the final form of the unification of knowledge: Does the systematic, sovereign unification of knowledge require reduction to a single vocabulary, or does it permit autonomous regions of knowledge at irreducibly different levels of description, or perhaps even apparently conflicting bodies of knowledge, which can be reconciled when each is situated in its own appropriate cultural or linguistic context?

Of course, the fourth and final question is ultimately the crucial one: What constitutes a legitimate exercise of sovereign epistemic judgment? What methods of epistemic adjudication could legitimately claim the force of right or truth? And here, as with the political theory of sovereignty, the question must be kept rigorously distinct from considerations of how judgments are actually made or enforced. For just as the sovereign power and its legislation are always in principle subordinate to the law, the reigning practices of adjudicating claims of knowledge must themselves be subjected to rational scrutiny.

The Dynamics of Scientific Knowledge

As I noted above, Foucault's principal arguments for rejecting the problematic of sovereignty within political theory were that many politically important phenomena could not be adequately understood in its terms, and indeed that those phenomena were in the end constitutive of the modern institutions and practices of sovereignty themselves. I want to suggest a similar line of argument against the way epistemological reflection has been shaped by the notion of sovereign knowledge. The practices of natural scientific research cannot be adequately understood in terms of the legitimation of a unified regime of knowledge, and indeed the actual certification of knowledge is shaped by those practices that outstrip analysis in terms of epistemically sovereign judgment. But the parallel goes deeper than just the basic structure of the argument. I want to suggest that many of the fundamental themes of Foucault's analytics of power will have analogs in a more adequate reflection upon scientific knowledge.

Foucault's analytics of power was supposed to go beyond the limits of the categories of political sovereignty by restricting inquiry to how power is exercised. Power was to be understood not as a thing possessed, but as a dynamic network of relations. My introduction of the notion of epistemic sovereignty was intended to suggest that we might take very seriously the intertwining of knowledge and political order, by considering in similar ways how we might escape the categorial limits of sovereign knowledge. What could it mean to ask about the "how" of knowledge in lieu of what knowledge is or why it is legitimately knowledge? How would such a refocusing of philosophical inquiry aid us in understanding the practices and achievements of the natural sciences?

My account will focus initially upon some of Foucault's more general remarks about power, as indications of what it would mean to bypass the issue of political sovereignty. However, I will not attempt an exposition of Foucault on power; instead, I will try to show how some of his central themes can be adapted to understand scientific knowledge. I shall consider six points:

Power is dynamic. It is not a commodity, an institution, a structure, or any other sort of *thing* (hence it is also not something possessed by agents, classes, or institutions), and it exists only through its exercise; if there is stability over time in power relations, it is because these relations are reenacted and reproduced: power "is the name that one attributes to a complex strategical situation in a particular society."[16]

Power is disseminated throughout the body politic. Power relations are material, and locally situated, and hence, as Foucault once put it, power is omnipresent "not because it has the privilege of consolidating everything under its invincible unity, but rather because it is produced from one moment to the next, at every point, or rather in every relation from one point to another. . . . It comes from everywhere."[17]

These power relations become linked or opposed tactically and strategically. "These relations find support in one another, thus forming a chain or a system, or on the contrary, disjunctions and contradictions which isolate them from one another."[18]

Power is always contested. "The existence of power relations depends upon a multiplicity of points of resistance [that] play the role of adversary, target, support, or handle, [and that] are present everywhere in the power network."[19]

Power needs an "analytics," not a theory. It does not constitute a self-contained domain, but is better understood as one way of examining the same phenomena that could also be seen in other terms; hence, "relations of power are not in a position of exteriority with respect to other types of relationships (economic processes, knowledge relationships, sexual relations), but are immanent in the latter."[20]

Power is productive. It does not merely tax, prohibit, or abolish various social goods, it helps produce or constitute them.

Stated so briefly, this list is undoubtedly oracular and cryptic, and could be dismissed as a repetition of one of the most infuriating aspects of Foucault's written style. I trust, however, that this litany will seem less obscure once we have considered how we might understand *knowledge* as likewise dynamic, disseminated, strategically linked, contested, analytical, and productive.

A dynamic understanding of knowledge may seem initially strange. Whatever one wants to say about power, surely knowledge *is* something possessed by a knower, and transmitted or exchanged through communicative interaction. Indeed, the content of knowledge (both the propositions known, and the evidence and reasoning that warrant them as knowledge) may seem to be independent of particular embodiments in texts, utterances, or thoughts, and of the specific history through which those propositions came to be known.

To understand scientific knowledge in this way, however, as an ideal, ahistorical content that a knower grasps or possesses, is to overlook the complex practical achievements through which scientific domains become accessible. Only within such a complex practical field, shaped by the availability of functional and reliable equipment, and a variety of subtle technical and theoretical skills, do electrons, viruses, tectonic plates, or quasars become possible objects of knowledge or discourse. Thus the propositions in which sovereign knowledge is supposedly expressed get their sense from a complex and heterogeneous field of practices and capabilities.

Foucault has discussed extensively how the body, the individual soul, and the population are constituted as possible objects of knowledge, and how sexuality and delinquency are organized as fields of knowledge. Individuals become knowable only through detailed practices of classification and documentation. Populations require different practices, most notably the tangled interconnection between categorization, counting, and statistics. Sexuality and delinquency have different kinds of history, shaped by patterns of association and strategies of intervention. Recall, for example, Foucault's claim that sexuality as a field of possible knowledge emerged from the nineteenth-century identification and association of "four privileged objects of knowledge, which were also targets and anchorage points for the ventures of knowledge: the hysterical woman, the masturbating child, the Malthusian couple, and the perverse adult. Each of them corresponded to [a] strategy."[21]

The natural sciences have their own histories of disclosure, through which domains of possible inquiry are also shaped by heterogeneous skills, practices, and equipment. The "gene," for example, becomes *available* for discussion in quite different ways, and hence as a different object of *knowledge*, successively through the hybridization studies of Mendel and his contemporaries,[22] the chromosome mappings initiated by Morgan and his colleagues, its molecular identification by Avery and Watson and Crick, and the sequencing and manipulation of genetic elements in recent molecular biology. We are all familiar with the retrospective reconstructions through which we have come to understand Mendel, de Vries, Morgan, McClintock, Avery, Crick, Berg, and Genentech to be talking about the same thing. And there is an important sense in which this reconstruction is not incorrect. But it required the excision or transformation of many of the forms of practice and know-how that at various points helped constitute knowledge of the gene.

In *Knowledge and Power* I talked about this point in terms of the importance of *local* knowledges.[23] "Genes" emerged as the objects of possible discourse through often-arduous accumulations of capabilities and insights in specific contexts (e.g., specific laboratories with their own projects, protocols, and materials; but also specific experimental systems such as *Drosophila*, maize, and bacteriophages). This knowledge cannot be extended to other locations, or related objects, without complex and subtle mutual adaptations. Laboratory practices and equipment themselves, and the knowledge they embody, must be standardized, simplified, and adapted to new purposes, while the working environment (both material and conceptual) to which they are extended must also be modified to accommodate them. I think it is useful to understand these gradual transformations, reproductions, extensions, and mutual alignments of local knowledges as strategic. Out of a confusing array of interacting projects, practices, and capabilities, there gradually emerges an overall pattern or direction (or rather, a plurality of these). This happens, however, not because this pattern was what was intended, however dimly, all along, but because some practices turn out to reinforce and strengthen one another, and are taken up, extended, and reproduced in various new contexts, while others remain isolated from or in conflict with these emergent strategies, and gradually become forgotten or isolated curiosities. Yet, I argued, these outcomes often have little to do with any intrinsic faults of the discarded practices.

An epistemological dynamics takes these strategic alignments to be constitutive of knowledge. Thus, knowledge is not a status that attaches to particular statements, skills, or models in isolation or instantaneously. Rather, their epistemic standing depends upon their relations to many other practices and capabilities, and especially upon the ways these relations are reproduced, transformed, and extended. Knowledge is temporally diffused or deferred: to take something as knowledge is to project its being taken up as a resource for various kinds of ongoing activity—whether in further research or in various applications of knowledge. In this sense, the word "applications" is somewhat misleading, since in the broadest sense we do not first gain knowledge, then apply it; something counts as knowledge only through the ways it is interpreted in use.

Knowledge in this sense *circulates*, and even the various points at which it is articulated, or even collected and assessed, are caught up in its circulation. What is proposed as possible new knowledge, whether in informal discussion or initial publication, has an element of tentativeness about it. What is gathered together in retrospective judgment is always oriented toward a further advance, and shaped by that projection.[24] What I would now conclude from my argument in *Knowledge and Power* and a subsequent paper[25] is thus that there is no place where epistemic sovereignty is actually located. The scientific literature itself is always continually reorganizing what is known as a resource for further investigation; it is also always contested. Yet philosophical attempts to stand outside or above the contested recycling of knowledge always verge upon irrelevance. As I have argued elsewhere,[26] if a judgment from a philosophical standpoint of supposed epistemic sovereignty were to conflict with the ways claims of knowledge are taken up and deployed in the course of research, it could be vindicated only within the contested strategic field in which claims of knowledge are transformed, reproduced, or left behind.

I need to say more about this claim that scientific knowledge is always contested, the parallel to Foucault's insistence that power always confronts resistance.[27] Once it is recognized that knowledge exists only through its reproduction and circulation, the importance of conflict becomes evident: conflict focuses and directs that circulation. Knowledge is developed in an agonistic field, and will typically be contested in very specific respects. And it is precisely in those respects that knowledge will be developed and articulated most extensively and precisely. Where there is (possible) resistance, new and more powerful techniques will be sought, more precise and careful measurement will be provided, and theoretical models will be refined to eliminate or bypass possible sources of inaccuracy or unrealistic assumption. These various refinements are themselves new knowledges, and often in turn provide further new directions or problems for research. Hence, around the specific points where knowledge is resisted, there emerges a whole cluster of new local capabilities and their extension into new contexts. But the contrary is also true: where knowledge goes unchallenged, where a claim "goes without saying," there is little or no articulation or development. And where previous resistance vanishes, knowledge also ceases to proliferate.

The forms taken by resistance to knowledge cannot be easily

reduced to traditional epistemic categories. Obviously, knowledge can be resisted because there are gaps in the data, dubious assumptions in the theoretical models, or countervailing evidence. But it can also be resisted because the procedures and capabilities for its articulation and development are too expensive, environmentally unsound, cruel to animals, politically sensitive, of too little or too much interest to the military, unprofitable, and so forth. Philosophers have often tried to separate considerations internal to knowledge from those that impinge upon it from the outside. The distinction is almost always bound up with a conception of epistemic sovereignty: only those issues that are codified in terms of method, the sovereign law in the realm of knowledge, count as "internal."

Yet the discovery of the local knowledges and their dynamics through which epistemic sovereignty is exercised undercuts any attempt to make such a distinction. The sense of a claim and the ways in which it is articulated and deployed in further research and development depend upon considerations that escape the boundaries that would constitute sovereignty in the realm of knowledge. All the small, local decisions about research materials, equipment, procedures, funding, personnel, skill development, and the like shape the actual development of the knowledges that invest and underwrite the sorts of knowledge claims that philosophers typically investigate. The actual justifications offered for these decisions typically interchange and balance supposedly internal and external considerations. Thus, a physicist may argue for a particular experimental strategy against its competitors by claiming that it is cheaper, provides a less diffuse particle beam, takes best advantage of the skills of available personnel, might interest new funding sources in the military, is more reliably established in the literature, would be adaptable to a variety of experiments, and would leak less radiation and thus counter the recently vocal objections of local environmental groups. These heterogeneous concerns and reasons function together in the shaping of knowledge.

This heterogeneity of knowledge and resistance to it thus points to the inadequacy of that aspect of epistemic sovereignty that presents knowledge as a distinct domain of investigation, which would be the object of a theory. Foucault's analytics of power was a recognition that power was not a more or less enclosed domain of objects, but a collection of strategies for codifying and intervening in things that could also be organized in overlapping and countervailing ways under other headings (economic, epistemic, sexual, etc.).[28] Knowledge should be similarly thought of as a strategic intervention rather than an isolable domain.[29] Understanding knowledge dynamically takes us into considerations that would "properly" belong to other domains if the world presented itself in such discrete bundles. But an adequate understanding of knowledge (even scientific knowledge) in all its local proliferation and heterogeneity regularly violates the boundaries of epistemic propriety. One could add, of course, that the same is true of other kinds of interventions. An investigation of economics or sexuality would have much to say about science, and could not be confined to considerations "external" even to knowledge narrowly construed. There is, after all, much to be said about the value and the pleasures of methodological rigor.

In concluding this introductory survey of a Foucaultian epistemological dynamics, I trust I do not need to say a great deal about the senses in which knowledge is productive. The ongoing practices in which knowledge is embodied are also increasingly the site of the production of health, wealth, military force, and so on. I use the word "production" advisedly. From within the circulation of the (re-)production of knowledge, there emerge new ways to be healthy (low cholesterol and high fiber, adequate T-cell count), new forms of wealth (most obviously in the form of access to and control of information), and new projections of destructive force. There also emerges, not accidentally, a proliferation of new knowledges. For the extension of Foucault's analytics of power to an epistemological dynamics shows more clearly how the continual *expansion* of scientific knowledge and its associated controls and constraints is not merely incidental, but is integral to the ways in which knowledges circulate and are validated.

Legitimation and the Specter of Relativism

I want to conclude with a brief reflection upon a likely source of resistance to a dynamic, nonsovereign epistemology. Attempts to situate scientific knowledge within the social and spatiotemporal circumstances of its production have often been linked with epistemological relativism. Relativism has frequently been attributed to

Kuhn or Feyerabend or both, and many more recent sociologists of science have actively embraced some version of relativism, at least methodologically. Foucault himself has also often been chastised as an arch-relativist, who denies any grounding to the legitimation or critique of power or knowledge.[30] My attempt to bypass or overcome the problematic of epistemic sovereignty may seem to suffer from the same, possibly self-defeating incapacity. A post-sovereign epistemology would presumably offer no standpoint, outside the contested domain in which conflicting and heterogeneous knowledge claims circulate, from which to assess what one *ought* to believe, including whether one ought to believe the assertions of post-sovereign epistemology.

There is, however, a slippery inference underlying this fear that the actions of tyrants and the beliefs of fools can no longer be effectively countered. This is the inference from there being *no* epistemic sovereignty, no privileged standpoint for legitimating knowledge, to *all* knowledge claims' being equally valid, however wacky or offensive they may be. But the latter sort of relativism is not only not entailed by the denial of epistemic sovereignty; the two are mutually inconsistent. Relativism is an assertion of epistemic sovereignty that proclaims the epistemic "rights" of all knowers or knowledges. The most fashionable forms of epistemic relativism today, which are also those frequently and mistakenly associated with Foucault, are those that dismiss all claims to objectivity or truth as merely masks for power. But such claims are the exact epistemological parallel to the radical critique of law as itself a form of violence, which Foucault insisted always "assumes that power must be exercised in accordance with a fundamental lawfulness."[31] To make this assumption, whether about power or about knowledge, is to remain committed to a conception of sovereignty, from which such fundamental lawfulness can be rightly assessed.

What, then, does a post-sovereign epistemology have to say about the legitimation of knowledge? The crucial point is not that there is no legitimacy, but rather that questions about legitimation are on the same "level" as any other epistemic conflict, and are part of a struggle for truth.[32] In the circulation of contested, heterogeneous knowledges, disputes about legitimacy, and the criteria for legitimacy, are part of the dynamics of that circulation. Understanding knowledge as "a strategical situation" rather than as a definitive outcome places epistemological reflection in the midst of ongoing struggles to legitimate (and delegitimate) various skills, practices, and assertions. Recognizing that the boundaries of science (or of knowledge) are what is being contested, epistemology is within those contested boundaries.

An example will clarify my point. What does it mean to say, about a recently prominent family of biological claims, that creationism has now been shown to be false? The standard epistemological interpretation of this claim is that it promises to stand above the myriad claims for and against creationism and assess which actually belong to the unitary regime of knowledge. A postsovereign epistemology would take this sentence to be a *commitment* to marshal available evidence sufficient to demonstrate to anyone audacious enough to challenge this claim that their challenge fails. Now the sovereign epistemologist will immediately ask about the standards of success and failure in any such ensuing contest. The challenger cites Scripture; the defender responds with data from fossils and breeding experiments. And the worry is that, without a sovereign standpoint to determine which appeals and standards are relevant and legitimate, Scripture is just as good as data.

But this conflict among competing standards will appear irresolvable only when one removes the conflict from any real setting, in which there are interested parties and something at stake. In any real conflict, there is a burden of proof,[33] which is sustained by a strategical alignment of people and things that can be relied upon to support and enforce that burden.[34] Epistemic conflict is always shaped by the goods, practices, and projects whose allocation and pursuit are at issue, and by the institutions and social networks that are organized around those pursuits. In such real contexts, there are constraints upon which arguments and which evidence will count as relevant and persuasive, based upon the need for support from others and for reliability from things. It matters what will count as persuasive to others who occupy strategic points in the circulation of knowledge and argument, and it also matters how things will manifest themselves in the contexts in which their behavior is recognized to be relevant.

It is crucial to recognize, however, that the alignment that determines the burden of proof and the standards that must be met is subject to challenge. Not long ago, creationism was readily dismissible as irrational, crackpot science, by appeal to standards that were recognized as relevant and decisive by a powerful social network that controlled access to the educational system, and to the other social goods to which adequate (or at least certified) knowledge of biology provided access. But this appeal was not "merely" a recourse to power; implicated in it was the history of the discovery and interpretation of fossils and other geological data, of political practices and goods surrounding religious life, of the institutional organization and placement of science, and so forth. This complexity and its dynamics are evident in the ways in which the proponents of creationism tried, with some success, to alter the epistemic and political alignment that denied their views any serious recognition.

Indeed, the recent prominence of this example testifies to the partial effectiveness of the new creationists' resistance to the dominant alignment. For they did not meekly accept their allotted place and vainly reassert the superiority of religious belief to science. They tried to claim the epistemic and political resources of science on multiple fronts: challenging the connection between fossil data and dominant interpretations of evolutionary theory, citing philosophical disputes about the demarcation of science, transforming their own views to resist attacks upon their "unscientific" character, and turning against their opponents their own objections to using political power to enforce scientific belief. Of course, they also mobilized a religiously committed political base and chose a sympathetic venue (legislatures and courts in states with strong ties to Christian fundamentalism and relatively weak ties to scientific and educational establishments). Their strategies were an interesting mix of trying to subvert or coopt elements of the dominant epistemic alignment that established and enforced the rationality of belief in Darwinian theory, and trying to create alternative alignments (Christian schools, creationist research and textbooks, etc.) that would enable them to bypass it.

Their initial successes in this strategy compelled defenders of mainstream biology to confront arguments and respond to strategies whose dismissal previously went virtually without saying. This response in the end was largely successful, as there were effective counters to many of the new creationist strategies and arguments. As a result, one still need not consider Scripture in most contexts in order to assess or advance biological knowledge. But new arguments, new knowledges (about the epistemic gaps in the Darwinian orthodoxy, about the creationist resistance, about religion, science, and politics), and new alignments were created in the course of these responses; resistance partially refocuses the organization of knowledge even when it substantially fails to overturn its target.

It is useful to compare this example with that of recent feminist challenges to sexist orthodoxies in many disciplines. Like creationism, feminist critiques could once be dismissed in most contexts without their having to be taken seriously. When articulated, the dismissal rejected not just their specific contributions to knowledge, but the very possibility of such contributions, which could only represent the illegitimate intrusion of religion or politics into knowledge. Like creationists, feminists combined a new program of research with resourceful appropriation and redirection of elements of orthodoxy, and the creation of alternative alignments that reduced their dependence upon unregenerate opponents (feminist presses and journals, professional associations and graduate programs, but also a variety of more straightforwardly political alliances).[35] Feminists have been rather more successful than have creationists, despite considerable remaining opposition. But this success must be understood in simultaneously epistemic and political terms. Their arguments were more persuasive. Sexism and gender have also shown themselves to be more resilient and readily manifest objects of inquiry than is biblical Creation. And the new feminist alignments made it more difficult to ignore or dismiss their arguments, avoid their vocabulary, or refuse to certify their achievements.

Were the feminists more rational, or more warranted in their claims, than were the creationists? Yes. But that judgment does not stand above the fray as an assessment of the unsituated rationality of each set of arguments and conclusions. That judgment is itself a move in ongoing epistemic struggles, which both draws upon and reinforces the successes and failures of each view. My own arguments, and allusions to arguments, were responses to the burden and standards of proof that I take to be effective in the context in which this is written; in their small way, they also aim to reinforce or transform the standards to which they appeal.

The moral for epistemology is, I hope, clear. The turn to a nonsovereign epistemological dynamics does not replace argument or a concern for truth with power and domination, even while insisting that argument and claims to knowledge are never politically innocent. The contested circulation of opposing knowledges, which cannot be consistently combined into a unitary framework of propositions, is a struggle for truth. Truth matters. Precisely because it matters, truth is often fiercely contested. And if we cannot stand outside that contest to assess it from a neutral standpoint, this does not mean that all claims to truth can be put forward on an equal basis. Knowledge claims are historically, socially, and materially situated in contexts that govern what can be intelligibly and seriously asserted, and how much or what kind of argument is necessary to support it. But such epistemic contexts are always in flux; their boundaries and configuration are continually challenged and partially reconstructed, as epistemic alignments shift. And these alignments are always intertwined with alignments of power and political resistance. To recognize this interconnection is not to devalue knowledge or science for political purposes, but to take seriously the stakes in struggles for knowledge and truth, and to place epistemology and philosophy of science squarely in their midst.

EVELYN FOX KELLER

The Dilemma of Scientific Subjectivity in Postvital Culture

Notions of historical arcs obviously rely on perceptions of continuity more than of discontinuity, and might accordingly appear directly antithetical to the spirit of this volume. Indeed, much in contemporary postmodern discourse, with its insistent emphasis on ruptures, fractures, and disunities, patently belies such notions. Equally antithetical might appear the invocation of "subjectivity." The language both of postmodernism and of postvital culture[1]—the one deriving from literary and philosophical discourse, and the other from scientific—reminds us of the artifactuality of subjectivity and of the illusoriness of an interior vital "self." Nonetheless, it seems to me useful to insert into this discussion some reminders of continuity in the historical evolution of the prosthetic subject of postmodern discourse—a subject (or antisubject) that can alternatively be seen as a product of epistemological rupture or of the convergence of what Freeman Dyson calls the "green" and "grey" revolutions of modern technology.[2] Reminders of continuity may be especially pertinent in the face of political crises that depend for their very meaning on the twin technologies of representation and intervention, on the proximity and interchangeability of simulated war games and enacted war, on an overdetermined synergy between the symbolic, political, and technological effacement of flesh-and-blood bodies consigned to immateriality. It was not long ago when, in response to the Gulf War, the very rallying cry of protest, "No blood for oil," called upon a resurrection of the distinction (premodern or otherwise) between the materiality of human bodies and that of hydrocarbons and silicon chips. In part, my effort here, as in a number of other papers,[3] is in frank support of the embodied subjects on which such a distinction depends. The lesson I take here from "historical arcs" is not as a recall of linear causal imperatives, but rather as a reminder of the power of cumulative amnesia. We are bound by the intertextual confines not only of contemporary texts, but also of the texts of the past we have imbibed along with our mothers' milk.

To be sure, more than one historical arc can be identified as leading to the present moment; and indeed, much of recent feminist theory can be seen as providing instructive alternatives to conventional narratives—ironically, thereby, as contributing to a revitalization of a history of ideas. The importance of "history," however, is now to be understood not as either moral or logical imperative, but precisely as political contingency. One particularly interesting attempt at retracing such a historical arc (which I read as being in manifest sympathy with feminist reworkings) is Brian Rotman's analysis of the logic of representation in *Signifying Nothing: The Semiotics of Zero*.[4] Rotman identifies a historical trajectory of progressive disembodiment in the semiotic logics of mathematics, painting, and money. What follows is an attempt to trace similar historical continuities in the logic of scientific representation; in this, I lean not only on Rotman, but also on recent work on the history of objectivity by Daston, Galison, Porter, and Dear.[5]

The history of scientific subjectivity is, as Daston and others have so eloquently argued,[6] enormously complex and layered, requiring many different kinds of readings. Ted Porter, for example, has suggested that the history of standardization—simultaneously of content and viewer—be read as a *political* history in which modern science provides the locus for the development of a "technology of distrust," appropriate not only to a dispersed and impersonal scientific practice but also to the needs of the "modern suspicious democratic political order."[7] For my purposes, the most relevant point of these arguments may be that once air pumps and telescopes became freely available as standardized instruments, neither the author of the text nor the original observer *needed* any longer to be identified. By the second half of the nineteenth century, the first-person narrator of the scientific text could be effectively replaced by the abstract "scientist"—then a newly coined word—who could speak for everyman but was no-man, in a double sense: not any particular man, and also a site for the not-man within each and every particular observer. Between the seventeenth and nineteenth centuries, a hollow place had been carved out in the mind of every actual or virtual witness into which a machine could vicariously be placed—the lacuna of classical perspective now filled in on the canvas, but emerging, instead, in the mind of the viewer.

The result of this semiotic progression was an enduring and final erasure: with the disappearance of all consciousness of the representation *qua* representation, both the presence of the knowing, doing subject behind representation and its corresponding absence *in* representation have been thoroughly occluded. The scientific subject arising in place of the embodied crafter, interpreter, and reporter of experiments is a classic instance of what Brian Rotman calls a "metasubject": invisible, autonomous, virtual—floating above the situated, dependent, and very real work that scientists actually engage in in the complex production of the scientific corpus, which can itself neither be seen on the canvas nor be noted for its absence. But if the original subject cannot be signified in the new representational scheme, the metasubject can. Its sign is the conspicuously replicable and representable mechanical detector—as it were, the machine in the ghost: a surrogate that stands in not for the presence of the knowing, doing subject, but for its absence.[8]

In *Signifying Nothing: The Semiotics of Zero*, Rotman tells a story of the progressive loss of the anteriority of things to signs throughout European culture, a story of the distantiation of metasubject from subject in art, in mathematics, and in finance. Each of his examples charts the same dilemma (or paradox) of subjectivity: in actuality, the viewing, acting, and doing subject—the one who paints, counts, or trades—remains as necessary as always for the actual production of art, mathematics, or financial transactions, but the new logic of representation allows only the metasubject to be signified, denying the subject behind the metasubject its signification even as an absence. The same dynamic, I suggest, can also be seen in science; indeed, I am suggesting that it is in the natural sciences that we see the dilemma of subjectivity in its most critical form. The man-made (or, more accurately, men-made) nature of scientific knowledge cannot be represented in the texts of science, for precisely what has been constructed is the illusion that this knowledge is *not* made by men—not crafted, articulated, or constructed, but discovered,[9] in a word, simply true. It is this erasure, this representational logic, that underlies (indeed, that guarantees) the representation of a single, unified spatiotemporal reality, of which the human subject could ultimately itself become part—re-presented in the only coinage available: as a machine among machines.

One face of the history of modern science is thus a history of semiotic repositioning—a repositioning that seems, at times, to point directly to twentieth-century realizations of Descartes's dream of presenting the viewer of the natural world as "an object or spectacle before his own vision."[10] Not surprisingly, Descartes's actual portrait of that object—his depiction of the processes of motion and emotion, of growth and reproduction, of waking and sleeping, even of perception and memory—has been transformed over the intervening centuries beyond recognition. Perhaps the only surprise is how stable has been his legacy of scientific subjectivity. Despite the radical transformation in representation of humans as objects, the subject that modern biology promises to reveal turns out after all to be in significant ways, still, a Cartesian object—merely knowable, never itself capable of knowing. Even where molecular determinants can be identified for memory and vision, for sleep and dreams, for depression and perhaps even for love, the agentic subjects responsible for this representation—still tacitly coded as the thinking searcher for the mechanical basis of thought—remain apart, off a canvas that has become coextensive with the world, outside a domain that denies it has an outside. After centuries of erasure, though invisible and unthinkable, the knowing subject is still there, resistantly anterior. The question is where, in an age of atheism, that anteriority, which had earlier been ceded to God but now can no longer be, is to be located. It is to this question, surely one of the most critical questions facing us in the late twentieth century, that I now wish to turn.

Consider, for a start, J. D. Bernal's own answer to this question in 1929, in his assessment of our progress in this long struggle (an assessment, incidentally, endorsed by Freeman Dyson in 1972).[11] Bernal called his treatise *The World, the Flesh and the Devil: An Enquiry into the Future of the Three Enemies of the Rational Soul.* As Bernal saw it, the greatest impediment that remains in the struggle of the rational soul against the inorganic forces of the world and the organic structure of our flesh is the endurance of the irrational forces of desire and fear: that is, the Devil. He wrote:[12]

We can abandon the world and subdue the flesh only if we first expel the devil, and the devil, for all that he has lost individuality, is still as powerful as ever: he is inside ourselves, we cannot see him. Our capacities, our desires, our inner confusions are almost impossible to understand or cope with in the present.

The goal awaiting us is to rout desire from its hiding place and bring it into line with our objective aims, "using and rendering innocuous the power of the id and leading to a life where a full adult sexuality would be balanced with objective activity"[13]—bringing "feeling, or at any rate, feeling-tones . . . under conscious control; . . . induced to favour the performance of a particular kind of operation."[14] Dimly aware of the absence of a subject in these pronouncements, Bernal, a Marxist, sought to relieve at least most of humanity from the dilemma of a still-unrepresented subjectivity. He speculated about a future "splitting of the human race— the one section developing a fully-balanced humanity, and the other groping unsteadily beyond it."[15] The latter, what Bernal himself described as "an aristocracy of scientific intelligence,"[16] would have[17]

a dual function: to keep the world going as an efficient food and comfort machine, and to worry out the secrets of nature for themselves. . . . A happy prosperous humanity enjoying their bodies, exercising the arts, patronizing the religions, may well be content to leave the machine, by which their desires are satisfied, in other and more efficient hands. Psychological and physiological discoveries will give the ruling powers the means of directing the masses in harmless occupations and of maintaining a perfect docility under the appearance of perfect freedom. But this cannot happen unless the ruling powers are the scientists themselves.

Perhaps, Bernal goes on to speculate (in a fashion close to Dyson's own heart), the colonization of space will offer a particularly convenient "solution." In that case,[18] mankind—the old mankind—would be left in undisputed possession of the earth, to be regarded by the inhabitants of the celestial spheres with a curious reverence. The world might, in fact, be transformed into a human zoo, a zoo so intelligently managed that its inhabitants are not aware that they are there merely for the purposes of observation and management.

Bernal's displacement of the scientific observer-manager to the celestial spheres is of course no solution at all to the problem of representing that observer's residual (albeit now fully rationalized) subjectivity, but it is at least evocative. The very phrase "celestial spheres," antiquated as it is, recalls an idea of proximity to God that Bernal would surely not have wanted to claim, but could nonetheless rely upon. Perhaps more to his point, it also evokes an ideal of ethereality, of disembodiment.

In Dyson's tribute to Bernal, he remarks that "the decisive change that has enabled us to see farther in 1972 than we could in 1929 is the advent of molecular biology."[19] And indeed, it is molecular biology that has provided us with the tools for identifying molecular determinants of perception, of emotion, and perhaps even of thought. Today, in the aperspectival genius of molecular biology, we seem to have come closer yet to Descartes's dream than even Bernal imagined. Only a few short decades after Watson and Crick's momentous discovery of "the secret of life," those whose business it is to "worry out the secrets of nature" can finally reveal, as Watson has put it, "what it really means to be human."

In the discourse of contemporary molecular biology, there is no talk of a scientific aristocracy, explicitly set apart from a malleable but content humanity. Here, the knowing subject asserts itself not by a residual identification with God, but by projection onto the canvas, through identification with those molecules—at once master architect and master builder, both encoding the blueprint for life and directing its production. (It is, Crick says, precisely through an identification with such molecules that we can restore our sense of unity with nature.)[20] Yet side by side with this willful identification with DNA, it is still possible to detect an anterior knowing subject—not removed to the celestial spheres, but right here on earth, behind the DNA, behind even (or perhaps especially behind) the authorial "we." One way in which this anteriority is exhibited is by an unwitting yet routine syntactical distancing of the authorial and subject pronoun (either singular or plural) from the represented object, be it nature, human beings, or society.

To see this, consider the use of pronouns in some of Francis Crick's informal asides. Take, for example, his remarks about "Nature's own analogue computer" in his 1966 book *Of Molecules and Men*: "The system itself—works so fantastically fast. Also, she knows the rules more precisely than we do. But we still hope, if not to beat her at her game, at least to understand her."[21] My point here is to note that the subject "we" in this sentence is not itself part of, but rather is on a par with, and, it is hoped, even better than Nature's own computer.

A similar distancing, but now from "people" or "society," can be detected in a series of comments Crick offered on the subject of eugenics and genetics in 1963. I quote three of these; first:[22]

Do people have the right to have children at all? . . . I think that if we can get across to people the idea that their children are not entirely their own business . . . , it would be an enormous step forward.

Here we have three subjects: the "I" who speculates, the "we" who knows, and finally, those to whom "we" must communicate our knowledge, namely "people."[23]

In a related remark, a similar syntactical distancing is expressed once again, but this time—perhaps because of the explicit message conveyed—yet further enhanced by invoking the even more abstract pronoun "one." Crick writes:[24]

The question . . . as to whether there is a drive for women to have children and whether [Crick's proposal of licensing] would lead to disturbances is very relevant. I would add, however, that there are techniques by which one can inconspicuously apply social pressure and thus reduce such disturbances.

And for a final, really embarrassing example that reads like a throwback to the earlier part of this century:[25]

We are likely to achieve a considerable improvement [in the human stock through genetics] . . . that is by simply taking the people with the qualities we like and letting them have more children.

The paradox of authorial representation in the natural sciences once found an exceedingly graphic image in the vanishing point of classical perspective. Over time, the scientific subject comes to be represented as progressively more abstract and more dispersed—until, in the nineteenth century, "the scientist" becomes a mere cipher, a depersonalized reporter of the recordings of a mechanical detector. All these moves can be seen as attempts to alleviate the particular difficulties that the paradox of authorship presented for the natural sciences. Here I have tried to show how, well into the twentieth century, we can still discern a semblance of the original paradox, although these days perhaps evident only in the informal discourse of actual scientists, in the locutory "we" or "one" in such off-the-cuff, unwitting remarks as Crick's above. Though still present, however, its days are clearly numbered. As molecular neurobiology extends its frontier ever deeper into the brain, promising to converge even on the problem of consciousness (Crick's own favorite), the scientist himself comes to be drawn ever deeper into the machine he has created. At such a point, the circle of scientific knowledge threatens to close in on itself, and the anchoring that has until now been provided by a residually anterior knowing subject finally disappears. Clearly, a new and radically different kind of subjectivity is called for. Indeed, we can already discern within the scientific project itself the outlines of the form this new, "postmodern," sensibility has, even as we speak, already begun to take.

In the late twentieth century, as Francis Crick boldly searches for the structure of consciousness in molecular arrangements, another kind of science, and with it another discourse, has developed to meet him at least halfway. Out of the science of computers has emerged a discourse that has as little need for the hypothetical "I" or "we" of molecular biology as the discourse of Newtonian physics earlier had for God. Instead it offers the promise of a mind that can indeed exist without the body, as Descartes once believed, but a mind that makes an ultimate mockery of all Descartes's hopes, a mind appropriated by its alter ego. This mind neither lacks extension nor meets God; it is spatial and banally corporeal: a bank of self-regulating knobs and switches that needs no outside agent to maintain it. Danny Hillis, designer of the "Connection Machine," anticipates the development of computers that would yield "not so much an artificial intelligence, but rather a human intelligence sustained within an artificial mind."[26] Hillis acknowledges:

Of course, I understand that this is just a dream, and I will admit that I am propelled more by hope than by the probability of success. But if this artificial mind can sustain itself and grow of its own accord, then for the first time human thought will live free of bones and flesh, giving this child of mind an earthly immortality denied to us.

Descartes too imagined that human thought could live "free of bones and flesh." In his Sixth Meditation he wrote, "I am truly distinct from my body, and . . . I can exist without it"; but he never dreamed that this marvelously autonomous "I" could itself be constituted of earthly matter.[27]

Only one twist is still missing from this apparently post-Cartesian vision: Hillis's artificial mind may be able to sustain itself, but for it to live and grow, it must be able to reproduce itself.[28] As it happens, thirty-seven years ago, just a year before Watson and Crick announced their discovery of "the secret of life," the mathematician John von Neumann showed us how this problem too could, in principle, be solved. Because of von Neumann, the technical culture of human robotics is emboldened to produce a dream auxiliary to Hillis's. For one of its more provocative expressions, I quote from the introduction to a work of Hans Moravec's called, appropriately enough, *Mind Children*:[29]

Unleashed from the plodding pace of biological evolution, the children of our minds will be free to grow to confront immense and fundamental challenges in the larger universe. We humans will benefit for a time from their labors, but sooner or later, like natural children, they will seek their own fortunes while we, their aged parents, silently fade away. Very little need be lost in this passing of the torch—it will be in our artificial offspring's power, and to their benefit, to remember almost everything about us, even, perhaps, the detailed workings of individual human minds.

In Moravec's vision, we, the subjects who have begotten these machines, have become entirely superfluous. Like the God who was once but is no longer thought necessary to account for our existence, our authorship too has become redundant—no longer required even for the intellectual and technical prowess that made such machines possible in the first place. Moravec and Hillis are the new magicians, presenting us in this ultimate vanishing act with a prosthetic subject that is fully autonomous. They paint a portrait, and a world, from which all residual signs of anteriority are utterly erased—a representation that is complete and unto itself, that needs neither the eye of God, nor the eye of the artist, nor the eye of an observer. It begets itself, it makes itself, it sees itself.

To the extent that this subjectivity—unmoored and floating free—is new, its apparently agentless and "denatured" autonomy bears a striking resemblance to what I might call the "antisubject" of much of poststructuralist theory. Replicated and dispersed as material (or immaterial) discourse, it offers the ultimately artifacted simulacrum as substitute for the classical subject, inviting us into an "ecstasy of communication" among radically disembodied and dislocated sites.[30] Indeed, such an "antisubject" might even be taken as providing ironic grounding for the "violent and reductive artifactualism" of postmodernism.[31] The irony derives from the fact that the subject Hillis and Moravec describe is less a departure than a culmination of over three hundred years of representational logic that has anchored the entire tradition of modernity. It is the end-point of a semiotic system in which, as Rotman puts it,[32]

the signs of the system become creative and autonomous. The things that are ultimately "real," that is numbers, visual scenes, and goods, [and now we add machines,] are precisely what the system allows to be represented as such. The system becomes both the source of reality, it articulates what is real, and provides the means of describing this reality as if it were some domain external and prior to itself.

Even at this late date, however, it may still be possible to articulate some resistance against this sensibility on behalf of the flesh-and-blood humans who must live out its consequences and even on behalf of those who have nurtured it and labored to embody it—that is, to remind ourselves of something like the facts of the matter. If so, they would have to be these: whatever lies in store for the future of robotics and artificial intelligence, the imaginings of Moravec and Hillis point at least as much to the past as to the future. Not only is it in fact not possible to construct such machines now, but the expectation that it ever will be possible in the future needs to be recognized as a particular, and very potent, fantasy—a fantasy that, like the machines that enliven it, grows out of, and speaks for, a particular and very potent historical development. The replacement of God's I/eye by a thinking and knowing machine may indeed mark a new way of speaking and thinking—perhaps neither more nor less veridical than the old—but it is one that emerges logically from the momentum of half a millennium of technoscientific and cultural history, in which particularly articulate and well-situated human subjects have increasingly sought particular kinds of representation in the interests of constructing particular kinds of products. We do, in fact, now know how to build parallel processors and computer viruses, just as we know how to construct nucleotide polymers and soon will undoubtedly be able to synthesize real viruses, and more. But these products of computer science and molecular biology do not arise autonomously; rather they emerge from the arts and artifacts of human industry, expressing certain human needs and desires. They are effected by historically contingent material and social practices that are themselves both enabled by and enabling of a tradition of representation that has sought for more than three hundred years the erasure of all evidence of the human agency behind both the practices and the representations, and even behind the allure of mechanical surrogacy that has provided such essential fuel for this search. In this sense, the vision that Hillis and Moravec offer us today clearly speaks to an ultramodern rather than a postmodern sensibility.

To insist that humanly meaningful texts—scientific or other—can no more be authored by silicon chips than they could have been authored by God is to resist erasure of the dependence of meaning on cultural and material history. It is to resist in the name of another sensibility—that is, on behalf of the variously situated, embodied, human subjects who have been destined to act or react in the terms made available to them by the culture; on behalf of those who, with their actions (be they manual, cognitive, or linguistic), have been most effective in creating these terms and acting out the meanings they inscribe; and on behalf of those, less well placed, whose options have been and remain more limited.

DONNA J. HARAWAY

Modest Witness: Feminist Diffractions in Science Studies

❖ I believe in the truth of what is perhaps
figurative, although Moshe Idel has found recipe
after recipe, precise as the instructions for
building a yurt or baking French bread, for
making golems.
—Marge Piercy, *He, She, and It*

A man whose narratives could be credited as
mirrors of reality was a *modest man*; his reports
ought to make that modesty visible.
—Steven Shapin and Simon Schaffer

The Modest Witness, a leading character in dramas of the Scientific
Revolution, is a figure in the stories of contemporary science studies, as well as of science. In literary practice indebted to the fraught
traditions of Christian Realism, figures collect up and reflect back
the hopes and beliefs of the community; they embody, or incarnate,
meanings. The Modest Witness is such a figure in science studies.
How could it be otherwise in the sacred secular stories of European
and Euro-American technoscience? S/he is about telling the truth,
giving reliable testimony, guaranteeing important things, providing good-enough grounding—while eschewing the addictive nar-
cotic of transcendental foundations—to enable compelling belief
and collective action.

The modest witness in the narrative field of this essay works to
refigure the subjects, objects, and communicative commerce of
technoscience into different kinds of knots.[1] I am consumed by the
project of materialized refiguration; I think that is what's happen-
ing in the worldly projects of technoscience and feminism. A figure

collects up the people; a figure embodies shared meanings in stories that inhabit their audiences. I take the term "modest witness" from the important book by Steven Shapin and Simon Schaffer, *Leviathan and the Air-Pump: Hobbes, Boyle, and the Experimental Life*. In order for the modesty referred to in the epigraph above to be visible, the man—the witness whose narratives mirror reality—must be invisible: that is, an inhabitant of the potent "unmarked category" that is constructed by the extraordinary conventions of self-invisibility. In Sharon Traweek's wonderfully suggestive terms, such a man must inhabit the space perceived by its inhabitants to be the "culture of no-culture."[2]

This is the culture within which contingent facts—the real case about the world—can be established with all the authority, but none of the considerable problems, of transcendental truth. This self-invisibility is the specifically modern, professional, European, masculine, scientific form of the virtue of modesty. This is the form of modesty that pays off its practitioners in the coin of epistemological and social power. This kind of modesty is one of the founding virtues of what we call modernity. This is the virtue that guarantees that the modest witness is the legitimate and authorized ventriloquist for the object world, adding nothing from his mere opinions, from his biasing embodiment. And so he is endowed with the remarkable power to establish the facts. He bears witness; he is objective; he guarantees the clarity and purity of objects. His subjectivity is his objectivity. His narratives have a magical power—they lose all trace of their history as stories, as products of partisan projects, as contestable representations, or as constructed documents in their potent capacity to define the facts. The narratives become clear mirrors, fully magical mirrors, without once appealing to the transcendental or the magical.

In what follows, I would like to queer the elaborately constructed and defended confidence of this civic man of reason, in order to help a more corporeal, inflected, and optically dense, if less elegant, kind of modest witness to matters of fact to emerge in the worlds of technoscience. I am less interested in the critical practice of reflection, of showing once again that the emperor has no clothes, than in finding a way to *diffract* critical inquiry in order to make difference patterns in a more worldly way. Reflection displaces the same elsewhere; diffraction patterns record the passage of difference, interaction, and interference. If, rooted etymologically in the Greek *trópos*, tropes are what make us swerve, what make us notice what we did not already know how to see, then refiguring the actors and actants in technoscience is a modest orthopedic exercise for becoming less literal-minded—a kind of aerobics for academics, perhaps. Experiments are tropic practices; they make their participants, human and nonhuman, swerve. In science studies, analytical and narrative choices are tropic choices. They determine whether we turn from the falsely self-evident or not. Immodestly, I like to figure feminist science studies as a practice of contributing to worldly diffractions.

Robert Boyle (1627–91) is memorialized in the narratives of the Scientific Revolution and of the Royal Society of London for Improving Natural Knowledge as a father of the experimental way of life. In a series of crucial developments in the 1650's and 1660's in post–Civil War, Restoration England, Boyle played a key role in forging the three constitutive technologies for such a new life-form: "a *material technology* embedded in the construction and operation of the air-pump; a *literary technology* by means of which the phenomena produced by the pump were made known to those who were not direct witnesses; and a *social technology* that incorporated the conventions experimental philosophers should use in dealing with each other and considering knowledge-claims."[3] Experimental philosophy—science—could spread only as its materialized practices spread. This was not a question of ideas, but of the apparatus of production of what could count as knowledge.[4]

At the center of this story is an instrument, the air pump. Embedded in the social and literary technologies of proper witnessing, and sustained by the subterranean labor of its building, maintenance, and operation, the air pump acquired the stunning power to establish matters of fact independent of the endless contentions of politics and religion. Such contingent matters of fact, such "situated knowledges," were constructed to have the earth-shaking capacity to ground social order *objectively*, literally. This separation of expert knowledge from mere opinion, as the legitimating knowledge for ways of life, without appeal to transcendent authority or to abstract certainty of any kind, is a founding gesture of what we call modernity. It is the founding gesture of the separation of the technical and the political. Much more than the existence or not of a vacuum was at stake in Boyle's demonstrations of the air pump. As Shapin and Schaffer put it, "the matter of fact can serve as the foundation of knowledge and secure assent insofar as it is not regarded as man-made. Each of Boyle's three technologies worked to achieve the appearance of matters of fact as *given* items. That is to say, each technology functioned as an *objectifying resource*."[5] The three technologies, metonymically integrated into the air pump itself, the neutral instrument, factored out human agency from the product. The experimental philosopher could say, "It is not I who say this; it is the machine."[6] "It was to be nature, not man, that enforced assent."[7] The world of subjects and objects was in place, and scientists were on the side of the objects. Acting as objects' transparent spokesmen, the scientists had the most powerful allies. As men whose only visible trait was their limpid modesty, they inhabited the culture of no-culture. Everybody else was left in the domain of culture and of society.

But there were conditions for being able to establish such facts credibly. To multiply its strength, witnessing should be public and collective. A public act must take place in a site that can be semiotically *accepted* as public, not private. But "public space" for the experimental way of life had to be rigorously defined; not everyone could come in, and not everyone could testify credibly. What counted as private and as public was very much in dispute in Boyle's society. His opponents, especially Thomas Hobbes (1588–1679), repudiated the experimental way of life precisely because its knowledge was dependent on a practice of *witnessing* by a special community, like that of clerics and lawyers. Hobbes saw the experimentalists to be part of private, or even secret, and not civil, public space. Boyle's "open laboratory" and its offspring evolved as a most peculiar "public space," with elaborate constraints on who legitimately occupies it: "What in fact resulted was, so to speak, a public space with restricted access."[8]

Indeed, it is even possible today, in special circumstances, to be working in a top-secret defense lab, communicating only to those with similar security clearances, and to be *epistemologically* in public, doing leading-edge science, nicely cordoned off from the venereal infections of politics. Since Boyle, only those who could disappear "modestly" could really witness with authority, rather than gawk curiously. The laboratory was to be open, to be a theater of persuasion, and at the same time it was constructed to be one of the culture of no-culture's most highly regulated spaces. Managing the public/private distinction has been critical to the credibility of the experimental way of life. This novel way of life *required* a special, bounded community. Restructuring that space—materially and epistemologically—is very much at the heart of late twentieth-century reconsiderations of what will count as the best science.[9]

Also, displaying the labor expended on stabilizing a matter of fact compromised its status. Unmasking the labor required to produce a fact showed the possibility of a rival account of the matter of fact itself—a point also not lost on Hobbes. Further, those actually, physically, present at a demonstration could never be as numerous as those virtually present by means of demonstration through the literary device of the written report. Thus, the rhetoric of the modest witness, the "naked way of writing," unadorned, factual, compelling, was crafted. Only through such naked writing could the facts shine through, unclouded by the flourishes of any human author. Both the facts and the witnesses inhabit the privileged zones of "objective" reality through a powerful writing technology. That technology is the Western scientific dream machine for the escape from the utter materiality of metaphoricity, the discrediting entrapments of troping. If language could become immaterial, in all senses of the term, figures and narratives could give way to explanations and facts. Ideas could not prevail until words were subdued. And, finally, only through the routinization and institutionalization of all three technologies for establishing matters of fact could the "transposition onto nature of experimental knowledge" be stably effected.[10]

All these criteria for credibility intersect with the question of modesty. Transparency is a peculiar sort of modesty. The philosopher of science Elizabeth Potter, of Mills College, gave me the key to this story in her wonderful paper "Making Gender/Making Science: Gender Ideology and Boyle's Experimental Philosophy."[11] Shapin and Schaffer attended to the submerging (literally, as represented by engravings of the regions under the room with the visible air pump) of the labor of the crucial artisans who built and tended the air pump—and without whom nothing happened—but they were totally silent on the structuring and meaning of the specific civil gendering of the modest witness. Like the stubbornly reproduced lacunae in the writing of many otherwise-innovative malestream science-studies scholars, the gap in their analysis seems to depend on the illusion that gender is a preformed, functionalist category, merely a question of preconstituted, race-neutral, "generic" men and women, beings resulting from either biological or social sexual difference and playing out "roles," but otherwise of no interest. And no self-respecting science-studies scholar has much to say about roles. The effect of the missing analysis is to treat race and gender, at best, as a question of empirical, preformed beings who are present or absent at the scene of action, but are not generically constituted in their doings in the new theaters of persuasion. This is a strange analytical aberration, to say the least, in a community of scholars who play games of epistemological chicken trying to beat each other in the game of showing how all the entities in techno-science are constituted *in* the action of knowledge production, not *before* the action starts.[12]

Elizabeth Potter, however, has a keen eye for how men became man in the practice of modest witnessing. Men in the making, not men, or women, already made, is her concern. Gender was *at stake* in the experimental way of life, not predetermined. To develop this topic, she turned to the burning early seventeenth-century English debates on the proliferation of genders in the practice of sexual cross-dressing. Her question is, How did Robert Boyle avoid the fate of being labeled *haec vir*, a feminine man, in his insistence on the virtue of modesty? How did the masculine practice of modesty, by appropriately civil men, enhance agency, epistemologically and socially, while modesty enforced on women of the same social class simply removed them from the scene of action? How did some men become transparent, self-invisible, legitimate witnesses to matters of fact, while most men and all women were made simply invisible, removed from the scene of action, either below the stage working the bellows that evacuated the pump, or offstage entirely? Women lost their security clearances very early in the stories of leading-edge science.

Women were, of course, literally offstage in early modern English drama, and the presence of men acting women's roles was the occasion for more than a little exploring and resetting of sexual and gender boundaries in the foundational settings of English drama in the sixteenth and seventeenth centuries. As the African-American literary scholar Margo Hendricks tells us, Englishness was also at stake in this period, for example in Shakespeare's *A Midsummer Night's Dream*.[13] And, she notes, the story of Englishness was part of the story of modern racial formations, rooted in this period in lineage, civility, and nation, rather than in color and physiognomy. But the discourses of "race" that were cooked in this cauldron, which melted states and bodies together in discourses on lineage, were found more than a little useful throughout the following centuries for marking the differentially sexualized bodies of "colored" peoples around the world, locally and globally, from the always unstably consolidated subject positions of self-invisible, civil inquirers. Gender and race never existed separately and never were about preformed subjects endowed with funny genitals and curious colors. Race and gender are about entwined, barely analytically separable, highly protean, *relational* categories. Racial, class, national, sexual, and gender formations (not essences) were, from the start, dangerous and rickety machines for guarding the chief fictions and powers of European civil manhood. To be unmanly is to be uncivil, to be dark is to be unruly: those metaphors have mattered enormously in the constitution of what may count as knowledge. The ongoing self-invisibility of these tropes in science and science studies has made us literal-minded. It is past time to swerve decisively from the smooth roads of such modest self-invisibility, perhaps less by means of unending, reflective self-vision than by passing through the rougher grids that produce worldly diffractions.

Let us attend to Elizabeth Potter's story. Modesty was not a medieval masculine virtue. Rather, noble, manly valor required patently heroic words and deeds. The modest man was a problematic figure for early modern Europeans, who still thought of nobility in terms of warlike battles of weapons and words. Potter insists that, in his literary and social technologies, Boyle helped to construct the new man and woman appropriate to the experimental way of life and its production of matters of fact: "The new man of science had to be a chaste, modest, heterosexual man who desires yet eschews a sexually dangerous yet chaste and modest woman."[14] Female modesty was of the body; the new masculine virtue was to be of the mind. This modesty was to be the key to his trustworthiness; he reported on the world, not on himself. "Masculine style" became English national style, a mark of the growing hegemony of the rising English nation. An unmarried man in a Puritan English moment that valued marriage highly, Boyle pursued his discourse on modesty in the context of the vexed *hic mulier/haec vir* (masculine woman/feminine man) controversies of the late sixteenth and early seventeenth centuries. In that anxious discourse, when gender characteristics were transferred from one sex to another, writers worried that third and fourth sexual kinds were created, proliferating outside all bounds of God and Nature. Boyle could not risk his modest witness's being a *haec vir*. God forbid the experimental way of life should have queer foundations.

The new science redeemed Boyle's man from gender confusion and made him a modest witness as the *type specimen* of modern heroic, masculine action—of the mind. Depleted of epistemological agency, modest women were to be invisible to others in the experimental way of life.[15] Women might watch a demonstration; they did not witness it. The definitive demonstrations of the working of the air pump had to take place in proper, civil public space, even if that meant holding a serious demonstration late at night to exclude women of his class, as Boyle did. Enhancing their agency thereby, modest men were to be self-invisible, transparent, so that their report might not be polluted by the body. Only in that way could they warrant their descriptions of other bodies and minimize critical attention to their own. This is a crucial epistemological move in the grounding of several centuries of race, sex, and class discourses as objective scientific reports.[16]

A central issue requires compressed comment here: the structure of heroic action in science. Several scholars have commented on the proliferation of violent misogynist imagery in many of the chief documents of the Scientific Revolution.[17] The modest man had at least a tropic taste for the rape of nature. Science made was nature undone, to embroider on Bruno Latour's metaphors in his important *Science in Action*. Nature's coy resistance was part of the story, and getting nature to reveal her secrets was the prize for manly valor—all, of course, merely valor of the mind. At the very least, the encounter of the modest witness with the world was a great trial of strength. In disrupting many conventional accounts of scientific objectivity, Latour and others have masterfully unveiled the self-invisible modest man. At the least, that is a nice twist on the

usual direction of discursive unveiling and heterosexual epistemological erotics.[18] Steve Woolgar would then keep the light relentlessly on this modest being, the "hardest case" or "hardened self"
that covertly guarantees the truth of a representation, which ceases
magically to have the status of a representation and emerges simply
as the fact of the matter.[19] That crucial emergence depends on many
kinds of transparency in the grand narratives of the experimental
way of life. Latour and others eschew Woolgar's relentless insistence on reflexivity, which, in perhaps ungenerously read versions,
seems not to be able to get beyond self-vision as the cure for self-
invisibility. The disease and the cure seem to be practically the same
thing, if what you are after is another kind of world and worldli-
ness. Diffraction, the production of difference patterns, should be a
more useful metaphor for the needed work than reflexivity.

Latour is generally less interested than his colleague in forcing
the Wizard of Oz to see himself as the linchpin in the technology
of scientific representation. Latour wants to follow the action in
science-in-the-making. He wants to make us swerve from the
unlively analytical habits of contemplating science already made.
Perversely, however, the structure of heroic action is only inten-
sified in this project—both in the narrative of science and in the
discourse of the science-studies scholar. For the author of *Science in
Action*, technoscience itself is war, the demiurge that makes and
unmakes worlds.[20] Privileging the younger face as science-in-the-
making, Latour adopts as the figure of his argument the double-
faced Roman god, Janus, who, seeing both ways, presides over the
beginnings of things. Janus is the keeper of the gate of heaven, and
the gates to his temple in the Roman Forum were always open in
times of war and closed in times of peace. War is the great creator and destroyer of worlds, the womb for the masculine birth of
time. The action in science-in-the-making is all trials and feats of
strength, amassing of allies, forging of worlds in the strength and
numbers of coerced allies. All action is agonistic; the creative abstraction is both breathtaking and numbingly conventional. Trials
of strength decide whether a representation holds or not. Period.
To compete, one must have the force equivalent of a counterlabo-
ratory capable of winning in these high-stakes trials of strength, or
give up dreams of making worlds. Victories and performances are
the action sketched in this all-too-seminal book: "The list of trials

becomes a thing; it is literally reified."[21] War is the paradigmatic rhetorical performance—the perfect technique of physical-spiritual persuasion. The word is performative. The word is spermatic; it is made flesh. This is an old figure in sacred secular technoscience narratives.

This powerful tropic system is like quicksand. *Science in Action* works by relentless, recursive mimesis. The story told is told by the same story. The object studied and the method of study mime each other. The analyst and the analysand both do the same thing, and the reader is sucked into the game. It is the only game imagined. The goal of the book is "penetrating science from the outside, following controversies and accompanying scientists up to the end, being slowly led out of science in the making."[22] The reader is taught how to resist both the scientist's and the false science-studies scholar's recruiting pitches. The prize is not getting stuck in the maze, but exiting the space of technoscience a victor, with the strongest story. No wonder Steven Shapin began his review of this book with the gladiator's salute: "Ave, Bruno, morituri te salutant."[23]

So, from the point of view of some of the best work in recent malestream science studies, "nature" is multiply the feat of the hero, more than it ever was for Boyle. First, nature is a materialized fantasy, a projection whose solidarity is guaranteed by the self-invisible representer. Unmasking this figure, he who would not be hoodwinked by the claims of philosophical realism and the ideologies of disembodied scientific objectivity fears to "go back" to nature, which was never anything but a projection in the first place.[24] The projection nonetheless tropically works as a dangerous female threatening manly knowers. Then, in Latour's work, another kind of nature is the result of trials of strength, also the fruit of the hero's action. Finally, the scholar too must work as a warrior, testing the strength of foes and forging bonds among allies, human and nonhuman, just as the scientist-hero does. The self-contained quality of all this is stunning. It is the self-contained power of the culture of no-culture itself, where all the world is in the sacred image of the Same. This narrative structure is at the heart of the potent modern story of European autochthony.[25]

What accounts for this intensified commitment to virile modesty? I have two suggestions. First, failing to draw from the understandings of semiotics, visual culture, and narrative practice coming specifically from feminist, postcolonial, and multicultural oppositional theory, many science-studies scholars leave their basic narratives and tropes essentially unexamined. In particular, the "self-birthing of man," "war as his reproductive organ," and "the optics of self-origination" narratives that are so deep in Western philosophy and science have been left in place, while so much else has been fruitfully scrutinized. Second, many contemporary science-studies scholars, like Latour, in their energizing refusal to appeal to Society to explain Nature, or vice versa, mistake other narratives of action about the production of scientific knowledge as functionalist accounts appealing in the tired old way to preformed categories of the Social, like gender, race, and class. Either critical scholars in antiracist, feminist cultural studies of science and technology have not been clear enough about racial formation, gender in the making, the forging of class, and the discursive production of sexuality *through the constitutive practices of technoscience production themselves*, or the science-studies scholars aren't reading or listening—or both. For the oppositional critical theorists, both the facts and the witnesses are constituted in the encounters that are technoscientific practice. Both the subjects and the objects of technoscience are forged and branded in the crucible of specific, located practices, some of which are global in their location. In the intensity of the fire, the subjects and objects regularly melt into each other. It is past time to end the failure of mainstream and oppositional science-studies scholars to engage each other's work. Immodestly, I think the failure to engage has not been symmetrical.

Let me close this meditation on figures who can give credible testimony to matters of fact by asking how to queer the modest witness this time around, so that s/he is constituted in the furnace of technoscientific practice as a self-aware, accountable, antiracist, worldly, hybrid, and mobile FemaleMan, one of the proliferating, uncivil children of the early modern *haec vir* and *hic mulier*. Like Latour, the feminist philosopher of science Sandra Harding is concerned with strength, but of a different order and in a different story. In *Whose Science? Whose Knowledge? Thinking from Women's Lives*,[26] Harding develops an argument for what she calls "strong objectivity" to replace the flaccid standards for establishing matters of fact insaturated by the literary, social, and material technologies inherited from Boyle.

Agreeing that science is the result of located practices at all levels, Harding concurs with Woolgar that reflexivity is a virtue the modest witness needs to cultivate. But her sense of reflexivity is closer to my sense of diffraction than it is to Woolgar's rigorous resistance to making strong knowledge claims. The point is to make a difference in the world, to cast our lot for some ways of life and not others. To do that, one must be in the action, be finite and dirty, not transcendent and clean. Knowledge-making technologies, including crafting subject positions and ways of inhabiting such positions, must be relentlessly made visible and open to critical intervention. Like Latour, Harding is committed to science-in-the-making. Unlike Latour, she does not mistake the constituted and constitutive practices that generate and reproduce systems of stratified inequality—and that issue in the protean, historically specific, marked bodies of race, sex, and class—as preformed, functionalist categories. I do not share her terminology of macrosociology and her all-too-self-evident identification of the social. But I think her basic argument is fundamental to a different kind of strong program in science studies, one that really does not flinch from an ambitious project of symmetry that is as much committed to knowing about the people and positions from which knowledge can come and for which it is targeted, as it is to the status of knowledge made.[27]

Critical reflexivity, or strong objectivity, does not dodge the world-making practices of forging knowledges with different chances of life and death built into them. All that critical reflexivity, diffraction, situated knowledge, or strong objectivity "dodges" is the double-faced, self-identical god of transcendent cultures of no-culture, on the one hand, and of subjects and objects exempt from the permanent finitude of engaged interpretation, on the other. No layer of the onion of practice that is technoscience is outside the reach of technologies of critical interpretation and critical inquiry about positioning and location; that is the condition of embodiment and mortality. The technical and the political are like the abstract and the concrete, the foreground and the background, the text and the context, the subject and the object. As Katie King teaches us, following Gregory Bateson, these are questions of pattern, not of ontological difference.[28] The terms pass into each other; they are shifting sedimentations of the one fundamental thing about the world—relationality. Oddly, embedded relationality is the prophylaxis for both relativism and transcendence. Nothing comes without its world, so trying to know those worlds is crucial. From the point of view of the culture of no-culture, where the wall between the political and the technical is maintained at all costs, and interpretation is assigned to one side and facts to the other, such worlds can never be investigated. Strong objectivity insists that both the objects and the subjects of knowledge-making practices must be *located*. Location is not a listing of adjectives or assigning of labels like race, sex, and class. Location is not the concrete to the abstract of decontextualization. Location is the always-partial, always-finite, always-fraught play of foreground and background, text and context, that constitutes critical inquiry. Above all, location is not self-evident.[29]

Location is also partial in the sense of being *for* some worlds and not others. There is no way around this polluting criterion for strong objectivity. In a wonderful paper called "Power, Technology, and the Phenomenology of Conventions: On Being Allergic to Onions,"[30] Susan Leigh Star explores taking sides in a way that is perhaps more readily heard by more science-studies scholars than Harding's more conventional philosophical vocabulary. Star is interested in taking sides with some people or other actors in the enrollments and alliance formations that constitute so much of the technoscientific action. Her points of departure are feminist and interactionist modes of inquiry that privilege the kind of witness possible from the point of view of those who suffer the trauma of not fitting into the standard. Not to fit the standard is another kind of transparency or invisibility; Star would like to see if this kind is conducive to crafting a better modest witness. She is compelled by the starting point of the monster, of what is exiled from the clean and light self; and so she suspects that the voices of those suffering from the abuses of technological power are among the most powerful analytically. Her own annoying but persistent allergy to onions, and the revealing difficulty of convincing service personnel in restaurants that such a condition is real, functions as Star's narrative wedge into the question of standardization. In order to address questions about power in science and technology, Star looks at how standards produce invisible work for some while clearing the way for others, and at how consolidated identities for some produce marginalized locations for others. She adopts what she calls a kind of "cyborg" point of view, where her "cyborg" is the "relationship between standardized technologies and local experience," where one falls "between the categories, yet in relationship to them."[31]

In short, Star thinks "that it is both more analytically interesting and more politically just to begin with the question, *cui bono*, than to begin with a celebration of the fact of human/nonhuman mingling."[32] She does not question the fact of the implosion of categorical opposites; she is interested in who lives and dies in the force fields generated. "Public" stability for some is "private" suffering for others; self-invisibility for some comes at the cost of public invisibility for others. I think that is the grammatical structure of "gender," "race," "class," and similar clumsy categorical attempts to name how the world is experienced by the nonstandard, who nonetheless are crucial to the technologies of standardization and the ease of fitting for others.

Star stresses that we are all members of many communities of practice. Multiplicity is in play with questions of standardization, and no one is standard or ill fitted in all communities of practice. Some kinds of standardization matter more than others, but all forms work by producing those who do not fit as well as those who do. Double vision is crucial to inquiring into the relations of power and standards that are at the heart of the subject-making and object-making processes of technoscience. Where to begin and where to be based are the fundamental questions in a world in which "power is about *whose* metaphor brings worlds together, and holds them there."[33] The point is to learn to remember that we might have been otherwise, and might yet be, as a matter of embodied fact. Being allergic to onions is a niggling tropic irritant to the scholarly drive to forget one's own complicity in apparatuses of exclusion that are constitutive to what may count as knowledge. Persistent, acute nausea and stomach pain can foster a keen appreciation of situated knowledges.

So I close this evocation of the figure of the modest witness in the narrative of science with the hope that the technologies for establishing what may count as the case about the world may be rebuilt to bring the technical and the political back into realignment, so that questions about possible livable worlds lie visibly at the heart of our best science—and science studies.

DAVID J. STUMP

Afterword:
New Directions in the
Philosophy of Science Studies

❖❖❖

In 1977, in the new afterword to the second edition of a collection of papers from a major conference on the state of the philosophy of science held in 1969, Frederick Suppe said that "philosophy of science is coalescing around a new movement or approach which espouses a hard-nosed metaphysical and epistemological realism that focuses much of its attention on 'rationality in the growth of scientific knowledge.'"[1] He was certainly right; realism and rationality have been important issues in postpositivist philosophy of science. I hope to be as accurate as Suppe when I claim that science studies is now coalescing around a new movement that dismisses questions of realism and relativism, and focuses on the multifaceted detail of scientific practice.

Rationality became an important issue in the philosophy of science in the wake of the skeptical challenges presented by Kuhn, Feyerabend, and others. At first, philosophers reacted by completely rejecting Kuhn and Feyerabend. They claimed that arguments about incommensurability were incoherent, that the logical analysis of science was still viable, and that the philosophical accounts of scientific methodology could be maintained. However, in order to respond convincingly to these challenges, philosophers were forced to take the history of science and current scientific practice far more seriously than they had previously, since the critics seemed to be able to present real cases of scientific practice that refuted philosophical views. Even if Kuhn's challenge was not the only reason that philosophers began to study the history of science, philosophers have had to pay some attention to historical cases and current scientific practice in order to make their case for the rationality of science. The new tool of philosophical analysis was an internal history of science, with which the philosopher could debate Kuhn on his own ground. During the same period, interdisciplinary programs in history and philosophy of science were formed, and philosophers considered various ways to naturalize their enterprise. Rather than attempting to make science fit into a logical structure, or to provide prescriptive methodological rules for scientists, the history of science and empirical study (of, say, cognitive psychology) were supposed to provide evidence for philosophical views about science.

Some philosophers and historians were aware at the time that the history of science cannot be seen as a database for philosophical theories and questioned whether philosophy and history of science really could be integrated. The relation of history of science and philosophy of science was widely discussed in the 1970's in terms of intimacy, marriage, and divorce.[2] However, it is unclear how much the discussion raised the sophistication of the philosophical use (and abuse) of history. Nevertheless, many saw integrating history and philosophy of science as absolutely necessary to maintaining the credibility of philosophical accounts of science and continued developing interdisciplinary programs.

We are now in the midst of a second round of philosophical reaction to skeptical challenges, and philosophers are again being forced to learn new methods of study and adopt new argumentative styles in response. Again philosophers are forced to consider more detail and take a closer look at science. This time the skeptical arguments come from sociologists and anthropologists of science and their allies, the cultural historians. While arguments about relativism and the rationality of science could be framed within internalist historical accounts of science in the first round, these accounts are inadequate to the task of responding to the social-contextual arguments of the second round. The new historiography of science studies rejects internal history before the debate can even get started. In order to engage in discussion with the skeptics, it will again be necessary to engage them on their own terms.

In order to drive home the point that philosophers cannot continue with business as usual, I want to block three possible philosophical attempts to avoid taking science studies seriously. First, it is impossible to escape the demand of the new historiography for the inclusion of heterogeneous elements of scientific practice by claiming that social factors play a role only in the discovery or the development of a scientific theory, not in its justification. A strong tradition in the philosophy of science argues that the process of discovery in science, whether method, madness, luck, or even social influence, has no effect on the ultimate result so long as theory testing is carried out by fully rational processes. One way to look at this philosophical response to the inclusion of social elements of scientific practice is to say that this tradition in the philosophy of science is simply at cross-purposes with science studies, and that the two look at different aspects of science; but attempting to separate discovery and justification does not address the current debate. The social constructivists have argued forcefully that justification, as well as discovery, of scientific theories involves social factors, so it is not so easy to slip away from the argument for the inclusion of social factors. The separation argument is bound to box philosophers into a corner by forcing them to an ultratraditionalist position that is still held by many philosophers who specialize in epistemology but that practically no one in philosophy of science would accept. The dismissal of discovery reduces philosophy of science to epistemology by making the task of philosophy of science that of certifying that science is knowledge (or that it is rational). According to the ultratraditionalists, methodology plays a foundational role. In order to ground science, methodology must be autonomous of science so as to avoid circularity, and it must be more certain. But in the move to naturalism, philosophers gave up both the autonomy of epistemology and the project of grounding science. Much of the disagreement between traditional epistemologists and naturalists has to do with a conception of philosophy as an autonomous intellectual discipline with a logically determined set of issues and a conception of philosophy as an interactive and interdisciplinary activity with a historically determined set of concerns. Accepting minimal standards of scholarship, the project of understanding science requires looking at heterogeneous influences on the local setting. By rejecting current historiographic standards, traditional epistemologists are put in the strange position of rejecting what is required to understand science in order to justify science, as if one could justify science without fully understanding it.

The science-studies counterargument to the separation of discovery and justification goes something like this: We want to account for the dynamics of scientific practice and to explain how changes and stabilizations of scientific practice take place. Therefore, we have to understand how the actors at the time saw the situation in order to see why they gave up or adopted certain practices (theories, methods, and aims; but also experimental set-ups, equipment, etc.). The Lakatosian project of coming back later and certifying the practice as justified or rational is of no use to the understanding of science as practiced. The traditionalists beg the question by claiming that only the epistemological status of the final product is interesting, and not the development of science. They cannot dismiss the development of science as unimportant by merely restating their commitment to foundational epistemology.

Second, it would be a mistake to dismiss science studies out of hand as mere relativism, or as idealism, or as an attack on science. One finds a full spectrum of views in science studies, and different sorts of arguments against traditional philosophical accounts of science can be found, as well as disputes among those writing in the field. What is required from philosophers is more help with developing a positive philosophy to supersede the failed views of the past—logical empiricism, realism, and Kuhnian relativism, as well as heroic biography and whiggism. The way to overcome the rhetorical excesses found in some works in the science-studies genre is for philosophers to become engaged positively in building the science-studies discipline. Philosophers can play a special role in that they bring an acquaintance with the philosophical tradition to the discussion of science-studies issues, but there is philosophical work to be done by all, not just by professional philosophers. After all, philosophical issues show up in many if not all disciplines, and interesting philosophical work is already being done by many in science studies.

Third, it is important that philosophers not ignore science studies and simply get lost in the technical detail of scientific subdisciplines without considering broader issues. The best philosophy of science has always engaged broad philosophical issues by being immersed in the technical aspects of cutting-edge science, just as the best history of science has always kept large issues in the forefront while documenting and analyzing the situated context in the most minute detail. In terms of institutional practice, philosophy of science has splintered into the philosophy of special sciences. There is a danger that philosophy of science will lose its generality, and this would be a real loss, since philosophers play the role of the generalists in an academia continually more dominated by specialists.

The historical and cultural contextualization of science presents new challenges to the philosophy of science. In this volume we have seen essays on what the unity and disunity of science have meant to philosophers. We have also seen an analysis of the epistemological and metaphysical implications of the new historiography of science studies and a clarification of the methodological position of science-studies researchers. I will mention six philosophical issues facing science studies, several of them already addressed in this volume. The list is by no means exhaustive, and there is some overlap between issues; it is meant to encourage further research.

Overcoming the dead-end debates over realism, the rationality of science, and relativism. The issues of realism, the rationality of science, and relativism cannot simply be dropped from the science-studies agenda, even if the majority of researchers in the field are quite weary of them. These issues will not go away until we have understood completely why the old debate flourished and have provided a new way to frame science-studies issues that will overcome both the realist/constructivist dichotomy and the rationalist/relativist one. Therefore, we will need a continuing debate on these topics, but a positive and consensus-building debate, not a hostile and divisive debate such as has occurred in the past. We must move toward the development of a new vocabulary, a new epistemology, and a new metaphysics that parties on both sides of the old debate can adopt. Some philosophical issues will be framed in a way that is fairly close to traditional philosophical issues, while others will be radically redefined.

Studying the philosophical implications of the contextualization of scientific practice. Recent studies of scientific practice integrate theoretical content, laboratory practice, materials, and instrumentation, and the social and political context in shaping science. The philosophical implications of the view of science that emerges from such studies still need further development. If contextualizing a scientific practice is not merely a way of saying that it is relative to a particular culture and time, what does the local nature of scientific practice imply? If close, detailed accounts of historical and contemporary scientific practice within a social context lead to neither realism nor relativism, where do they lead? We need to provide an account of the process of contextualization that shows why contextualization does not amount to delegitimation. We also need to reconsider the role of the analyst. One can easily think up arguments and positions that the scientists themselves did not use, a practice that has a long tradition in philosophy as rational reconstruction, and in the social construction of knowledge as opening up black boxes. Given that these arguments are not historical, what is their status? Can they be legitimate analyses of the situations in which scientists find themselves in a particular setting, or do they always illegitimately import our current knowledge, a taken-for-granted social context, or some other misleading factor into an analysis? Finally, we need to provide an account of how knowledge and skills are disseminated out of a specified context despite their being local.

Rethinking our position as science-studies researchers. In the process of developing a new vocabulary, epistemology, and metaphysics for science studies, we must be aware of the potential political and social effects of our studies. Thus, we must avoid the danger of adopting language and metaphors that are associated with traditions that we want to reject. We must clarify our goals as science-studies researchers and discuss the political content of our work. The role of philosophy in science studies also needs to be clarified, and the relation of the philosophy of science to its own history is an important part of this process. In recent years philosophers have actively reevaluated the logical positivists and found a movement that is quite different from the one that has been represented as canonical. Of special interest in this volume is a reevaluation of the logical positivists' unity of science program, since it seems to be in sharp contrast to the current view of science as disunified. Philosophical foundational studies of special sciences will also have to be seen in a new light. What is the relation of foundational studies to science studies? Has the philosophy of science itself become disunified because of specialization?

Studying the philosophical implications of the unity and disunity of science. In accord with the rest of science studies, a central theme of recent philosophy of science has been the view that the methods of science themselves are not unitary. This can be seen in Dudley Shapere's argument that science has developed into independent domains, in the new philosophy of experiment advocated by Ian Hacking and Peter Galison, in which experiment has a life of its own independent of theory, in Arthur Fine's advocating of an anti-essentialist view of science as a way of getting beyond debates over realism, in John Dupré's pluralistic metaphysics, and in Thomas Nickles's claim that scientists concern themselves only with do-main-specific methods.[3] The disunity of science follows from its contextualization—that is, from studies of the influence of inde-pendent domains on the development of science. The disunity of science is crucial to overcoming the realism/relativism debate, and may be our guide in developing a view of science that is radically different from both traditional and relativist views of science.

Rethinking explanation of scientific practice. Whig histories take current knowledge for granted when analyzing the development of a scientific practice; some constructivist analyses take the social context for granted instead. In explaining how a scientific practice develops, the social context is just as much in need of explanation as scientific knowledge. The story of how a particular social context was shaped by the influence of a scientific practice may prove as interesting as the story of how a given scientific practice was shaped by its social and political context. Can we do both? The notion of explanation itself requires rethinking. Is explanation of a phenome-non tantamount to a reduction of one domain to another? I think that recent accounts of explanation as responding to why-questions take us much farther from the tradition than most discussions would suggest.[4] On this view, there is no such thing as an explana-tion *per se*; and there is no such thing as the explanatory power of a theory. (Imagine a discussion of the answering power of a theory. Immediately, we would say that it depends on what kinds of ques-tions are asked.) Realists are committed to there being only one kind of objective story (sequence of events), namely the causal sequence; but science-studies researchers will not want to be sad-dled with such a commitment.

Expressing the temporality of scientific practice. Whig histories, realist explanation, and logical analysis or scientific practice are all manifestly atemporal, whereas the material practice, the institutional organization, the content of theories, the methods of analysis, the aims of inquiry, and the ontology associated with scientific theories all change through time. A major goal of science studies must be to account for the temporality of scientific practice, and a rethinking of philosophy is especially important in this area since traditional philosophy has never been able to deal adequately with temporal change. Emphasis on temporal changes can also begin to dissolve the dualist representational account of knowledge if we deny that there is an *a priori* atemporal distinction between nature and mind or society.

In sum, the philosophy of science is in a period of transition. We must follow contemporary historiography when gathering evidence about the way scientific practice functions and redefine our own enterprise, as well as help those in our sister disciplines. The beginning of a real integration of philosophy into science studies has been the aim of this volume. I will end with "A Little Fable" for those who do not see a new direction in the science-studies literature, those who would rather maintain the traditional and comfortable debate on the social versus the rational nature of science:[5]

"Alas," said the mouse, "the world is growing smaller every day. At the beginning it was so big that I was afraid, I kept running and running, and I was glad when at last I saw walls far away to the right and left, but these long walls have narrowed so quickly that I am in the last chamber already, and there in the corner stands the trap that I must run into." "You only need to change your direction," said the cat, and ate it up.

Reference Matter

Notes

Galison, Introduction

1. Evelyn Fox Keller, *A Feeling for the Organism* (San Francisco: Freeman, 1983).

2. For more on unity of science and the "limits of science" debate, see Keith Anderton, "The Limits of Science: A Social, Political, and Moral Agenda for Epistemology in Nineteenth-Century Germany" (diss., Harvard, 1993), 115ff., 200ff.; Anderton's thesis also contains extensive citations to primary and secondary literature on Helmholtz, Dubois-Reymond, and Virchow.

3. Ibid.

4. Carl G. Hempel, pers. comm., 1986.

5. John Dewey, "Unity of Science as a Social Problem," in R. Carnap, O. Neurath, and C. W. Morris, eds., *Encyclopedia of Unified Science* (Chicago: University of Chicago Press, 1938), 32–33.

6. Victor Lenzen, "Procedures of Empirical Science," in R. Carnap, O. Neurath, and C. W. Morris, eds., *Encyclopedia of Unified Science* (Chicago: University of Chicago Press, 1938), 338.

7. P. Oppenheim and Hilary Putnam, "Unity of Science as a Working Hypothesis," in H. Feigl, M. Scriven, and G. Maxwell, eds., *Concepts, Theories, and the Mind-Body Problem*, Minnesota Studies in the Philosophy of Science 2 (Minneapolis: University of Minnesota Press, 1958), 3–36.

8. See, e.g., Timothy Lenoir, "The Politics of Vision" (in press), and "Social Interests and the Organic Physics of 1847," in Edna Ullmann-Margalit, ed., *Science in Reflection: Boston Studies in the Philosophy of Science*, vol. 3 (Dordrecht: Kluwer, 1988), 169–81.

9. Steven Weinberg, *Dreams of a Final Theory: The Search for the Fundamental Laws of Nature* (New York: Pantheon, 1992).

10. Howard Georgi, "A Treasure at Need," talk presented at Cabot House, Harvard University, 1994.

11. P. Suppes, "The Plurality of Science," in P. D. Asquith and I. Hacking, eds., *PSA 1978*, vol. 2 (East Lansing: Philosophy of Science Association, 1978), 3–16.

12. See Ian Hacking, "Language, Truth, and Reason," in M. Hollis and S. Lukes, eds., *Rationality and Relativism* (Cambridge, Mass.: MIT Press, 1982), 48–66.

13. Foucault develops the notion of positivity in the following crucial passage from *The Order of Things* (New York: Vintage, 1970), xxi–xxii: "I am concerned to show . . . in what way, as one traces . . . language as it has been spoken, natural creatures as they have been perceived and grouped together, and exchanges as they have been practised; in what way, then, our culture has made manifest the existence of order, and how, to the modalities of that order, the exchanges owed their laws, the living beings their constants, the words their sequence and their representative value. . . . [This] is an inquiry whose aim is to rediscover on what basis knowledge and theory became possible; within what space of order knowledge was constituted; on the basis of what historical *a priori*, and in the element of what positivity, ideas could appear, sciences be established, experience be reflected in philosophies, rationalities be formed, only, perhaps, to dissolve and vanish soon afterwards." Hacking focuses his attention particularly on the candidacy for truth or falsehood; Foucault, I would argue, takes positivity (in this quotation) to set the evaluational framework a larger task than the determination of what propositions have truth values. The episteme will fix, *inter alia*, the associations, exchanges, order, and identifying features of things.

14. Ian Hacking, *Representing and Intervening* (Cambridge: Cambridge University Press, 1983), 70–71.

15. Arnold Davidson, "Closing Up the Corpses," in G. Boulos, ed., *Meaning and Method* (Cambridge: Cambridge University Press, 1992), 295–325.

16. While this book was in production, Alexander Rosenberg's volume *Instrumental Biology; or, The Disunity of Science* (Chicago: University of Chicago Press, 1994) has appeared with a very interesting variation on Dupré's thesis. Rosenberg contends that it is the nature of biology itself that demands that it be disunified from the physical sciences. In particular, building on David Hull's work, Rosenberg begins with the observation that there may well be many molecular structures that lead to a single phenotype, just as there may be many phenotypes that can issue from the same underlying molecular structure. This problem of the "many and the many" (according to Rosenberg) underscores the difficulty we face in

trying to bridge the gap between the molecular and the Mendelian. Luckily for our scientific understanding, natural selection acts on function, and function can be studied more or less independently of the underlying physicochemical structure. Rosenberg concludes that it is *our* problem, *our* inability as humans to reduce biological phenomena to molecules, that forces us to take an instrumental stance toward elephants and egrets. Physics may aim for prediction and irreducible, invariable laws; biology, according to Rosenberg, aims at the "sharpening of tools for interacting with the biosphere, but not telling the whole truth about it," insofar as the biological accounts may not now or ever be reducible to fundamental theories. Thus Dupré and Rosenberg agree that biology is necessarily disunified. But while Dupré argues that "the failure of reductionism [in the practice of biology] implies the falsity of causal completeness," Rosenberg contends that it is not metaphysics but the limit of our human cognitive capacities that lies at the root of the disunity of science. (Dupré cited in Rosenberg, p. 12; Rosenberg cited from p. 38.)

17. Peter Galison, "History, Philosophy and the Central Metaphor," *Science in Context* 2 (1988): 197–212; the notion of the trading zone is developed in Peter Galison, "Contexts and Constraints," in J. Buchwald and I. Hacking, eds., *Theories of Practice/Stories of Practice* (Chicago: University of Chicago Press; forthcoming); and Peter Galison, *Image and Logic: The Material Culture of Twentieth-Century Physics* (forthcoming), chap. 9; "The Trading Zone: Coordinating Action and Belief"; this chapter of *Image and Logic* has been circulated as a preprint of the same title cited in several essays in this volume as a separate preprint presented at the UCLA Center for Cultural History of Science and Technology Conference, "TECH-KNOW Workshops on Places of Knowledge, Their Technologies, and Economies," 2 December 1989; and Virginia Polytechnic Institute and State University Department of Philosophy Conference, "The Role of Experiment in Scientific Change," 31 March 1990.

18. E.g., Barry Barnes, *T. S. Kuhn and Social Sciences* (New York: Columbia University Press, 1982), 87: "In actual historical cases, as a new paradigm arises so there is an associated transformation in the entire conceptual fabric. What has to be evaluated are two alternative frameworks of discourse and activity. There is a reconstruction of the whole pattern of both. Terms connect to other terms differently, and they connect to nature differently."

19. Thomas Uebel in *Noûs* (forthcoming).

20. Galison, "History, Philosophy, and the Central Metaphor" and "The Trading Zone," both cited above.

21. Bruno Latour, *The Pasteurization of France*, tr. A. Sheridan and J. Law (Cambridge, Mass.: Harvard University Press, 1988), 35.

22. Simon Schaffer, "The Eighteenth Brumaire of Bruno Latour," *Studies in History and Philosophy of Science* 22 (1991): 174–92.

23. While Biagioli shares a great deal with evolutionary epistemologists, he does not argue that the selection process is any guarantee of the rationality of the process of theory selection.

24. S. Shapin and S. Schaffer, *Leviathan and the Air-Pump: Hobbes, Boyle, and the Experimental Life* (Princeton: Princeton University Press, 1985).

25. P. Galison, "Aufbau/Bauhaus: Logical Positivism and Architectural Modernism," *Critical Inquiry* 16 (1990): 709–52.

26. Donna J. Haraway, "A Cyborg Manifesto: Science, Technology, and Socialist-Feminism in the Late Twentieth Century," in *Simians, Cyborgs, and Women* (London: Routledge, Chapman and Hall, 1991), 149.

27. T. Nagel, *The View from Nowhere* (Oxford: Oxford University Press, 1986).

Hacking, The Disunities of the Sciences

1. William Whewell, *The Philosophy of the Inductive Sciences, Founded upon Their History*, vol. 1 (London: Parker, 1840), 1.

2. Ibid., 2.

3. Auguste Comte, *Cours de philosophie positive*, vol. 1 (Paris, 1830 [reprint: Paris: Schleicher, 1907]), lesson 1, 28–30.

4. The connection between the unity of science and the modern professionalization of the philosophy of science is quite strong. Historians measure professionalization by organizations and publications. The original *Synthese*, founded in Holland in 1936, bears its intention in its name. In the early years it regularly included special issues edited by the Forum for Unified Science and the Institute for the Unification of Science. *Philosophy of Science* began in 1934. The name may suggest "science" rather than the sciences, but this journal was always an eclectic organ. Its first issue opened with papers by Feigl and Carnap, but the editor, William Malisoff, welcomed all points of view and added many a spritely and nondoctrinaire touch of his own. The grim professionalism of today had not yet taken hold of the subject.

5. Otto Neurath, "Unified Science as Encyclopedic Integration," in O. Neurath, R. Carnap, and C. W. Morris, eds., *International Encyclopedia of Unified Science*, vol. 1, no. 1 (Chicago: University of Chicago Press, 1938), 1.

6. I owe the distinction to a talk on the soul by Elizabeth Anscombe, in Toronto, 28 September 1989.

7. I did it very differently in a talk, "Disunified Science," given at the

25th Nobel Conference at Gustavus Adolphus College, October 1989; published in R. J. Elvee, ed., *The End of Science?* (Lanham: University Press of America, 1991), 33–52.

8. In 1876, to the bishop of Gloucester and Bristol: L. Campbell and J. Garnett, *The Life of James Clerk Maxwell with a Selection from His Correspondence and Occasional Writings, and a Sketch of His Contributions to Science* (London: Macmillan, 1882), 394.

9. Carlos Castaneda, *The Teachings of Don Juan: A Yaqui Way of Knowledge* (Berkeley and Los Angeles: University of California Press, 1968), and *A Separate Reality: Further Conversations with Don Juan* (New York: Simon & Schuster, 1971).

10. For more on this idea, see n. 39 below.

11. William Blake, in a letter to Thomas Butts, 22 November 1802 (G. Keynes, ed., *The Complete Writings of William Blake* [London: Oxford University Press, 1966], 816). The first line of the poem is "With happiness stretch'd across the hills."

12. Or so I argue in "A Tradition of Natural Kinds," *Philosophical Studies* 61 (1991): 109–26.

13. P. A. M. Dirac, "Methods in Theoretical Physics," reprinted in Abdus Salam, *Unification of Fundamental Forces: The First of the 1988 Dirac Memorial Lectures* (Cambridge: Cambridge University Press, 1990), 128. The lecture was originally published in 1968 as one of a series of evening talks at the International Centre for Theoretical Physics, Trieste.

14. Ibid. 129.

15. Abdus Salam, *Unification of Fundamental Forces: The First of the 1988 Dirac Memorial Lectures* (Cambridge: Cambridge University Press, 1990), 5.

16. Freeman Dyson, *Infinite in All Directions* (New York: Harper and Row, 1985), 45.

17. Ibid., 47.

18. I owe the word "vicarious" to a conversation with a notable unifier, Sheldon Glashow.

19. See Margaret Morrison, "A Study in Theory Unification: The Case of Maxwell's Electromagnetic Theory," *Studies in History and Philosophy of Science* 23 (1992): 103–45.

20. Alexander Rosenberg, *The Structure of Biological Science* (Cambridge: Cambridge University Press, 1985).

21. See Peter Galison, "Aufbau/Bauhaus: Logical Positivism and Architectural Modernism," *Critical Inquiry* 16 (1990): 709–52.

22. Paul Oppenheim and Hilary Putnam, "Unity of Science as a Working Hypothesis," in H. Feigl, M. Scriven, and G. Maxwell, eds., *Concepts, Theories and the Mind-Body Problem*, Minnesota Studies in the

Philosophy of Science 2 (Minneapolis: University of Minnesota Press, 1958), 3–36; the quotations that follow are from pp. 3–4. All emphases and capitals are in the original.

23. The levels are familiar, but better argued for here than elsewhere in a debate that is well over a century old: "Social groups, (multicellular) living things, cells, molecules, atoms, elementary particles" are provided as an example of a sequence that satisfies some carefully stated "conditions of adequacy": ibid., 9.

24. Patrick Suppes, *Probabilistic Metaphysics* (Oxford: Blackwell, 1984), 121. The chapter is based on "The Plurality of Science," in Peter Asquith and Ian Hacking, eds., *PSA 1978* (East Lansing: Philosophy of Science Association, 1978), vol. 2, pp. 3–16.

25. Dyson, *Infinite*, 5.

26. Campbell and Garnett, *Life*, 243.

27. "I myself prefer an Argentinian fantasy. God did not write a Book of Nature of the sort that the old Europeans imagined. He wrote a Borgesian library, each book of which is as brief as possible, yet each book of which is inconsistent with every other. No book is redundant. For every book, there is some humanly accessible bit of Nature such that that book, and no other, makes possible the comprehension, prediction and influencing of what is going on. Far from being untidy, this is New World Leibnizianism. Leibniz said that God chose a world which maximized the variety of phenomena while choosing the simplest laws. Exactly so: but the best way to maximize phenomena and have simplest laws is to have the laws inconsistent with each other, each applying to this or that but none applying to all." (Ian Hacking, *Representing and Intervening* [Cambridge: Cambridge University Press, 1983], 219.)

28. The Holy Quran (XXXI, 27), quoted in Salam, *Unification*, 80.

29. *The New York Times*, 12 Sept. 1989, section A, p. 13.

30. Steven Weinberg, "The Forces of Nature," *Bulletin of the American Academy of Arts and Sciences* 29 (1976): 28. Weinberg attributes the specific phrase "Galilean style of reasoning" to Husserl, but I don't think Husserl himself used it; one has to go back to Spengler for talk of *Stil* in this sense, and then, perhaps (contemporaneous with Husserl's discussion of Galileo), to Ludwik Fleck's *Denkstil*.

31. Noam Chomsky, *Rules and Representations* (New York: Columbia University Press, 1980), 9.

32. For the list of styles, see A. C. Crombie, "Philosophical Presuppositions and Shifting Interpretations of Galileo," in J. Hintikka et al., eds., *Theory Change, Ancient Axiomatics and Galileo's Methodology* (Dordrecht: Reidel, 1981), 284. Crombie's work has now come to fruition in a three-volume study, *Styles of Scientific Thinking in the European Tradition* (London: Duckworth, 1994).

33. Dyson, *Infinite*, 13.

34. Alistair C. Crombie, "Designed in the Mind: Western Visions of Science, Nature and Humankind," *History of Science* 24 (1988): 1–12.

35. The uses for historians and for philosophers of the sciences are different; cf. Ian Hacking, "'Style' for Historians and Philosophers," *Studies in History and Philosophy of Science* 23 (1992): 1–20.

36. The first statement was in Ian Hacking, "Language Truth and Reason," in M. Hollis and S. Lukes, eds., *Rationality and Relativism* (Oxford: Blackwell, 1982), 48–66.

37. Campbell and Garnett, *Life*, 361–62.

38. Richard Rorty, "Is Science a Natural Kind?" in E. McMullin, ed., *Construction and Constraint: The Shaping of Scientific Rationality* (Notre Dame: University of Notre Dame Press, 1988), 49–74.

39. Bernard Williams, "The Scientific and the Ethical," in S. C. Brown, ed., *Objectivity and Cultural Divergence* (Cambridge: Cambridge University Press, 1984), 214; cf. idem., *Ethics and the Limits of Philosophy* (Cambridge, Mass.: Harvard University Press, 1985), 138–39.

40. I have made a start at this in a series of papers: "Statistical Language, Statistical Truth and Statistical Reason: The Self-Authentication of a Style of Reasoning," in E. McMullin, ed., *Social Dimensions of Science* (Notre Dame: University of Notre Dame Press, 1992), 130–55; "The Self-Vindication of Laboratory Science," to appear in A. Pickering, ed., *Science as Practice and Culture* (Chicago: The University of Chicago Press, 1992), 29–64; "Radically Constructivist Theories of Mathematical Progress," forthcoming in an issue of *Iride* edited by Alessandro Pagnini.

41. A. Pickering, "Living in the Material World," in D. Gooding, T. Pinch, and S. Schaffer, eds., *The Uses of Experiment: Studies in the Natural Sciences* (Cambridge: Cambridge University Press, 1989), 275–98.

Davidson, Styles of Reasoning

Portions of the second half of this essay have appeared, in somewhat different form, in "Conceptual Analysis and Conceptual History: Foucault and Philosophy," *Stanford French Review* 8.1 (June 1984): 105–22, and in "Sex and the Emergence of Sexuality," *Critical Inquiry* 14.1 (autumn 1987): 16–48. Much of this essay was originally written in 1983–84, and over the years it has been read at a number of different universities. I have made no attempt to "update" its content, or to alter its oral style of presentation. In rereading it, I discovered that Michel Foucault is referred to in the present tense, and I have left this temporal designation as an indication of my gratitude for the conversations we had on some of the topics of this essay. I am also indebted to Ian Hacking and Peter Galison for many discussions of the basic ideas.

1. Woody Allen, "My Philosophy," in *Getting Even* (New York: Vintage, 1978), 21. The quotation from Kierkegaard is found in *The Sickness unto Death*, in *Fear and Trembling and The Sickness unto Death* (Princeton: Princeton University Press, 1941), 146.

2. A. C. Crombie, *Styles of Scientific Thinking in the European Tradition*, 3 vols. (London: Duckworth, 1994).

3. J. L. Austin, "Intelligent Behaviour: A Critical Review of *The Concept of Mind*," in Oscar P. Wood and George Pitcher, eds., *Ryle: A Collection of Critical Essays* (New York: Anchor, 1970), 51.

4. Kennedy Fraser, "Style," in *The Fashionable Mind* (New York: Knopf, 1981), 73, 77.

5. I should mention that an analysis of fashion that is methodologically closer to what I am attempting here can be found in Roland Barthes, *Système de la mode* (Paris: Seuil, 1967).

6. Michel Foucault, "Truth and Power," in *Power/Knowledge* (New York: Pantheon, 1980), 133.

7. Veyne's remark occurs in a discussion of Foucault by some of France's leading historians: *Magazine Littéraire*, April 1977, 21.

8. Ian Hacking, "Language, Truth and Reason," in Martin Hollis and Steven Lukes, eds., *Rationality and Relativism* (Cambridge, Mass.: MIT Press, 1982), 64–65.

9. Ibid., 60.

10. Meyer Schapiro, "Style," in Morris Philipson and Paul J. Gudel, eds., *Aesthetics Today*, rev. ed. (New York: New American Library, 1980).

11. W. L. Wisan, "Galileo and the Emergence of a New Scientific Style," in Jaakko Hintikka, David Gruedner, and Evandro Agazzi, eds., *Theory Change, Ancient Axiomatics, and Galileo's Methodology: Proceedings of the 1978 Pisa Conference on the History and Philosophy of Science* (Dordrecht: Reidel, 1981), 314.

12. Ibid., 326–27.   13. Ibid.
14. Ibid., 325.   15. Ibid.

16. For some general remarks, see section 2 of Schapiro's "Style."

17. Meyer Schapiro, *Words and Pictures: On the Literal and the Symbolic in the Illustration of a Text* (The Hague: Mouton, 1973), 41.

18. Ibid., 42–43.

19. Ibid., 43.

20. Ibid., 44.

21. Heinrich Wölfflin, "Preface to the Sixth Edition," in *Principles of Art History: The Problem of the Development of Style in Later Art*, 7th ed. (New York: Dover, 1950), vii.

22. Some of the problems with Wölfflin's account are discussed in section 5 of Schapiro's "Style."

23. Wölfflin, *Principles*, 12.

24. Ibid.

25. Ibid., 227.

26. Paul Frankl, *Principles of Architectural History: The Four Phases of Architectural Style, 1420–1900* (Cambridge, Mass.: MIT Press, 1968).

27. Wölfflin, *Principles*, ix. The thesis is repeated on p. 11.

28. Ibid., 228.

29. Michael Baxandall, *Painting and Experience in Fifteenth-Century Italy* (Oxford: Oxford University Press, 1972). See also Michael Baxandall, *The Limewood Sculptures of Renaissance Germany* (New Haven: Yale University Press, 1981); especially relevant is chap. 6.

30. Baxandall, *Painting and Experience*, 152.

31. Ibid., 29–30.

32. Hacking, *The Emergence of Probability* (Cambridge: Cambridge University Press, 1975), 9.

33. Ian Hacking, "Proof and Eternal Truths: Descartes and Leibniz," in S. Gaukroger, ed., *Descartes: Philosophy, Mathematics and Physics* (Barnes and Noble, 1980), 169.

34. Ian Hacking, *Emergence*, chap. 19.

35. Michel Foucault, *The Archeology of Knowledge* (New York: Pantheon, 1972), 183.

36. Arnold I. Davidson, "Closing Up the Corpses," in G. Boulos, ed., *Meaning and Method* (Cambridge: Cambridge University Press, 1990). More detailed historical documentation for my claims can be found in that chapter.

37. Paul Moreau (du Tours), *Des aberrations du sens génésique* (Paris: Asselin, 1880), 2.

38. Ibid., 3.

39. *Oxford English Dictionary* (Oxford: Clarendon Press, 1933) s.v., vol. 7, p. 739.

40. Richard von Krafft-Ebing, *Textbook on Insanity* (Philadelphia: Davis, 1904), 79. Krafft-Ebing considers abolition to be the extreme case of diminution.

41. Ibid., 77–81.

42. Ibid., 81. This same classification is given in Richard von Krafft-Ebing, *Psychopathia Sexualis* (New York: Stein and Day, 1965), 34.

43. Krafft-Ebing, *Textbook*, 83–86; *Psychopathia*, 34–36.

44. Krafft-Ebing, *Psychopathia*, 16, 62–63; see also *Textbook*, 81. For other representative statements, see Albert Moll, *Perversions of the Sex Instinct* (Newark: Julian Press, 1931), 172, 182 (originally published in German in 1891); and Dr. Laupts (pseudonym of G. Saint-Paul), *L'homosex-*

ualité et les types homosexuels: Nouvelle édition de perversion et perversités sexuelles (Paris: Vigot, 1910).

45. Thomas Nagel, "Sexual Perversion," in *Mortal Questions* (Cambridge: Cambridge University Press, 1979), 44.

46. Ibid., 47.   47. Ibid., 46.
48. Ibid., 48.   49. Ibid., 39.
50. Ibid., 48 (my emphasis).   51. Ibid., 50.

52. Thomas Nagel, "Sexual Perversion," *Journal of Philosophy*, 16 January 1969, 15.

53. Ibid., 16.

54. Nagel, *Mortal Questions*, 50–51.

55. Hacking, "Proof," 179.

56. See n. 36.

57. A. C. Crombie, "Philosophical Presuppositions and Shifting Interpretations of Galileo," in *Theory Change* (above, n. 11), 283.

58. Michel Foucault, *Herculine Barbin, Being the Recently Discovered Memoirs of a Nineteenth-Century French Hermaphrodite* (New York: Pantheon, 1980), vii–viii.

59. Ibid., viii.

60. Ibid., 127–28 (my emphasis).

61. Ibid., 135–36.

62. Ibid., 123. Tardieu's book was published in 1874. Parts of it had previously appeared in the *Annales d'hygiène publique* in 1872. A fuller discussion of questions of sexual identity would have to consider this document in detail.

63. Havelock Ellis, "Sexo-Aesthetic Inversion," *Alienist and Neurologist* 34 (1913): 156–67.

64. Ibid., 156.

65. Ibid., 159.

66. Ibid., 158, n. 7.

67. American Psychiatric Association, *Diagnostic and Statistical Manual of Mental Disorders*, 3d ed. (Washington, D.C.: American Psychiatric Association, 1980), 261.

68. D. M. Rozier, *Des habitudes secrètes ou des maladies produites par l'onanisme chez les femmes* (Paris: Audin, 1830). For a discussion of the changing iconography of the insane, see Sander L. Gilman, *Seeing the Insane* (New York: Wiley, 1982).

69. Stanley Cavell, *The Claim of Reason: Wittgenstein, Skepticism, Morality, and Tragedy* (Oxford: Clarendon Press, 1979), 121.

70. Lieberman's remark, "Nonsense is nonsense, but the history of nonsense is a very important science," is made with reference to Gnosti-

cism: "How Much Greek in Jewish Palestine?" in *Texts and Studies* (New York: Ktav, 1974), 228.

Dupré, Metaphysical Disorder

1. The metaphysical views advocated in this paper are worked out in much greater detail in my book *The Disorder of Things: Metaphysical Foundations of the Disunity of Science* (Cambridge, Mass.: Harvard University Press, 1993).

2. The diversity of the ontology of particle physics suggests the fascinating possibility that this homogeneity is at a maximum at one particular level, but not the lowest. If so, it would make more attractive the natural speculation that perhaps there is no lowest level. This in turn might suggest that the apparent homogeneity of the atomic level was to some considerable extent an illusion grounded in the peculiar importance of that level of the kinds of structures that are of central concern to us.

3. For many kinds of organisms it is not even clear what constitutes an individual. For some biological purposes an entire clone of a vegetatively reproducing plant, or the entire community of a social insect, may be the appropriate individual.

4. See B. Barnes and D. Bloor,"Relativism, Rationality and the Sociology of Knowledge," in M. Hollis and S. Lukes, eds., *Rationality and Relativism* (Cambridge: MIT Press, 1982), 21–47.

5. See D. Davidson, "Mental Events," in *Essays on Actions and Events* (Oxford: Oxford University Press, 1980), 207–25.

6. That supervenience entails a kind of reductionism has been argued by J. Kim, "Supervenience and Nomological Incommensurables," *American Philosophical Quarterly* 15 (1978): 149–56.

7. Davidson's arguments for supervenience (Davidson, *Actions and Events*) are actually dependent on a causal thesis of this kind. Davidson's views on causality are laid out in greater detail in his essay "Causal Relations," in *Actions and Events*, 149–62.

8. I pass over here special problems that might arise in notorious cases such as raw feels, moral properties, and suchlike. If the general pluralism that I wish to defend is established, the special difficulties with such cases will be considerably alleviated.

9. Nancy Cartwright, *How the Laws of Physics Lie* (Oxford: Oxford University Press, 1983).

10. An excellent paradigm, because largely free of broad and therefore controversial theoretical conclusions, is Anne Fausto-Sterling's assault on most of the main areas of scientific theorizing about sexual difference: *Myths of Gender* (Cambridge, Mass.: Harvard University Press, 1986).

Galison, Computer Simulations

I would like to thank Nicholas Metropolis for his assistance at many stages of this work. My gratitude goes as well to Nancy Cartwright, Cuthbert Hurd, Kenneth Ford, and Caroline Jones for helpful discussions, and to the Library of Congress, Manuscript Division, and the archives of the American Philosophical Library in Philadelphia. Portions of this text are excerpted from my *Image and Logic: The Material Culture of Twentieth-Century Physics* (forthcoming).

1. Von Neumann to Huxley, 28 Mar. 1946, John von Neumann Correspondence, Manuscript Division, Library of Congress, hereafter referred to as JvNC.

2. See, for example: Herman H. Goldstine, *The Computer: From Pascal to von Neumann* (Princeton: Princeton University Press, 1972); William Aspray, "The Mathematical Reception of the Modern Computer: John von Neumann and the Institute for Advanced Study Computer," in Esther Phillips, ed., *Studies in the History of Mathematics* (Washington, D.C.: Mathematical Association of America, 1987), 166–94; William Aspray, *John von Neumann and the Origins of Modern Computing* (Cambridge, Mass.: MIT Press, 1990); Alice R. Burks and Arthur W. Burks, *The First Electronic Computer: The Atanasoff Story* (Ann Arbor: University of Michigan Press, 1988); and Nancy Stern, *From ENIAC to UNIVAC: An Appraisal of the Early Eckert-Mauchly Computers*, Digital Press History of Computing Series (Bedford, Mass.: Digital Press, 1981).

3. Egon Bretscher, Stanley P. Frankel, Darol K. Froman, Nicholas C. Metropolis, Philip Morrison, L. W. Mordheim, Edward Teller, Anthony Turkovich, and John von Neumann, "Report of Conference on the Super," 16 Feb. 1950, Los Alamos National Laboratory–575, 4.

4. Ibid., 3.

5. Ibid., 20.

6. John von Neumann, "Proposal and Analysis of a New Numerical Method for the Treatment of Hydrodynamical Shock Problems," OSRD-3617, 1944; reprinted in John von Neumann, *Collected Works*, ed. A. H. Taub, vol. 6 (London: Pergamon Press, 1963), pp. 361–69, on 367.

7. Ulam (and Tuck?), draft of "Preliminary Jet Experiments of Interest for Thermonuclear Study," 1: Stanislaw Ulam Papers, American Philosophical Library, hereafter referred to as SUP; handwritten note indicates that this document is a draft of LA-560 (Ulam and Tuck, "Possibility of Initiating a Thermonuclear Reaction . . .").

8. Nicholas Metropolis and S. Ulam, "The Monte Carlo Method," *Journal of the American Statistical Association* 44 (1949): 335–41, on 336.

9. Householder to von Neumann, 15 Jan. 1948, JvNC.

10. Von Neumann to Householder, 3 Feb. 1948, JvNC.

11. See Householder to von Neumann, 15 Jan. 1948, and von Neumann to Householder, 3 Feb. 1948, JvNC. Ulam also endorsed this "reasonably random" notion in his presentation to a conference in September 1949: "It may seem strange that the machine can simulate the production of a series of random numbers, but this is indeed possible. In fact, it suffices to produce a sequence of numbers between 0 and 1 which have a uniform distribution in this interval but otherwise are uncorrelated." (Published as Metropolis and Ulam, "The Monte Carlo Method," on 339.)

12. John von Neumann, "Various Techniques Used in Connection with Random Digits," in A. S. Householder, G. E. Forsythe, and H. H. Germond, eds., *Monte Carlo Method*, National Bureau of Standards, Applied Mathematics Series, no. 12 (Washington, D.C.: USGPO, 1951), 36–38, on 36; reprinted in von Neumann, *John von Neumann Collected Works*, ed. A. H. Taub, vol. 5 (New York: Macmillan, 1963), pp. 768–70, on 768.

13. Von Neumann to Richtmyer, 11 Mar. 1947, in von Neumann, *Collected Works*, vol. 5, pp. 751–62, on 753.

14. Richtmyer to von Neumann, 2 Apr. 1947; in von Neumann, *Collected Works*, vol. 5, pp. 763–64.

15. Von Neumann to Ulam, 17 Dec. 1947, SUP.

16. Von Neumann to Ulam, 28 Mar. 1949, JvNC.

17. Von Neumann added: "I think that I have now come near to a procedure, which is reasonably economic at least as a first try: I will treat the Ph[otons?]. by the old "blackjack" method, which seems very well suited to the peculiarities of this species in this milieu; the other species [presumably neutrons] by the normal MC, and, of course, one of them [presumably charged particles such as helium nuclei] by 'local deposition'. [That is, all their energy was deposited at once.]" I do not know (nor does Metropolis remember) what the "blackjack method" was—though I suspect it was a simple method of generating random collisions.

18. Ulam to von Neumann, 17 Mar. 1950, JvNC, box 22.

19. Ulam to von Neumann, 27 Jan. 1950, JvNC, box 22.

20. Von Neumann to Ulam, 7 Feb. 1950, JvNC, box 22.

21. Ulam to von Neumann, 24 Mar. 1950, JvNC, box 22.

22. S. Ulam, "Introduction" to E. Fermi, J. Pasta, and S. Ulam, "Studies of Non-linear Problems," in *Enrico Fermi: Collected Papers*, ed. E. Segrè, vol. 2 (Chicago: University of Chicago Press, 1965), pp. 977–78, on 978.

23. Handwritten diaries in SUP.

24. Ibid.

25. Ibid.

26. K. V. Roberts, "Computers and Physics," in International Center for Theoretical Physics, *Computing as a Language of Physics* (Vienna: International Atomic Energy Agency, 1972), 3–26, on 7.

27. Ibid.

28. Metropolis and Ulam, "The Monte Carlo Method," 339.

29. J. M. Hammersley and D. C. Handscomb, *Monte Carlo Methods* (London: Methuen, 1964; reprint, 1979), 1.

30. Evans Hayward and John Hubbell, "The Albedo of Various Materials for 1-Mev Photons," *Physical Review* 93 (1954): 955–56, on 955.

31. L. Kowarski, "The Impact of Computers on Nuclear Science," in International Center for Theoretical Physics, *Computing as a Language of Physics*, 27–37, on 35.

32. Ibid., 29.

33. Ibid., 36.

34. On replicability, see H. M. Collins, *Changing Order: Replication and Induction in Scientific Practice* (London: Sage Publications, 1985); S. Shapin and S. Schaffer, *Leviathan and the Air-Pump: Hobbes, Boyle, and the Experimental Life* (Princeton: Princeton University Press, 1985); and S. Schaffer, "Glass Works: Newton's Prisms and the Uses of Experiment," in D. Gooding, T. Pinch, and S. Schaffer, eds., *The Uses of Experiment: Studies in the Natural Sciences* (Cambridge: Cambridge University Press, 1989), 67–104.

35. Roberts, "Computers and Physics," 17.

36. Ibid., 19.

37. Ibid., 20.

38. Kowarski, "The Impact of Computers," 35.

39. Ibid., 36. Sharon Traweek has put this nicely in her *Beamtimes and Lifetimes: The World of High Energy Physicists* (Cambridge, Mass.: Harvard University Press, 1988), on 158, where she writes that detectors are where "two kinds of cosmological time converge," one of which is the time of particle interactions and the other "experiential" time.

40. G. Goertzel and H. Kahn, "Monte Carlo Methods for Shield Computation," Summer Shielding Session, Oak Ridge National Laboratory, ORNL-429, 19 Dec. 1949, 10.

41. Ibid., 8.

42. S. Ulam, "John von Neumann, 1903–1957," *Bulletin of the American Mathematical Society* 64 (1958): 1–49, on 34.

43. Gilbert W. King, "Further Remarks on Stochastic Methods in Quantum Mechanics," in Cuthbert C. Hurd, ed., *Computation Seminar December 1949* (New York: International Business Machines Corporation, 1951), 92–94, on 92.

44. Gilbert W. King, "Monte Carlo Method for Solving Diffusion Problems," *Industrial and Engineering Chemistry* 43 (1951): 2475–78, on 2475.

45. Ibid., 2476.

46. Ibid. (emphasis added).

47. From the discussion following Gilbert W. King, "Stochastic Method in Quantum Mechanics," in Cuthbert C. Hurd, ed., *Seminar on Scientific Computation Proceedings, 16–18 November 1949* (New York: International Business Machines Corporation, 1950), 42–48, on 48.

48. Robert R. Wilson, "Showers Produced by Low Energy Electrons and Photons," in Householder et al., *Monte Carlo Method*, 1–3.

49. Robert R. Wilson, "Monte Carlo Study of Production," *Physical Review* 86 (1952): 262–69.

50. A. S. Householder, "Neutron Age Calculations in Water, Graphite, and Tissue," in Householder et al., *Monte Carlo Method*, 6–8, on 8; emphasis added.

51. Nancy M. Dismuke, "Monte Carlo Computations," in Herbert A. Meyer, ed., *Symposium on Monte Carlo Methods Held at the University of Florida, March 16 and 17, 1954* (New York: Wiley, 1956), 52–62, on 52 (emphasis added).

52. A. W. Marshall, "An Introductory Note," in Meyer, ed., *Symposium on Monte Carlo Methods*, 1–14, on 4–5.

53. Dr. Howlett, in E. C. Fieller et al., "Discussion on Symposium on Monte Carlo Methods," *Journal of the Royal Statistical Society* B 16 (1954): 61–75, on 63.

54. Both quotations from the discussion following M. D. Donsker and Mark Kac, "The Monte Carlo Method and Its Applications," in Hurd, *Computation Seminar December 1949*, 74–81, on 81.

55. F. James, "Monte Carlo Theory and Practice," *Reports on Progress in Physics* 43 (1980): 1145–89; reprinted in Thomas Ferbel, ed., *Experimental Techniques in High Energy Physics* (Menlo Park: Addison-Wesley, 1987), 627–78, on 628 (emphasis added).

56. John Hammersley and Keith Morton, in Fieller et al., "Discussion on Symposium on Monte Carlo Methods," 61–75, on 74.

57. William Gibson's trilogy consists of *Neuromancer* (New York: Ace, 1984), *Count Zero* (New York: Arbor, 1986), and *Mona Lisa Overdrive* (New York: Bantam, 1988).

58. K. Ford, "H-Bomb Interview" (with author and with P. Hogan [New York: Media Transcripts, Inc., Phillip Burton Productions, 1987]), SR 21, 53.

59. Ulam to Kahn, 23 Mar. 1954, SUP.

60. H. H. Germond, "Round Table Discussion: Summary," in Householder et al., *Monte Carlo Method*, 39–42, on 39.

Creath, The Unity of Science

1. While both Carnap and Neurath were active in the Vienna Circle, by the time that the unity of science movement was under way Carnap had moved to Prague. Eventually Neurath fled to England, and Carnap moved to Chicago and to Los Angeles. Even so it is easiest to treat them as from Vienna. That is, after all, the city where they became famous and with which they are still appropriately identified.

2. Rudolf Carnap, "Logical Foundations of the Unity of Science," in Otto Neurath, Rudolf Carnap, and Charles W. Morris, eds., *International Encyclopedia of Unified Science* (Chicago: University of Chicago Press, 1938), vol. 1, no. 1, pp. 42–62.

3. Ibid.

4. See especially Carnap's remarks in Paul A. Schilpp, ed., *The Philosophy of Rudolf Carnap* (LaSalle: Open Court Press, 1963), 23, 52, 880.

5. Contrast this with the view of Paul Oppenheim and Hilary Putnam, "Unity of Science as a Working Hypothesis," in H. Feigl, M. Scriven, and G. Maxwell, eds., *Concepts, Theories, and the Mind-Body Problem*, Minnesota Studies in the Philosophy of Science 2 (Minneapolis: University of Minnesota Press, 1958), 3–36.

6. For more on the substantive and priority issues involved, see an extremely interesting paper by Thomas Uebel, "Physicalism in Wittgenstein and the Vienna Circle," forthcoming.

7. E.g., Oppenheim and Putnam, "Unity of Science as a Working Hypothesis."

8. See especially his "Computer Simulations and the Trading Zone" in this volume; and "The Trading Zone: Coordinating Action and Belief," forthcoming in *Image and Logic: The Material Culture of Modern Physics*.

Fuller, Talking Metaphysical Turkey

An earlier version of this essay was delivered at the "Epistemological Chicken" session at the joint meeting of the Society for Social Studies of Science (4S) and European Association for the Study of Science and Technology (EASST) in Göteborg, Sweden, August 1992. Many thanks to my interlocutor Harry Collins and the large and lively audience that attended.

The epigraph for this essay is quoted from Bruno Latour, "One More Turn after the Social Turn," in E. McMullin, ed., *The Social Dimen-*

sions of Science (Notre Dame: University of Notre Dame Press, 1992), 287–88.

1. Harry Collins and Steven Yearley, "Epistemological Chicken," in A. Pickering, ed., *Science as Practice and Culture* (Chicago: University of Chicago Press, 1992), 301–26.

2. Thomas Kuhn, "Reflections on My Critics," in I. Lakatos and A. Musgrave, eds., *Criticism and the Growth of Knowledge* (Cambridge: Cambridge University Press, 1970), 231–78.

3. Jean Aitchison, *Language Change: Progress or Decay?* (New York: Universal Books, 1981), 208–25.

4. Peter Galison and I have independently been exploring the use of trade languages for bridging problems of incommensurability for some time now, though the reader will see that our emphases are somewhat different. My own interest arose from trying to reconcile genuine cultural diversity with a nonrelativistic ethical stance (see Steve Fuller, "Towards Objectivism and Relativism," *Social Epistemology* 1 [1987]: 351–62, esp. 358). In *Philosophy of Science and Its Discontents* (Boulder: Westview Press, 1989), 144 (2d ed. [New York: Guilford Press, 1993], 189–90) I suggested that the emergence of trade languages could explain the possibility of communication between disciplines that have some overlapping concerns but otherwise operate with different theoretical languages. Here my views coincided with those of Galison in *Image and Logic: The Material Culture of Twentieth-Century Physics* (forthcoming). However, since that time I have become more impressed by the potential of trade languages to develop into full-fledged, all-purpose languages, while Galison has continued to emphasize their more typically partial character.

5. Jonathan Culler, *Structuralist Poetics* (Ithaca: Cornell University Press, 1975).

6. Steve Woolgar, *Science: The Very Idea!* (London: Tavistock, 1988).

7. Bruno Latour, *Science in Action* (Cambridge, Mass.: Harvard University Press, 1987).

8. Cf., e.g., Simon Schaffer, "The Eighteenth Brumaire of Bruno Latour," *Studies in History and Philosophy of Science* 22 (1991): 174–92.

9. Michel Serres, *Hermes* (Baltimore: The Johns Hopkins University Press, 1982).

10. There is an interesting theory currently popular in primatology that supports something like the Hegelian view that reason is domesticated, or "civilized," once it must be exercised in the presence of others who are taken to be acting similarly. This is the Machiavellian Intelligence hypothesis, which argues that the cognitive complexity of primates is a direct function of the complexity of their social organization: in Latourian terms, how many "obligatory passage points" need to be passed before a

license is given for action? See Robert Byrne and Andrew Whiten, *Machiavellian Intelligence* (Oxford: Oxford University Press, 1987).

11. There would seem to be profound lessons here for animal-rights activists who presume that animals can be included as persons without attenuating the property of "being a person." As Yves Gingras pointed out during discussion of this paper at 4S/EASST in Göteborg, it may be that animal-rights activists are simply mobilizing animals as resources to consolidate a power base that they otherwise would not have. After all, who says that the whales need someone to speak on their behalf? Perhaps the test of this Latourian hypothesis would be whether it is possible to construct a narrative of animal-rights activism in which the animals resist their spokespersons. According to Latour (pers. comm., 19 Oct. 1992), such things happen whenever the animals survive even when the activists fail to liberate them!

12. Cf. Steve Fuller, *Social Epistemology* (Bloomington: Indiana University Press, 1988), 36–45.

13. Steve Woolgar, ed., *Knowledge and Reflexivity* (London: Sage, 1988); and Malcolm Ashmore, *The Reflexive Thesis* (Chicago: University of Chicago Press, 1989).

14. Michel Callon, "Some Elements of a Sociology of Translation," in J. Law, ed., *Power, Action, and Belief* (London: Routledge and Kegan Paul, 1986), 196–229.

15. Umberto Eco, *A Theory of Semiotics* (Bloomington: Indiana University Press, 1976).

16. Bruno Latour, "The Politics of Explanation," in Woolgar, ed., *Knowledge and Reflexivity*, 155–76.

17. Fuller, *Social Epistemology* (1988) and *Philosophy of Science and Its Discontents* (1989, 1993).

18. Latour, *Science in Action*, esp. 215–57.

19. Jean Piaget, *Psychology and Epistemology* (Harmondsworth: Penguin, 1976).

20. Cf. Nicholas Georgescu-Roegen, *The Entropy Law and the Economic Process* (Cambridge, Mass.: Harvard University Press, 1971), 114–40.

21. Cf. Jonathan Culler, *On Deconstruction* (Ithaca: Cornell University Press, 1982), 86–88; John Searle, "The World Turned Upside Down," *New York Review of Books* 30, no. 16 (1983): 74–79.

22. Cf. Fuller, *Social Epistemology* (1988), 128–38.

23. Peter Galison, "The Trading Zone: Coordinating Action and Belief," in *Image and Logic: The Material Culture of Modern Physics*, chap. 9 (forthcoming); Donald McCloskey, *If You're So Smart* (Chicago: University of Chicago Press, 1991).

24. Cf. Galison, "The Trading Zone" (forthcoming).

25. Kenneth Boulding, *Beyond Economics* (Ann Arbor: University of Michigan Press, 1968), 145–47.

26. Steve Fuller, *Philosophy, Rhetoric, and the End of Knowledge: The Coming of Science and Technology Studies* (Madison: University of Wisconsin Press, 1993).

Biagioli, From Relativism to Contingentism

1. David Oldroyd, "Why Not a Whiggish Social Studies of Science?" *Social Epistemology* 3 (1989): 355–60 (a review of *Leviathan and the Air-Pump*); George Bowker and Bruno Latour, "A Booming Discipline Short of Discipline: (Social) Studies of Science in France," *Social Studies of Science* 17 (1987): 724–26.

In "Scientific Discoveries and the End of Natural Philosophy," *Social Studies of Science* 16 (1986): 397, Simon Schaffer argues: "We can say that the research of the early nineteenth century produced the discovery of photosynthesis in the late 1770s. Without some form of *teleology*, there is no reconciliation available. It seems simultaneously unnecessary, ill-mannered, and impossible to find a mark of discovery separate and superior to the locally generated rules of communities of natural philosophers" (emphasis mine). However, nothing in Schaffer's analysis indicates that historians of science are exempt from this "teleogy" (definitely a whiggish category) when they reproduce that discovery historiographically.

My "The Anthropology of Incommensurability" (*Studies in History and Philosophy of Science* 21 [1990]: 183–209) displays the same aporia. In fact, while describing the debate on buoyancy as undecidable, I also said that Galileo was right without explaining how the two apparently opposed stances could be reconciled. Some colleagues have read those claims about Galileo's being "right" as signs of my hidden whiggish agenda. Although I insisted on denying charges of whiggism, I had to admit that I had not made explicit my system of reference.

During a TECH-KNOW Workshop at UCLA in November 1989, a number of participants discussed reconsidering the taboo about whig history of science.

2. David Hull, "In Defense of Presentism," *History and Theory* 18 (1979): 1–15; and A. Wilson and T. G. Ashplant, "Whig History and Present-Centred History," *Historical Journal* 31 (1988): 1–16; idem, "Present-Centred History and the Problem of Historical Knowledge," *Historical Journal* 31 (1988): 253–74. See also an earlier piece by George W. Stocking, Jr., "On the Limits of 'Presentism' and 'Historicism' in the Historiography of the Behavioral Sciences," *Journal of the History of the Behavioral Sciences* 1 (1965): 211–18.

3. Some anthropologists have begun to look at the way ethnographic records are produced as *texts* and the ways in which the *dialogue* between the ethnographer and the informant is usually effaced in the final narrative in order to create a sense of scientific distance between the ethnographer and the foreign culture being described: George E. Marcus and D. Cushman, "Ethnographies as Texts," *Annual Review of Anthropology* 11 (1982): 25–69; George W. Stocking, Jr., ed., *Observers Observed* (Madison: University of Wisconsin Press, 1983); James Clifford and George E. Marcus, eds., *Writing Culture* (Berkeley and Los Angeles: University of California Press, 1986); Renato Rosaldo, *Culture and Truth* (Boston: Beacon Press, 1989). Similarly, in the wake of Hayden White's argument that modern historiographical discourse is inexorably emplotted by literary tropes, a number of authors have analyzed the limits and "hidden agendas" of historical interpretations: Hayden White, *Metahistory* (Baltimore: The Johns Hopkins University Press, 1973); Michel De Certeau, *L'écriture de l'histoire* (Paris: Gallimard, 1975); idem, "Writing vs. Time: History and Anthropology in the Works of Lafitau," *Yale French Studies* 59 (1980): 37–64; Sande Cohen, *Historical Culture* (Berkeley and Los Angeles: University of California Press, 1986); Michael Taussig, "History as Sorcery," *Representations* 7 (1984): 87–109.

4. The main exception is Steve Woolgar, who has repeatedly called for a reflexive sociology of science. His most articulated statement is perhaps "Reflexivity Is the Ethnographer of the Text," in Steve Woolgar, ed., *Knowledge and Reflexivity* (London: Sage, 1988), 14–34. The ethnomethodologists have also stressed the role of reflexivity in their method. However, the meaning they attribute to reflexivity is quite different from the one adopted by most participants in the reflexivity debate. To the ethnomethodologists, reflexivity is not a "problem" or an issue that would question or undermine the scientificity of the social sciences. Quite to the contrary, they see it as a necessary component of their "scientific" methodology.

5. For a good summary of the positions on reflexivity by a number of sociologists of science, see Malcolm Ashmore, *The Reflexive Thesis* (Chicago: University of Chicago Press, 1990), 26–86. On reflexivity within the sociology of scientific knowledge, see also Woolgar, ed., *Knowledge and Reflexivity*. For reflexivity as the predicament of postmodern philosophy, see Hilary Lawson, *Reflexivity* (La Salle: Open Court, 1985).

6. Although Yehuda Elkana has been discussing "reflexivity" extensively, the meaning he attributes to that term is quite different from the one it has assumed in the contemporary debate on reflexivity in the social sciences.

7. Augustine Brannigan, *The Social Basis of Scientific Discovery* (Cam-

bridge: Cambridge University Press, 1981), is a good example of this trend. See Simon Schaffer's "Scientific Discoveries and the End of Natural Philosophy," *Social Studies of Science* 16 (1986): 387–420, for a review of some of the more recent literature on scientific discoveries.

8. Feyerabend and Kuhn have used history as a tool for philosophical critique, that is, to attack the separation between the contexts of discovery and justification. At the same time, their introduction of the notion of incommensurability has given their critics a good resource to undermine their statements by pointing to the fact that incommensurability would jeopardize their ability to interpret past science and, therefore, to use history to undermine received rationalistic philosophies of science: see Hilary Putnam, *Reason, Truth and History* (Cambridge: Cambridge University Press, 1981), 113–19. Feyerabend's response to Putnam is reprinted in his *Farewell to Reason* (London: Verso, 1987), 265–72. The anthropologist Dan Sperber has presented a critique of relativist anthropology that is structually analogous to Putnam's argument against relativist philosophy of science: "Apparently Irrational Beliefs," in M. Hollis and S. Lukes, eds., *Rationality and Relativism* (Cambridge, Mass.: MIT Press, 1982), 180.

9. Thomas S. Kuhn, *The Essential Tension* (Chicago: University of Chicago Press, 1978), xi–xii; idem., "What Are Scientific Revolutions?" in Lorenz Kruger, Lorraine J. Daston, and Michael Heidelberger, eds., *The Probabilistic Revolution* (Cambridge, Mass.: MIT Press, 1987), vol. 1, p. 9.

10. This point cannot be fully articulated here, but will become clearer in the later part of the essay. See also note n. 27 below.

11. Of course, not all academic social scientists and humanists are relativists, though the vast majority of the younger ones are. Academic relativism comes in all forms and shapes. For instance, many academics are relativists only when they discuss social, cultural, or political issues but often display a less tolerant "Enlightenment gestalt" and a sanguine veneration for Rationality in their discussions of the production and acceptance of scientific knowledge. However, as shown by recent reforms in the teaching of Western Civilization (as at Stanford and Berkeley), it seems that cultural relativism has become accepted (actually or rhetorically) by many academic institutions in the United States—especially those having an ethnically and culturally diverse student population.

12. For instance, relativism helps to represent the claims of minorities as local and bound to the specific views and needs of their subcultures even when, instead, they address issues of a much more general currency. In short, minorities are nominally entitled to have their cultures and claims, but they are also kept in check by the representation of their claims as those of a "special-interest group."

13. Donna Haraway, "Situated Knowledges: The Science Question in Feminism and the Privilege of Partial Perspective," *Feminist Studies* 14 (1988): 575–99; reprinted in revised form in Donna Haraway, *Simians, Cyborgs, and Women* (London: Routledge, Chapman and Hall, 1991), 183–201.

14. Haraway, "Situated Knowledges" 284. Haraway's claims echo Renato Rosaldo's critique of North American social scientists who "pretend to speak either from a position of omniscience or from no position at all": *Culture and Truth*, 204.

15. Things have changed since 1990, when this essay was initially written. As shown by Paul R. Gross and Norman Levitt, *Higher Superstition: The Academic Left and Its Quarrels with Science* (Baltimore: Johns Hopkins University Press, 1994), we may be witnessing the beginning of an attack on relativistic science studies as part of a reaction against postmodernism (construed by its critics in extremely broad terms). I believe these reactions reflect precisely the widespread success of relativistic approaches within the humanities and social sciences.

16. During the "Which Way L.A.?" program.

17. In particular, I do not endorse the rationalists' critique of relativism, and share most of Clifford Geertz's argument in "Anti Anti-Relativism," *American Anthropologist* 88 (1984): 263–78.

18. Biagioli, "The Anthropology of Incommensurability." This article has been revised and expanded into chaps. 3 and 4 of *Galileo Courtier* (Chicago: University of Chicago Press, 1993).

19. The view of scientific change presented in my "Anthropology of Incommensurability" is quite different from what goes under the label of "evolutionary epistemology"—an approach (first proposed by Donald Campbell) based on a Popperian interpretation of Darwin. "Evolutionary epistemology" draws a close analogy between Darwinian natural selection and Popperian falsification and ends up seeing natural selection as a "rational" process of "error elimination." As the rest of the article shows, I do not share this Popperian reading of Darwin and, in particular, do not see "natural selection" as a "rational" process, and do not use categories such as progress or "directed evolution." On evolutionary epistemology, see Donald T. Campbell, "Evolutionary Epistemology," in Paul A. Schilpp, ed., *The Philosophy of Karl Popper* (La Salle: Open Court, 1974), vol. 1, pp. 413–63; and Kai Hahlweg and C. A. Hooker, eds., *Issues in Evolutionary Epistemology* (Albany: SUNY Press, 1989). A much more interesting (and empirically articulated) evolutionary view of science available is found in David Hull's *Science as a Process* (Chicago: University of Chicago Press, 1988).

20. Although in this essay I keep talking about "groups" and "tribes,"

I do so for the sake of simplicity. My claims can, I think, be transferred to sets of historical actors who are connected through networks rather than by sharing the same physicogeographical niche. Also, later in the essay I talk about scientific species, tribes, theories, and paradigms. Although the nomenclature may not be apparently consistent, what I refer to through all these terms is a set of related scientists holding a certain theory or set of theories. So when I talk about tribes and scientific species, I also mean their theories. Vice versa, when I mention theories or paradigms, I also mean to refer to the people that hold them.

21. Although incommensurability should not be confused with the various noncommunicative behaviors we may observe during scientific controversies, it is, I think, connected to them. At an early stage of the process, it is quite probable that the subgroup or "variety's" linguistic grid may be still largely commensurable with that of the original group. However, the subgroup's cohesion and socioprofessional identity, maintained through these strategies of noncommunication, would commit its members to develop their new worldview and lexical structure so that, eventually, it may become linguistically incommensurable with the old one. Linguistic "sterility" between the subgroup and the rest would then intervene, indicating that the scientific "variety" had turned into a "species."

22. The view on the evolutionary process I am proposing here is closer to that of Stephen J. Gould, as articulated in *Wonderful Life: The Burgess Shale and the Nature of History* (New York: Norton, 1989), than to many "progressivist" interpretations of Darwin's views.

23. Because of this, I strongly disagree with the evolutionary epistemologists' association of natural selection and Popperian falsification.

24. Actually, one may try to think of "rationality" as a result rather than a cause of knowledge production: that is, as a set of protocols that, at some point, were developed by a culture looking back at the process through which it had produced its own knowledge. In short, rationality may be perceived as a set of guidelines (rather than prescriptions) developed *a posteriori*. "Rationality" may be simply seen as a method—one that, at some point and to some people, seemed to describe how knowledge had been successfully produced.

25. The "binarity" of "fit" is one reason why I think it wrong-headed to try to assess the (degree of) "closeness" between a theory and "reality."

26. However, this does not suggest that the distinction between internal and external features of science is meaningless. In fact, here I am simply talking about "socionatural selection" and I am not discussing the processes of "scientific variation."

27. In trying to convey the basic core of the argument, I have glossed

over a number of important issues. One of them is the role of bilingualism in the interpretation of old, alien, and possibly incommensurable world-views. In "The Anthropology of Incommensurability," I discussed the ways in which processes of identity preservation tend to prevent members of a scientific tribe from becoming bilingual—that is, from learning the language of the "other." I believe these considerations apply to historians as well.

The "exemplary" historian of science described by Kuhn or Feyerabend is somebody who can become bilingual. However, neither Kuhn nor Feyerabend has analyzed the conditions regulating historians' access to bilingualism. I would argue that when historians encounter very alien systems of belief in the process of doing history, they respond to that encounter in ways that reflect their socioprofessional identity, that is, they may or may not decide to try to become bilingual. However, historians' socioprofessional identity is quite different from scientists', and this allows historians to become bilingual more easily. In fact, because historians' socioprofessional identity does not need to be linked to the theories of the scientists they are studying, they may not feel threatened (as the past scientists may have) in learning the language of the "other." In short, the academic historians' option to become bilingual is not merely a result of his or her "objective distance" from the events he or she studies. More simply, it derives from the historian's having a socioprofessional identity that is centered on beliefs that are different from those held by the scientists. In short, a historian's potential ability to become bilingual is not a matter of *distance* but of *difference*. It is not that the historian is "objective" by virtue of not having high stakes, but simply that the stakes may lie elsewhere.

However, even bilingualism does not avoid the incommensurability between the historian's culture and the past scientist's theories. Bilingualism allows the historian to detect incommensurability, but not to solve it. Consequently, the "defeated" scientific theory does not have much chance to be fully understood. In fact, finding himself or herself in a situation of undecidability, the bilingual historian would "lean" on the side to which he or she is genealogically connected. This is so for at least two reasons. First, the historian's identity is more or less (but inextricably) tied to that of the "winner." Second, the "other" theory stands there isolated, without a comprehensive picture of what it could have turned into. In short, because of the way scientific change takes place, we are bound to some extent to reinforce the memory of the winners even when we think that we are "really" understanding the "other."

28. An eloquent statement against the institution of a canon in science studies (and the dangers that that move may entail) is Sharon Traweek,

"Border Crossings: Narrative Strategies in Science Studies and among Physicists in Tsukuba Science City, Japan," in Andrew Pickering, ed., *Science as Practice and Culture* (Chicago: University of Chicago Press, 1992), 429–65.

Schaffer, Contextualizing the Canon

I thank Mario Biagioli and Peter Galison for their generous help. I am also grateful to the participants in a meeting on the history of the human sciences organized by Roger Smith at Lancaster University in September 1990, where this paper was first presented.

The epigraph sources for this essay are as follows: Antonio Gramsci, *Selections from the Prison Notebooks*, ed. Q. Hoare and G. N. Smith (London: Lawrence and Wishart, 1971), 325; Thomas Hobbes, *Leviathan* (London: Crooke, 1651), 199; Walter Benjamin, *Illuminations*, ed. Hannah Arendt (New York: Schocken, 1969), 263; Christopher Marlowe, *Doctor Faustus*, act 1, scene 1, lines 118–20.

1. Mario Biagioli, "From Relativism to Contingentism" (in this volume), text following n. 20. For the historical relation between the human and the natural sciences, see John Christie, "The Human Sciences: Origins and Histories," *History of the Human Sciences* 6 (1993): 1–12.

2. Biagioli, "From Relativism to Contingentism," n. 24. For the patchiness and temporality of scientific practice, see Andrew Pickering, "From Science as Knowledge to Science as Practice," in A. Pickering, ed., *Science as Practice and Culture* (Chicago: University of Chicago Press, 1992), 1–26, esp. 8–9.

3. Steve Woolgar, *Science: The Very Idea* (London: Tavistock, 1988), 30–37, 91–94.

4. E. P. Thompson, *Customs in Common* (London: Merlin, 1991), 1–15, 185–89.

5. Barry Barnes, "On the Conventional Character of Knowledge and Cognition," in Karin Knorr Cetina and Michael Mulkay, eds., *Science Observed: Perspectives on the Social Studies of Science* (Beverly Hills: Sage, 1983), 19–51, esp. 30–31.

6. For "shadow histories," see Richard Watson, "Shadow History in Philosophy," *Journal of the History of Philosophy* 31 (1993): 95–123. For the noncanonical meaning of *Leviathan* itself, see Quentin Skinner, "The Ideological Context of Hobbes' Political Thought," *Historical Journal* 9 (1966): 286–317; and "Thomas Hobbes and the Engagement Controversy," in G. E. Aylmer, ed., *The Interregnum* (London: Macmillan, 1972), 79–98.

7. Steven Shapin and Simon Schaffer, *Leviathan and the Air-Pump:*

Hobbes, Boyle, and the Experimental Life (Princeton: Princeton University Press, 1985), 92–107; Bruno Latour, *Nous n'avons jamais été modernes* (Paris: La Découverte, 1991), 30–33.

8. Hobbes, *Leviathan*, 395; Shapin and Schaffer, *Leviathan and the Air-Pump*, 311.

9. Michael Oakeshott, "The Activity of Being an Historian," in *Rationalism in Politics* (London: Methuen, 1962), 137–64; and "Introduction," in Thomas Hobbes, *Leviathan* (Oxford: Blackwell, 1948), v–lxvi, esp. viii, xi.

10. Steven Lukes, *Emile Durkheim* (Harmondsworth: Peregrine Books, 1975), 278, 299.

11. Emile Durkheim, *Montesquieu and Rousseau: Founders of Sociology* (Ann Arbor: Ann Arbor Books, 1960), 2.

12. David Bloor, *Knowledge and Social Imagery* (London: Routledge, 1976), 40–47.

13. Joseph Ben-David, "Rationality and Scientific Research in the Social Sciences," in S. J. Doorman, ed., *Images of Science* (Aldershot: Gower Books, 1989), 121–34.

14. Michael McKeon, *The Origins of the English Novel* (London: Radius, 1987), 19.

15. Richard Ashcraft, *Revolutionary Politics and Locke's Two Treatises of Government* (Princeton: Princeton University Press, 1986), 7.

16. Ibid., 589–90. Compare J. G. A. Pocock, *The Ancient Constitution and the Feudal Law*, 2d ed. (Cambridge: Cambridge University Press, 1987), 356.

17. Bruce Kuklick, "Seven Thinkers and How They Grew," in Richard Rorty et al., eds., *Philosophy in History* (Cambridge: Cambridge University Press, 1984), 125–40, esp. 128; Paul B. Wood, "The Hagiography of Common Sense: Dugald Stewart's Account of the Life and Writings of Thomas Reid," in A. J. Holland, ed., *Philosophy: Its History and Historiography* (Dordrecht: Reidel, 1985), 305–22, esp. 318.

18. Kuklick, "Seven Thinkers," 128.

19. Nicholas Jardine, *The Scenes of Inquiry* (Oxford: Clarendon Press, 1991), 124–30, for strategies and genres of historical legitimation; Jonathan Rée, *Philosophical Tales* (London: Methuen, 1987), 42–43, 54–55, for the anticanon of philosophy.

20. Latour, *Nous n'avons jamais été modernes*, 96–97: "The asymmetry between nature and culture becomes an asymmetry between the past and the future." For "anticonquest," see Mary Louise Pratt, *Imperial Eyes: Travel Writing and Transculturation* (London: Routledge, 1992), 7, 78.

21. Roland Barthes, *Image Music Text*, ed. Stephen Heath (London:

Fontana, 1977), 165. Compare, for example, Paul Forman, "The Discovery of the Diffraction of X-Rays by Crystals: A Critique of the Myths," *Archive for History of the Exact Sciences* 6 (1969): 38–71; Augustine Branigan, "The Reification of Mendel," *Social Studies of Science* 9 (1979): 423–54; F. A. J. L. James, "The Creation of a Victorian Myth: The Historiography of Spectroscopy," *History of Science* 23 (1985): 1–24.

22. Barthes, *Image Music Text*, 169.

23. Raymond Williams, *The Country and the City* (St. Alban's: Paladin, 1975), 363–66. For the modernist role of Marx's dictum (*Communist Manifesto* [Harmondsworth: Penguin, 1967], 82–83), see Marshall Berman, *All That Is Solid Melts into Air: The Experience of Modernity* (London: Verso, 1983), 111–29.

24. Williams, *The Country and the City*, 18–22, 348.

25. Ibid., 348–49.

26. Eric Hobsbawm, "Inventing Traditions," in Eric Hobsbawm and Terence Ranger, eds., *The Invention of Tradition* (Cambridge: Cambridge University Press, 1983), 1–14; Walter Benjamin, "A Small History of Photography" (1931), in *One-Way Street and Other Writings*, with an introduction by Susan Sontag (London: Verso, 1985), 250.

27. Eric Hobsbawm, "Mass-Producing Traditions, Europe 1870–1914," in Hobsbawm and Ranger, eds., *Invention of Tradition*, 263–307.

28. Martin J. Wiener, *English Culture and the Decline of the Industrial Spirit, 1850–1980* (Cambridge: Cambridge University Press, 1981), 159, 166; Keith Joseph's new version of Smiles's *Self-Help* is discussed in Allen McLaurin, "Reworking 'Work' in Some Victorian Writing and Visual Art," in Eric Sigsworth, ed., *In Search of Victorian Values* (Manchester: Manchester University Press, 1988), 27–41, esp. 31–33. For a survey of the historiography, see Harvey Kaye, "The Use and Abuse of the Past: The New Right and the Crisis of History," in *Socialist Register 1987* (London: Merlin Press, 1987), 332–64.

29. James Walvin, *Victorian Values* (London: Sphere Books, 1988), 4; Patrick Wright, *On Living in an Old Country: The National Past in Contemporary Britain* (London: Verso, 1985). Wright's epigraph is from Walter Benjamin's sixth thesis on the philosophy of history (spring 1940), in *Illuminations*, ed. Hannah Arendt (New York: Schocken, 1969), 255.

30. George Marcus, "A Broad(er) Side to the Canon," *Cultural Anthropology* 6 (1991): 385–405, esp. 401–2 on Bruno Latour's comments on the absence of a purely literary canon in the sciences. Mario Biagioli kindly lent me a copy of this paper. See also Steven Lukes, "Political Ritual and Social Integration," *Sociology* 9 (1975): 289–308, esp. 302–3.

31. Claude Lévi-Strauss, *Tristes tropiques* (New York: Atheneum, 1974), 75–76.

32. Stephen Greenblatt, *Learning to Curse: Essays in Early Modern Culture* (London: Routledge, 1990), 163–70.

33. Stephen Greenblatt, *Renaissance Self-Fashioning from More to Shakespeare* (Chicago: University of Chicago Press, 1980), 193–200.

34. Edward Said, *Orientalism* (New York: Vintage, 1979), 20–28.

35. Martin Bernal, *Black Athena: The Afroasiatic Roots of Classical Civilization*, vol. 1 (London: Free Association, 1987), p. 442; see Iain Boal, "Facing the Past," *Before Columbus Review* 1 (1990): 7–8, 28–31.

36. Jardine, *The Scenes of Inquiry*, 74–75.

37. Marcus, "A Broad(er) Side to the Canon," 385.

38. The principal sources for the standard interpretation are presented in J. C. Beaglehole, ed., *The Journals of Captain James Cook on His Voyages of Discovery*, vol. 3 (Cambridge: Cambridge University Press, 1967), pp. cxliv, clxxi–ccxvii; and, for Hawaiian myth, David Malo, *Hawaiian Antiquities*, 2d ed., trans. N. B. Emerson (Honolulu: Bishop Museum, 1951). The comparison with Mexico is suggested in Marshall Sahlins, "The Apotheosis of Captain Cook," in Michel Izard and Pierre Smith, eds., *Between Belief and Transgression: Structuralist Essays in Religion, Myth and History* (Chicago: University of Chicago Press, 1982), 73–102, esp. 74 n. 1.

39. Sahlins, "The Apotheosis of Captain Cook," 80–82, described as an "expanded" version of a 1977 lecture.

40. Marshall Sahlins, *Historical Metaphors and Mythical Realities: Structure in the Early History of the Sandwich Islands Kingdom* (Ann Arbor: University of Michigan Press, 1981), 17–28; Marshall Sahlins, *Islands of History* (London: Tavistock, 1987), 104–35.

41. Steen Bergendorff, Ulla Hasager, and Peter Henriques, "Mythopraxis and History: On the Interpretation of the Makahiki," *Journal of the Polynesian Society* 97 (1988): 391–408.

42. Gananath Obeyesekere, "'British Cannibals': Contemplation of an Event in the Death and Resurrection of James Cook, Explorer," *Critical Inquiry* 18 (1992): 630–54. This is a preliminary study for Obeyesekere's own *The Apotheosis of Captain Cook: European Mythmaking in the Pacific* (Princeton: Princeton University Press, 1992), 138–39. Sahlins mentions the context of the priests' question in debates about cannibalism at *Islands of History*, 122 n. 9.

43. Pratt, *Imperial Eyes*, 7, 102.

44. The term "evidential context" is used by Trevor Pinch to describe the agreed zone of relevance of some experiment: "Toward an Analysis of Scientific Observation: The Externality and Evidential Significance of Observations Reports in Physics," *Social Studies of Science* 15 (1985): 3–35.

45. Sahlins, *Islands of History*, 120 n. 8. Compare J. G. Frazer, *The*

Golden Bough, abr. ed. (London: Macmillan, 1963), chap. 24, "The Killing of the Divine King." The novelist Anthony Powell chose Frazer's term "Temporary Kings" ironically to describe the condition of participants at academic conferences.

Fine, Science Made Up

Part of the work for this paper was done during the tenure of a fellowship at the Center for Advanced Study in the Behavioral Sciences at Stanford. I am grateful for financial support provided by the National Science Foundation, Grant BNS-8011494, and for the assistance of the staff of the Center. I also want to thank David Bloor, Stephen Downes, David Hull, and Andy Pickering for offering good advice and criticism, some of which I have heeded.

The epigraph to this essay is quoted from Robert Boyle, as cited in R. Sargent, "Explaining the Success of Science," in A. Fine and J. Leplin, eds., *PSA 1988*, vol. 1 (East Lansing: Philosophy of Science Association, 1988), 55–63.

1. Later developments in this line include ethnomethodology. See H. Garfinkel, *Studies in Ethnomethodology* (Englewood Cliffs: Prentice-Hall, 1967); and W. Sharrock and R. J. Anderson, *The Ethnomethodologists* (London: Tavistock, 1986).

2. R. K. Merton, *The Sociology of Science: Theoretical and Empirical Investigations* (Chicago: University of Chicago Press, 1973).

3. S. Woolgar, ed., *Knowledge and Reflexivity* (London: Sage, 1988); and S. Woolgar, *Science: The Very Idea* (London: Tavistock, 1988).

4. K. D. Knorr Cetina, *The Manufacture of Knowledge: An Essay on the Constructivist and Contextual Nature of Science* (Oxford: Pergamon Press, 1981); B. Latour and S. Woolgar, *Laboratory Life: The Social Construction of Scientific Facts* (London: Sage, 1979 [2d ed. as *Laboratory Life: The Construction of Scientific Facts* (Princeton: Princeton University Press, 1986)]); A. Pickering, *Constructing Quarks: A Sociological History of Particle Physics* (Chicago: University of Chicago Press, 1984).

5. For a sample of some especially sympathetic admirers, see H. Lawson and L. Appignanesi, eds., *Dismantling Truth: Science in Postmodern Times* (London: Weidenfeld and Nicolson, 1988).

6. S. Shapin, review of Andrew E. Benjamin et al., *The Figural and the Literal*, *Isis* 79 (1988): 127–28, on 127.

7. L. Laudan, "The Pseudo-science of Science?" *Philosophy of the Social Sciences* 11 (1981): 173–98; D. Bloor, "The Strengths of the Strong Programme," *Philosophy of the Social Sciences* 11 (1981): 199–213. Of course it is not only analytic philosophers who take a combative stance toward

constructivism. Critical theorists do too. See, for example, T. McCarthy, "Scientific Rationality and the 'Strong Program' in the Sociology of Knowledge," in E. McMullin, ed., *Construction and Constraint* (Notre Dame, Indiana: University of Notre Dame Press, 1988), 73–96.

8. I will discuss relativism only in passing. For more on that topic, see McCarthy, "Scientific Rationality"; and M. Hollis and S. Lukes, eds., *Rationality and Relativism* (Oxford: Blackwell, 1982).

9. The platform below is an amalgam that constitutes an "ideal type" description of constructivism. I believe it captures important features of the school (at least in many of its stages). Nevertheless I have no doubt that each member of the school will find some things in my mix with which to quarrel. In any case, I have drawn the amalgam from the social-interest Edinburgh school, especially B. Barnes, *Interests and the Growth of Knowledge* (London: Routledge and Kegan Paul, 1977); D. Bloor, *Knowledge and Social Imagery* (London: Routledge and Kegan Paul, 1976 [2d ed. 1991]); Knorr Cetina, *The Manufacture of Knowledge*; Pickering, *Constructing Quarks*; the historical sociology of knowledge promoted by S. Shapin, "The Sociology of Science," *History of Science* 20 (1982): 157–211; and the actor–network analysis pioneered by Latour and Woolgar, *Laboratory Life*, and developed further in B. Latour, *Science in Action* (Cambridge, Mass.: Harvard University Press, 1987).

10. My claim here (namely, that a loose consensus theory of truth binds the planks together into a constructivist platform) might be questioned—or even denied—by some constructivists. Thus Bloor writes: "The question may be pressed: does acceptance of a theory by a social group make it true? The only answer that can be given is that it does not" (*Knowledge and Social Imagery*, 2d ed., 43). Yet in the afterword Bloor tells us that truths form not a natural but a social kind, in the sense that their membership in the class "is a result of how they are treated by other people" (p. 174), and he goes on to hedge as to whether this conception of truth amounts to idealism or, more accurately, he qualifies the *kind* of idealism it does amount to (p. 175). (A decade earlier one finds a similar superposition of realist and idealist conceptions in a seminal paper of Barnes and Bloor, where their well-known relativist tribesman "might claim that not even the fact that his own tribe believes something is, *in itself*, sufficient to make it true. But he would then have to mend the damage [*sic*] of this admission by adding that it just was a fact that what his tribe believed *was* true": B. Barnes and D. Bloor, "Relativism, Rationalism and the Sociology of Knowledge," in Hollis and Lukes, eds., *Rationality and Relativism*, 30. My case here is that constructivists *need* a consensus theory to make their planks hold together as constructivism, and that, in fact, they use consensus for this very purpose. Below, how-

ever, I argue that this conception of truth undermines their core insight into the openness of science.

11. For a review, see A. Wylie, "Explaining Confirmation Practice," *Philosophy of Science* 55 (1988): 292–303.

12. Richard Miller, *Fact and Method* (Princeton: Princeton University Press, 1987), advances this theme and identifies the "worship of generality" as the philosophical legacy of neopositivism. That may be too strong a historical claim, for there are after all other sources of that worship, starting with Socrates and carrying on, for example, with Descartes and especially with Kant, who was himself an important influence on neopositivism. Still, Miller's critique of specific generalizing programs is insightful, and his context-sensitive realist alternative is very interesting. See "The Challenge to Realism" below for some discussion; and A. Fine, "Piecemeal Realism," *Philosophical Studies* 61 (1991): 79–96.

13. A. Fine, "Unnatural Attitudes: Realist and Instrumentalist Attachments to Science," *Mind* 95 (1986): 149–79.

14. The essays in J. C. Traynham, ed., *Essays on the History of Organic Chemistry* (Baton Rouge: Louisiana State University Press, 1987), support the instrumentalist claim, especially the article by Alan J. Rocke, "Convention Versus Ontology in Nineteenth-Century Organic Chemistry," 1–20.

15. Pickering, *Constructing Quarks*, 413.

16. Woolgar, *Science*, 65, 73.

17. Ibid., 13.

18. This idealizing option has been revived in Putnam's internal realism (H. Putnam, *Reason, Truth and History* [Cambridge: Cambridge University Press, 1981]), which he now prefers to call "pragmatic" realism (*The Many Faces of Realism* [LaSalle: Open Court, 1987], 17). James's way was part of Rorty's epistemological behaviorism, a part to which he no longer subscribes: R. Rorty, *Philosophy and the Mirror of Nature* (Princeton: Princeton University Press, 1979). See especially the introduction to his essays on pragmatism: R. Rorty, *Consequences of Pragmatism* (Minneapolis: University of Minnesota Press, 1982). I suggest a general line of criticism of these pragmatic theories, under the label "truthmongers," in my *The Shaky Game: Einstein, Realism and the Quantum Theory*, rev. ed. (Chicago: University of Chicago Press, 1988), and in "Truthmongering: Less is True," *Canadian Journal of Philosophy* 19 (1989): 611–16.

19. See Fine, "Unnatural Attitudes," for more on this theme.

20. Fine, "Unnatural Attitudes" and *Shaky Game*.

21. Shapin, "Sociology of Science," 164.

22. Woolgar, *Science*, 49.

23. This bypasses the question of which community by supposing in

effect that there is only one relevant community. Thus, as promised, I ignore the problem of relativism. For the constructivists, relativism stands or falls with the consensus theory of truth; and, as we shall see, it falls.

24. The vocabulary of production also seems tinged with a Marxism not only consonant with the sociological idea of knowledge as a superstructure, but also with an antiestablishment (i.e., anticapitalist) politics.

25. E. Husserl, *The Crisis of the European Sciences and Transcendental Phenomenology*, tr. David Carr (Evanston: Northwestern University Press, 1970).

26. See Fine, "Unnatural Attitudes" and *Shaky Game*.

27. See Fine, *Shaky Game*, esp. chap. 7, "The Natural Ontological Attitude," 112–35; and "Unnatural Attitudes," esp. Metatheorem 2, for the details.

28. See preceding note. Miller, *Fact and Method*, 448–61, attacks the explanationist argument over the "success" theme, showing just how carefully this has to be qualified.

29. Pickering, *Constructing Quarks*.

30. For example, P. Galison, *How Experiments End* (Chicago: University of Chicago Press, 1987), who is also sensitive to social factors in science, takes issue with Pickering's account of the neutral-current episode (*Constructing Quarks*, 257–60). The verdict on this is not yet in. For a negative assessment of what constructivism can do in explaining scientific success, see R. Sargent, "Explaining the Success of Science," in A. Fine and J. Leplin, eds., *PSA 1988* (East Lansing: Philosophy of Science Association, 1988), vol. 1, 55–63. I am more optimistic.

31. I. Hacking, *Representing and Intervening* (Cambridge: Cambridge University Press, 1983), 22–23.

32. Miller, *Fact and Method*.

33. Ibid., 10.

Stump, Epistemology and Metaphysics

An early version of this paper was presented under a different title at Stanford University on 1 April 1991. This material is based upon work supported by the National Science Foundation under a grant awarded in 1990. I thank the NSF, Stanford University, and the Science Studies program at UCSD for support, and David Bloor, Michael A. Dennis, Arthur Fine, Peter Galison, David Gooding, Bruno Latour, Michael Lynch, Andrew Pickering, Joseph Rouse, and V. B. Smocovitis for helpful discussions and/or correspondence on this material. The term "concrete" in the title not only expresses the idea of forming a strong compound of heterogeneous elements, but also resonates with Jean Arp's nonrepresentational art and literature: "Since this art doesn't have the slightest trace of abstraction, we name it: concrete art. Works of concrete art should not be signed by the artist. These paintings, sculptures—these objects—should remain anonymous in the huge studio of nature, like clouds, mountains, seas, animals, men. Yes! Men should go back to nature!" (J. Arp, "Concrete Art," in *Arp on Arp: Poems, Essays, Memories*, ed. M. Jean [New York: Viking, 1972], 139–40, on 139).

The epigraphs for this essay are quoted from William James, "The Meaning of the Word 'Truth,'" remarks at the meeting of the American Philosophical Association, Cornell University, December 1907, published in H. S. Thayer, ed., *Pragmatism: The Classic Writings* (Indianapolis: Hackett, 1982), 249–50, on 249; and from Friedrich Nietzsche, "On Truth and Lies in a Non-Moral Sense," in *Philosophy and Truth: Selections from Nietzsche's Notebooks of the Early 1870s*, ed. and trans. D. Breazeale (Atlantic Highlands, N.J.: Humanities Press, 1979), 79–97, on 92.

1. D. Shapere, "Notes towards a Post-Positivist Interpretation of Science," in P. Achinstein and S. F. Barker, eds., *The Legacy of Logical Positivism* (Baltimore: Johns Hopkins University Press, 1969), esp. 149–56 and 158–59.

2. M. Gardner, "Realism and Instrumentalism in Nineteenth-Century Atomism," *Philosophy of Science* 46 (1979): 1–34; and "Realism and Instrumentalism in Pre-Newtonian Astronomy," in J. Earman, ed., *Testing Scientific Theories*, Minnesota Studies in the Philosophy of Science 10 (Minneapolis: University of Minnesota Press, 1983), 201–65.

3. Gardner, "Realism" (1983), 201–2.

4. Ibid., 244.

5. Ibid.

6. See L. Laudan, "A Confutation of Convergent Realism," *Philosophy of Science* 48 (1981): 218–49; reprinted in J. Leplin, ed., *Scientific Realism* (Berkeley and Los Angeles: University of California Press, 1984), 226.

7. M. Friedman, *Foundations of Space-Time Theories: Relativistic Physics and Philosophy of Science* (Princeton: Princeton University Press, 1983).

8. Ibid., 251.   9. Ibid., 3–4.

10. Ibid., 236.   11. Ibid., 242.

12. Discussion of Laudan's program in naturalized philosophy of science is outside the scope of this paper; the reader is referred to the thoughtful reviews by Gooding and Nickles: see L. Laudan, "Progress or Rationality? The Prospects for Normative Naturalism," *American Philosophical Quarterly* 24 (1987): 19–31; and "Normative Naturalism," *Philoso-*

phy of Science 57 (1990): 44–59; D. Gooding, "How to Be a Good Empiricist," *British Journal for the History of Science* 22 (1989): 419–27; T. Nickles's review of A. Donovan, L. Laudan, and R. Laudan, eds., *Scrutinizing Science, Isis* 80 (1989): 665–69.

13. R. Miller, *Fact and Method: Explanation, Confirmation and Reality in the Natural and the Social Sciences* (Princeton: Princeton University Press, 1987), 8.

14. A. Fine, "Piecemeal Realism," *Philosophical Studies* 61 (1991): 79–96, on 94.

15. I do not accept van Fraassen's constructive empiricism (which I am calling "instrumentalism"), but rather only his rejection of abduction. I agree with Fine's view—we should not add any philosophical interpretation at all to scientifically accepted theories. The most important issue here is that scientific realism is not debated by citing cases, but rather by general principles. Miller's critique of van Fraassen follows this pattern, for he moves from specific cases to a general argument about when antirealist interpretations of theory lack rationally compelling force (*Fact and Method*, 383).

16. Ibid., 4. Miller does seem to know that van Fraassen is not a framework relativist; see ibid., 372–73, 379.

17. See R. M. Chisholm, *The Problem of the Criterion*, Aquinas Lecture 1973 (Milwaukee: Marquette University Press, 1973), for one classic discussion.

18. Michael Lynch has pointed out that Karl Mannheim made a similar point about relativism long ago: M. Lynch, "Allan Franklin's Transcendental Physics," in A. Fine, M. Forbes, and L. Wessels, eds., *PSA 1990*, vol. 2 (East Lansing: Philosophy of Science Association, 1991), 471–586, on 474–75.

19. P. Galison, "Multiple Constraints, Simultaneous Solutions," in A. Fine and J. Leplin, eds., *PSA 1988*, vol. 2 (East Lansing: Philosophy of Science Association, 1989), 157–63; P. Galison, *How Experiments End* (Chicago: University of Chicago Press, 1987); P. Galison, "History, Philosophy, and the Central Metaphor," *Science in Context* 2 (1988): 197–212; I. Hacking, *Representing and Intervening* (Cambridge: Cambridge University Press, 1983); P. Kosso, "Dimensions of Observability," *British Journal for the Philosophy of Science* 39 (1988): 449–67; P. Kosso, "Objectivity," *Journal of Philosophy* 86 (1989): 245–57; P. Kosso, *Observability and Observation in Physical Science* (Dordrecht: Kluwer, 1989).

20. See D. Bloor's review of Peter Galison, *How Experiments End*, *Social Studies of Science* 21 (1991): 186–89; 188 for the complaint. Examples of the charge include L. Laudan, "The Pseudo-science of Science?" *Philosophy of the Social Sciences* 11 (1981): 41–73; and P. Roth and R. Barrett,

"Deconstructing Quarks," *Social Studies of Science* 20 (1990): 579–632. The direct replies are D. Bloor, "The Strengths of the Strong Programme," *Philosophy of the Social Sciences* 11 (1981): 199–213; and A. Pickering, "Mere Construction: A Reply to Roth and Barnett," *Social Studies of Science* 20 (1990): 682–729. I will note here that I do not count the symmetry thesis to be a form of relativism. The symmetry thesis is a demand for what counts as a good explanation and is clearly compatible with the historiography that I am highlighting here. However, Barnes and Bloor's further claim that arguments always "run in a circle" shows their use of the traditional argument for relativism; see B. Barnes and D. Bloor, "Relativism, Rationalism and the Sociology of Knowledge," in M. Hollis and S. Lukes, eds., *Rationality and Relativism* (Oxford: Blackwell, 1982), 27.

21. Lynch, "Allan Franklin's Transcendental Physics," 474–75.

22. Philosophical interest in justification of theories might be thought to keep the social out in much the same way that the distinction between the context of discovery and the context of justification kept the actual historical development of science out of philosophical accounts, but social analyses have been concerned with how scientific theories are justified, as well as how they develop.

23. B. Latour, *The Pasteurization of France* (Cambridge, Mass.: Harvard University Press, 1988), part 2, "Irreductions," 153–236; and B. Latour, "One More Turn after the Social Turn," in Ernan McMullin, ed., *The Social Dimensions of Science* (Notre Dame: University of Notre Dame Press, 1992), 272–94.

24. As with all rejections of dichotomies, we are left with a continuum. Just as some beliefs are more theory-laden than others, some practices are more purely scientific, and others more cultural. Still, a separation cannot be maintained: the "purely scientific" aspects must be given a concrete content in a specific scientific setting, and the "cultural elements" must be given a concrete function as part of local scientific practice.

25. A. Pickering, "Living in the Material World," in D. Gooding, T. Pinch, and S. Schaffer, eds., *The Uses of Experiment* (Cambridge: Cambridge University Press, 1989), 275–98.

26. A. Pickering, "Openness and Closure: On the Goals of Scientific Practice," in H. E. Le Grand, ed., *Experimental Inquiries* (Dordrecht: Kluwer, 1990), 215–39, on 231–32.

27. See Latour, "One More Turn."

28. When I say that social factors always play a role, I mean that one can always tell a story of the social/cultural context of a scientific practice. This is important work, but there is no justification for reducing scientific

practice to a particular cultural setting. The results of embodied experimental practice are not built into the cultural setting and cannot be fully explained by it.

29. A. Pickering, "Beyond Constraint: The Temporality of Practice and the Historicity of Knowledge," unpublished talk given at the conference "Philosophical and Historiographical Problems about Small-Scale Experiments," University of Toronto, 23–25 March 1990, ms. p. 2.

30. A. Pickering, "Objectivity and the Mangle of Practice," *Deconstructing and Reconstructing Objectivity*, Special Issue, *Annals of Scholarship* 8 (1991): 409–25, on 418.

31. Pickering, "Beyond Constraint."

32. Galison, "History, Philosophy, and the Central Metaphor."

33. I do not mean to gloss over the differences in Galison's and Pickering's accounts; the senses in which the results are not built in are different. Nevertheless, the two accounts are more compatible than they are generally thought to be. For another important account that overcomes the debate over the social construction of science, see M. Rudwick, *The Great Devonian Controversy* (Chicago: University of Chicago Press, 1985), esp. 450–56.

34. G. Doppelt, "Kuhn's Epistemological Relativism: An Interpretation and Defense," *Inquiry* 21 (1978): 33–86; reprinted in M. Krausz and J. Meiland, eds., *Relativism: Cognitive and Moral* (Notre Dame: University of Notre Dame Press, 1982), 109–51; D. Stump, "Fallibilism, Naturalism and the Traditional Requirements for Knowledge," *Studies in History and Philosophy of Science* 22 (1991): 451–69.

35. H. Putnam, *Reason, Truth and History* (Cambridge: Cambridge University Press, 1981), 238.

36. See, e.g., D. Papineau, *Reality and Representation* (Oxford: Blackwell, 1987), 2–3.

37. Horwich claims that the minimal theory does explain the truth of sentences in general, but this claim seems implausible because he provides no account of the independence of facts from our beliefs: P. Horwich, *Truth* (Oxford: Blackwell, 1990), 110–12. I am suggesting here that the epistemic disunity of science can provide such an account of the correspondence intuition on an everyday level. I claimed above that no theory, not even the correspondence theory, satisfies the demand for a general explanation of what makes sentences true. Advocates of the minimal account of truth accept the realist view that truth should be "nonepistemic," but are unwilling to accept the skepticism that many realists embrace. For example, Friedman has gone so far as to argue that the access problem is one of the great virtues of realism: M. Friedman, "Truth and Confirmation," *Journal of Philosophy* 76 (1979): 361–82; reprinted in Hilary

Kornblith, ed., *Naturalizing Epistemology* (Cambridge, Mass.: MIT Press, 1983), 147–68.

38. Pickering, "Objectivity and the Mangle," 418.

39. I also do not see why Pickering is tempted to claim that historically contingent events are "relative" ("Beyond Constraint," ms. p. 8). He does emphasize the difference between relativism and historicism, but I think the point needs clarification regarding contingency. The only people who could be bothered by contingency are those who look for some kind of teleology. Unless we hold that they are given by God, all scientific laws are contingent, but they are still as objective and factual as we can get. The fact that we cannot predict what will contingently happen does not take away the objective nature of the events that do in fact occur. It is a contingent fact (indeed, highly improbable!) that I was born rather than one of my possible siblings, but there is no historical relativism here; rather, this is what we mean by a historical fact, by something that "just happened." Rationalists tend to dismiss such events as "mere facts," but science does not live up to the rationalist vision.

40. Pickering, "Openness and Closure," 233.

41. See Fine, this volume.

42. S. Woolgar, *Science: The Very Idea* (London: Tavistock, 1988), 53–54; M. Ashmore, *The Reflexive Thesis: Wrighting Sociology of Scientific Knowledge* (Chicago: University of Chicago Press, 1989); B. Barnes, *About Science* (Oxford: Blackwell, 1985); H. M. Collins, "Stages in the Empirical Programme of Relativism," *Social Studies of Science* 11 (1981): 3–10; H. M. Collins, *Changing Order: Replication and Induction in Scientific Practice* (London: Sage, 1985); T. Pinch, *Confronting Nature: The Sociology of Solar-Neutrino Detection* (Dordrecht: Reidel, 1986).

43. See H. Field, "The Deflationary Conception of Truth," in G. MacDonald and C. Wright, eds., *Fact, Science and Morality* (Oxford: Blackwell, 1987); A. Fine, *The Shaky Game: Einstein, Realism, and the Quantum Theory* (Chicago: University of Chicago Press, 1986), 136–50; Fine, this volume; Horwich, *Truth*; J. Rouse, *Knowledge and Power: Towards a Political Philosophy of Science* (Ithaca: Cornell University Press, 1987), 127–65; and M. Williams, "Do We (Epistemologists) Need A Theory of Truth?" *Philosophical Topics* 14 (1986): 223–42. Donald Davidson is also associated with a deflationary or minimal account of truth, but he now distances himself from the previously mentioned authors: see D. Davidson, "The Structure and Content of Truth," *Journal of Philosophy* 87 (1990): 279–328.

44. For example, Descartes was a "structural realist." He held that ordered sets of numbers can adequately represent geometric objects, even though they clearly do not "copy" geometric objects. They are not similar

(a common requirement of the modern correspondence theory), but they are isomorphic structures, which is compatible with ontological relativity.

45. We do not have direct access to the facts to which statements are supposed to correspond, so the correspondence relation, which was supposed to explain so much, turns mysterious. The first problem is defining a fact, an issue that exercised nineteenth-century philosophers greatly. Tarski seemed to solve the problem by showing how to eliminate facts. Truth can be defined in a formal language in terms of reference and predicate satisfaction without any mysterious facts, but the debate over correspondence theories of truth then shifted to the problem of reference, which has greatly exercised twentieth-century philosophers: see H. Putnam, "Why Reason Can't Be Naturalized," Howison Lecture, University of California, 1981; reprinted in *Realism and Reason: Philosophical Papers*, vol. 3 (Cambridge: Cambridge University Press, 1983), 229–47, for a standard criticism of theories of reference. The old problem of access has never been solved.

46. C. S. Peirce, *Selected Writings*, ed. P. Wiener (New York: Dover, 1958), 379; W. James, "Pragmatism's Conception of Truth," in *Pragmaticism* (New York: Hafner, 1907), 160–61.

47. See Fine's regress argument, *Shaky Game*, 141.

48. Here we have an ironic use of a typical skeptical argument against induction, the curve-fitting problem, applied against an antirealist definition of truth.

49. Russell says "criterion of truth" and "meaning of truth"; see B. Russell, "William James's Conception of Truth" (1910), reprinted in *Philosophical Essays* (London: Allen and Unwin), 127–49. These arguments are rehearsed in contemporary debates over truth; for example, see Putnam, *Reason, Truth and History*, 54–55; or Davidson, "Structure and Content," §2, and the many references there. James anticipates this argument against his view, but his reply shows that he does not stay consistently on the everyday level and takes a verificationist stand that nothing exists outside our practices.

50. See Fine, *Shaky Game*, 140.

51. I first used this formulation in a paper on Ludwik Fleck: D. Stump, "The Role of Skill in Experimentation: Reading Ludwik Fleck's Study of the Wassermann Reaction as an Example of Ian Hacking's Experimental Realism," in A. Fine and J. Leplin, eds., *PSA 1988*, vol. 1 (East Lansing: Philosophy of Science Association, 1988), 302–8, on 305.

52. Horwich, *Truth*, 5–6.

53. Ibid., §35.

54. Ibid.; Williams, "Do We (Epistemologists) Need a Theory of Truth?"

55. Of course, Hegel attempts to overcome the distinction between the internal and the external, but it is not clear that this attempt succeeds.

56. No doubt many in sociology will deny that they adopt any theory of truth. I started this section by pointing out three kinds of arguments within the social-construction literature, so it should be clear that I am not painting everyone with the same brush.

57. W. James, "The Meaning of the Word 'Truth,'" in H. S. Thayer, ed., *Pragmatism: The Classic Writings* (Indianapolis: Hackett, 1982), 249–50, on 249.

58. Fine, this volume, 231–54.

59. Stump, "Role of Skill."

60. L. Fleck, *Genesis and Development of a Scientific Fact*, tr. F. Bradley and T. J. Trenn (Chicago: University of Chicago Press, 1979), 155.

61. It is somewhat misleading to say that we can no longer explain science using social categories, as Latour sometimes does ("One More Turn"). It is more correct to say that we cannot reduce scientific practice to a social category (emphasized in Latour, *Pasteurization*, part 2). A broader notion of explanation could help here, and I think that the pragmatics of explanation—the study of explanation as answering why-questions—provides the outlines of such a theory. See B. van Fraassen, *The Scientific Image* (Oxford: Clarendon Press, 1980). The central claim of such a theory is that there are no explanations *per se*; explanations are responses to specific why-questions. Alan Garfinkel's application of the pragmatics of explanation to an antireductionist account of various social sciences is especially helpful in the current context: A. Garfinkel, *Forms of Explanation: Rethinking the Questions in Social Theory* (New Haven: Yale University Press, 1981). See P. Kitcher and W. Salmon, "Van Fraassen on Explanation," *Journal of Philosophy* 84 (1987): 315–30; and P. Kitcher, "Explanatory Unification and the Causal Structure of the World," in P. Kitcher and W. Salmon, eds., *Scientific Explanation*, Minnesota Studies in the Philosophy of Science 8 (Minneapolis: University of Minnesota Press, 1989), 410–505, for a critique of the pragmatics of explanation as an incomplete account of explanation; and A. Richardson, "Explanation: Pragmatics and Asymmetry," unpublished talk given at the Pacific Division Meeting of the American Philosophical Association, Portland, 27 March 1992, for a defense and further discussion.

62. Latour, "One More Turn," 283.

63. E.g., B. Latour, "Mixing Humans and Nonhumans Together: The Sociology of a Door-Closer," *Social Problems* 35 (1988): 298–310.

64. A. Pickering, *Constructing Quarks: A Sociological History of Particle Physics* (Chicago: University of Chicago Press, 1984), 7–8.

65. Latour, *Pasteurization*, 185.

66. Latour, "One More Turn," 280.

67. Latour, *Pasteurization*, 93. See M. Lynch, *Art and Artifact in Laboratory Science: A Study of Shop Work and Shop Talk in a Research Laboratory* (London: Routledge and Kegan Paul, 1985).

68. D. Gooding, *Experiment and the Making of Meaning* (Dordrecht: Kluwer, 1990), 216.

69. Ibid., 187.   70. Ibid., 255.
71. Ibid., 160.   72. Ibid., 165.
73. Ibid., 179.   74. Ibid., 186.
75. Ibid., 13.

76. Fine, *Shaky Game*, 112–50; and Fine, this volume; Rouse, *Knowledge and Power*, 127–65.

77. Some of the issues here are dealt with in a different way in J. Rouse, "Arguing for the Natural Ontological Attitude," in A. Fine and J. Leplin, eds., *PSA 1988*, vol. 1 (East Lansing: Philosophy of Science Association, 1988), 294–301.

78. I argue against this tendency in D. Stump, "Naturalized Philosophy of Science with a Plurality of Methods," *Philosophy of Science* 59 (1992): 456–60. Since science itself is not unitary, the attempt to privilege one scientific discipline as the method of naturalized philosophy of science has left naturalism as a form of reduction that distorts the philosophy of science just as badly as traditional philosophy of science distorted science.

79. Thomas McCarthy also suggests interest models for science studies, and Donna Haraway once adopted such a view: T. McCarthy, "Scientific Rationality and the 'Strong Program' in the Sociology of Knowledge," in Ernan McMullin, ed., *Construction and Constraint: The Shaping of Scientific Rationality* (Notre Dame: University of Notre Dame Press, 1988), 75–95; D. Haraway, "Situated Knowledges: The Science Question in Feminism and the Privilege of Partial Perspective," *Feminist Studies* 14 (1988): 575–99.

80. R. Rorty, "Solidarity or Objectivity?" in J. Rajchman and C. West, eds., *Post-Analytic Philosophy* (New York: Columbia University Press, 1985), 3–19; and R. Rorty, "Is Natural Science a Natural Kind?" in McMullin, ed., *Construction and Constraint*, 49–74, on 55.

81. N. Rescher, *Methodological Pragmatism: A Systems-Theoretic Approach to the Theory of Knowledge* (Oxford: Blackwell, 1977); L. Laudan, *Science and Values: The Aims of Science and Their Role in the Scientific Debate* (Berkeley and Los Angeles: University of California Press, 1984); D. Shapere, *Reason and the Search for Knowledge* (Dordrecht: Reidel, 1984); and "The Universe of Modern Science and Its Philosophical Exploration," in E. Agazzi and A. Cordero, eds., *Philosophy and the Origin and Evolution of the Universe* (Dordrecht: Kluwer, 1991), 87–202.

82. See J. Maffie, "Recent Work on Naturalized Epistemology," *American Philosophical Quarterly* 27 (1990): 281–93, for an overview of epistemological naturalism; and "Realism, Relativism and Naturalized Meta-epistemology," *Metaphilosophy* 24 (1993): 1–13, for a realist attempt to distinguish sharply interest models and naturalized epistemology.

83. C. S. Peirce, "What Pragmatism Is," *The Monist* 15 (1905): 161–81; reprinted in Thayer, ed., *Pragmatism*, 107.

84. Recently John Worrall and Harvey Siegel have taken such a position, and I criticize them in my "Fallibilism, Naturalism." Naturalists in the philosophy of science also accept a version of the traditional hierarchy when they claim that there must be a unified metamethod in science studies.

85. Fine, *Shaky Game*, 148.

86. This charge is often made against naturalized philosophy of science as well. For a discussion of the issues in a political context and for a different defense of the possibility of criticism, see N. Fraser and L. Nicholson, "Social Criticism without Philosophy: An Encounter between Feminism and Postmodernism," in A. Cohen and M. Dascal, eds., *The Institution of Philosophy: A Discipline in Crisis?* (La Salle: Open Court, 1989), 284–302. A public discussion between Bruno Latour and Donna Haraway on 13 May 1992 at the University of California, San Diego, was helpful in clarifying the issues in this section.

87. M. Callon and B. Latour, "Unscrewing the Big Leviathan: How Actors Macro-structure Reality and How Sociologists Help Them Do So," in K. Knorr Cetina and A. V. Cicourel, eds., *Advances in Social Theory and Methodology: Toward an Integration of Micro- and Macro-sociologies* (Boston: Routledge and Kegan Paul, 1981), 277–303.

Knorr Cetina, The Care of the Self

I am extremely grateful to the many scientists who made this research possible through the advice they offered, through their patience, and through their indulgence. I also thank the Deutsche Forschungsgemeinschaft for financing and the Center for Science Studies in Bielefeld for facilitating this research.

1. The first laboratory studies were conducted in the late 1970's and led to a new understanding of scientific research: see K. Knorr Cetina, "Laboratory Studies: The Cultural Approach to the Study of Science," in J. C. Petersen, G. E. Markle, S. Jasanoff, and T. J. Pinch, eds., *Science, Technology and Society Handbook* (London: Sage, 1994). Examples for such laboratory studies are: K. D. Knorr Cetina, "Producing and Reproducing Knowledge: Descriptive or Constructive? Toward a Model of Research

Production," *Social Science Information* 16 (1977): 669–96; K. Knorr Cetina, *The Manufacture of Knowledge: An Essay on the Constructivist and Contextual Nature of Science* (Oxford: Pergamon Press, 1981); B. Latour and S. Woolgar, *Laboratory Life: The Social Construction of Scientific Facts* (London: Sage, 1979); M. Lynch, *Art and Artifact in Laboratory Science: A Study of Shop Work and Shop Talk in a Research Laboratory* (London: Routledge and Kegan Paul, 1985); and S. Traweek, *Beamtimes and Lifetimes: The World of High Energy Physics* (Cambridge, Mass.: Harvard University Press, 1988).

2. H. Maturana and F. Varela, *Autopoeisis and Cognition: The Realization of the Living* (Boston: D. Reidel, 1980).

3. One physicist described them to me in German as "irreale Gegenstände," as irrational objects (somewhat like irrational numbers).

4. The quote is from the physicist Andy Paches in the experiment UA2 at CERN.

5. See V. D. Barger and R. J. N. Phillips, *Collider Physics* (Redwood City: Addison-Wesley, 1987).

6. The quote is from a physicist who conducted the Alpha S analysis in the experiment UA2.

7. See n. 6.

8. M. Foucault, "Of Other Spaces," trans. J. Milkowiec, *Diacritics* 16 (1986): 22–27.

9. The quote is from a postdoc who paraphrased "understanding the behavior of the detector" as indicated.

10. For an example of this attitude, see K. Knorr Cetina, *Epistemic Cultures: How Scientists Make Sense* (forthcoming).

11. Victor Turner uses the term to characterize periods during which the status of ritual subjects is ambiguous, as in rites of transition performed by native tribes: see *The Ritual Process* (Chicago: Aldine, 1969).

12. See Knorr Cetina, *Epistemic Cultures*, chap. 3.

13. For a detailed analysis of the experiential regime see Knorr Cetina, *Epistemic Cultures*, chap. 4.

14. Ian Hacking, *Representing and Intervening* (Cambridge: Cambridge University Press, 1983).

15. H. Collins and S. Yearley, "Epistemological Chicken," in A. Pickering, ed., *Science as Practice and Culture* (Chicago: University of Chicago Press, 1992).

16. G. H. Mead, *Mind, Self, and Society* (Chicago: University of Chicago Press, 1967).

17. T. Kuhn, "The Trouble with the Historical Philosophy of Science," Robert and Maureen Rothschild Distinguished Lecture, Department of the History of Science, Harvard University, 1992.

Wylie, Gender Politics and Science

I gratefully acknowledge the support of research funding from the Social Sciences and Humanities Research Council of Canada, and sabbatical and research leave affiliations that I held with the Department of Anthropology at the University of California at Berkeley and at Clare Hall, Cambridge, during the period I developed the research that is the basis for this paper.

1. Roland B. Dixon, "Some Aspects of North American Archaeology," *American Anthropologist* 15 (1913): 565.

2. Lewis R. Binford, "Archaeological Perspectives," in *An Archaeological Perspective* (New York: Seminar Press, 1972), 86.

3. Lewis R. Binford, *Working at Archaeology* (New York: Academic Press, 1983).

4. John Fritz, "Archaeological Systems for Indirect Observation of the Past," in Mark P. Leone, ed., *Contemporary Archaeology* (Carbondale: Southern Illinois University Press, 1972), 135–57.

5. Clyde Kluckhohn, "The Conceptual Structure of Middle American Studies," in C. L. Hay, R. L. Linton, S. K. Lothrop, H. L. Shapiro, and G. C. Vaillant, eds., *The Maya and Their Neighbours* (New York: Appleton-Century, 1940), 41–51.

6. Berthold Laufer, "Comments On 'Some Aspects of North American Archaeology,'" *American Anthropologist* 15 (1913): 576.

7. Ibid.

8. Ibid., 577.

9. W. D. Strong, "Anthropological Theory and Archaeological Fact," in Robert H. Lowie, ed., *Essays in Anthropology* (Berkeley: University of California Press, 1936), 365.

10. Ibid.

11. Ibid., 364.

12. See, for example, Albert C. Spaulding, "Statistical Techniques for the Discovery of Artifact Types," *American Antiquity* 18 (1953): 305–13. In fact, the architect of one of the most influential typological schemes in the 1930's describes his Midwest taxonomy as a "tool," a "method of discovering order in the world": W. C. McKern, "The Midwest Taxonomic Method as an Aid to Archaeological Culture Study," *American Antiquity* 4 (1939): 304. In a similar vein, Cole and Deuel endorsed the system as "begin[ning] to furnish us with materials on which a chronology may perhaps be based"; they regarded such systematization as the necessary basis for any reconstruction of cultural events, contacts, interactions, or developments such as might comprise a culture history: Fay-Cooper Cole and Thorne Deuel, *Rediscovering Illinois: Archaeological Explorations in and around Fulton County* (Chicago: University of Chicago Press, 1937), 200.

13. Strong, "Anthropological Theory and Archaeological Fact," 365.

14. Dixon, "Some Aspects of North American Archaeology," 565.

15. Ibid., 563.

16. Clark Wissler, "The New Archaeology," *American Museum Journal* 17 (1917): 100. Several articles along similar lines had appeared in the previous decade: Edgar L. Hewett, "The Groundwork of American Archaeology," *American Anthropologist* 10 (1908): 591–95; Harlan I. Smith, "Archaeological Evidence as Determined by Method and Selection," *American Anthropologist* 13 (1911): 445–48. Laufer's defense of a more narrowly empirical approach was, in fact, formulated in reaction against these proposals for a "new" archaeology.

17. Wissler, "The New Archaeology," 100.

18. Clyde Kluckhohn, "The Place of Theory in Anthropological Studies," *Philosophy of Science* 6 (1939): 328–44.

19. Ibid.

20. Kluckhohn identifies these approaches with a "narrow empiricism" or, alternatively, a "*simpliste* mechanistic-positivistic philosophy": Kluckhohn, "The Conceptual Structure of Middle American Studies," 46.

21. Ibid., 46.

22. Ibid., 47. Kluckhohn concludes that to insist that data collection can go forward on its own prior to engaging theoretical and methodological questions about its evidential significance is to reason enthymematically. As he puts the point, "the alternative is not . . . between theory and no theory or a minimum of theory, but between adequate and inadequate theories": Kluckhohn, "The Place of Theory in Anthropological Studies," 330. In a similar vein, Steward and Setzler are adamant that data collection and systematization (and any improvement in the techniques deployed in these connections) can proceed effectively "only with reference to their purpose, which involves the question of research objectives": Julian H. Steward and Frank M. Setzler, "Function and Configuration in Archaeology," *American Antiquity* 4 (1938): 3. They maintain that, so long as researchers proceed blindly, not only will they "overlook" data that might prove useful in addressing the questions of ultimate concern; they will also miss interpretive possibilities: "no one in the future will be able to interpret the data one tenth as well as the persons now immersed in them," so it is imperative that those actually recovering and analyzing the primary data do so with a clear problem orientation and a sound conceptual framework (ibid., 7). Writing in reaction against McKern's taxonomic proposals, Steward generalizes on these observations, arguing that "facts are totally without significance and may even be said not to exist without reference to theory. It is wholly impossible to collect bare facts. . . . It is

equally impossible merely to give significant order to facts without reference to some theory or problem": Julian Steward, "RE Archaeological Tools and Jobs," *American Antiquity* 10 (1944): 99. See also John W. Bennett, "Empiricist and Experimentalist Trends in Eastern Archaeology," *American Antiquity* 11 (1946): 198–200.

23. Bennett, "Empiricist and Experimentalist Trends in Eastern Archaeology," 198.

24. Ibid., 200.

25. J. A. Ford, "Spaulding's Review of Ford," *American Anthropologist* 56 (1954): 109.

26. Spaulding, "Statistical Techniques for the Discovery of Artifact Types."

27. Albert C. Spaulding, "Review of *Measurements of Some Prehistoric Design Developments in the Southeastern States*," *American Anthropologist* 55 (1953): 590.

28. Raymond Thompson, *Modern Yucatecan Pottery Making*, Society for American Archaeology, Memoir 15 (Salt Lake City: SAA, 1958). See also Raymond Thompson, "The Subjective Element in Archaeological Inference," *Southwestern Journal of Anthropology* 12 (1956): 1–2.

29. Thompson, *Modern Yucatecan Pottery Making*, 8.

30. M. A. Smith, "The Limitations of Inference in Archaeology," *The Archaeological Newsletter* 6 (1955): 6.

31. There is virtually no recognition by the New Archaeologists of these antecedents, despite a number of striking similarities between the program they proposed in the 1960's and 1970's and the arguments for problem-oriented, interpretively rich modes of practice that appeared in the second decade of the century and then, again, in the late 1930's and 1950's.

32. Kent V. Flannery, "Cultural History v. Cultural Process: A Debate in American Archaeology," *Science* 217 (1967): 120.

33. Binford, "Archaeological Perspectives," 90.

34. Ibid., 86. See also Lewis R. Binford and Jeremy A. Sabloff, "Paradigms, Systematics and Archaeology," *Journal of Anthropological Research* 38 (1982): 137–53.

35. There are antecedents for this position. In the late 1940's, critics of "empiricist" approaches, such as Bennett, and Steward and Setzler, took the interdependence between archaeological "facts" and "theory" to be cause for optimism. It meant that archaeology need not be confined to "the 'Baconian observation' of empirical detail" simply because it necessarily deals with tangible, "sense-perceivable data": Bennett, "Empiricist and Experimentalist Trends in Eastern Archaeology," 200. Archaeological data need not be assumed to have any "intrinsic qualities" that "pro-

hibit" their interpretive analysis as cultural material; Steward and Setzler, "Function and Configuration in Archaeology," 7.

36. Binford, *Working at Archaeology*, 12.

37. Binford and Sabloff, "Paradigms, Systematics and Archaeology."

38. Michael Shanks and Christopher Tilley, *Re-constructing Archaeology* (Cambridge: Cambridge University Press, 1987), 15.

39. Ian Hodder, "Archaeology in 1984," *Antiquity* 58 (1984): 26. For a more detailed presentation of the analysis that follows, see Alison Wylie, "On 'Heavily Decomposing Red Herrings': Scientific Method in Archaeology and the Ladening of Evidence with Theory," in Lester Embree, ed., *Metaarchaeology: Reflections by Archaeologists and Philosophers*, Boston Studies in the Philosophy of Science (Dordrecht: Kluwer, 1992), 269–88.

40. Hodder, "Archaeology in 1984," 26.

41. Ibid., 27.

42. Ibid., 26.

43. Shanks and Tilley, *Re-constructing Archaeology*, 111.

44. Ibid., 20–21.

45. Ibid., chap. 3, "Facts and Values in Archaeology," 46–67.

46. Ibid., 103.

47. Ian Hodder, "Archaeology, Ideology and Contemporary Society," *Royal Anthropological Institute News* 56 (1983): 7.

48. Shanks and Tilley, *Re-constructing Archaeology*, 246; Ian Hodder, *Reading The Past: Current Approaches to Interpretation in Archaeology* (Cambridge: Cambridge University Press, 1986), 159–61; Ian Hodder, "Interpretive Archaeology and Its Role," *American Antiquity* 56 (1991): 7.

49. The account that follows of how and why archaeological research on gender has emerged in the last few years is developed in more detail in Alison Wylie, "Gender Theory and the Archaeological Record: Why Is There No Archaeology of Gender?" in Dale Walde and Noreen Willows, eds., *The Archaeology of Gender: Proceedings of the 22nd Annual Chacmool Conference* (Calgary: Archaeological Association of the University of Calgary, 1991), 31–54. See also "The Interplay of Evidential Constraints and Political Interests: Recent Archaeological Research on Gender," *American Antiquity* 57 (1992): 15–34; and "Evidential Constraints: Pragmatic Empiricism in Archaeology," in Lee McIntyre and Michael Martin, eds., *Readings in the Philosophy of Social Science* (Cambridge, Mass.: MIT Press, 1993), 747–65.

50. Margaret W. Conkey and Janet D. Spector, "Archaeology and the Study of Gender," in Michael B. Schiffer, ed., *Advances in Archaeological Method and Theory*, vol. 7 (New York: Academic Press, 1984), 1–38.

51. Joan M. Gero and Margaret W. Conkey, eds., *Engendering Archaeology: Women and Prehistory* (Oxford: Blackwell, 1991). There had been

some exploration of feminist themes in other connections. For example, a number of essays that dealt centrally with questions about gender appeared in a collection published by the Archaeology Division of the American Anthropology Association in 1990: Sarah M. Nelson and Alice B. Kehoe, eds., *Powers of Observation: Alternative Views in Archaeology*, Archaeological Papers of the American Anthropological Association, no. 2 (Washington, D.C.: AAA, 1990). Several more narrowly focused discussions along these lines had also appeared in problem- or region-specific literatures where a developed anthropological and historical interest in the status and roles of women was extended to archaeology. See, for example, contributions to Patricia Albers and Beatrice Medicine, eds., *The Hidden Half: Studies of Plains Indian Women* (Washington, D.C.: University Press of America, 1983): Alice Kehoe, "The Shackles of Tradition," 53–73; Janet D. Spector, "Male/Female Task Differentiation among the Hidatsa: Toward the Development of an Archaeological Approach to the Study of Gender," 77–99. See also Anne Barstow, "The Uses of Archeology for Women's History: James Mellaart's Work on the Neolithic Goddess at Çatal Hüyük," *Feminist Studies* 4.3 (1978): 7–17. In addition, a number of analyses of the status of women in archaeology (equity studies) had appeared, some of them undertaken by "critical archaeologists" who also repudiated the scientism of the New Archaeology: Joan M. Gero, "Gender Bias in Archaeology: A Cross-Cultural Perspective," in Joan Gero, David M. Lacy, and Michael L. Blakey, eds., *The Socio-Politics of Archaeology*, Research Report no. 23 (Amherst: University of Massachusetts Press, 1983); Joan M. Gero, "Socio-Politics and the Woman-at-Home Ideology," *American Antiquity* 50 (1985): 342–50. Finally, although feminist approaches did not become a central research focus for postprocessualists, the influence of feminist perspectives is to be seen in some of their work, for example in contributions to Daniel Miller and Christopher Tilley, eds., *Ideology, Power, and Prehistory* (Cambridge: Cambridge University Press, 1984): Ian Hodder, "Burials, Houses, Women and Men in the European Neolithic," 51–68; Mary Braithwaite, "Ritual and Prestige in the Prehistory of Wessex c. 2200–1400 BC: A New Dimension to the Archaeological Evidence," 93–110.

Two annotated bibliographies have recently appeared that provide quite comprehensive summaries of conference presentations and publications on the subject of archaeology and gender: Cheryl Claassen, "Bibliography of Archaeology and Gender: Papers Delivered at Archaeology Conferences, 1964–1992," *Annotated Bibliographies for Anthropologists* 1.2 (1992); Elisabeth A. Bacus, Alex W. Barker, Jeffrey D. Bonevich, Sandra L. Dunavan, J. Benjamin Fitzhugh, Debra L. Gold, Nurit S. Goldman-Finn, William Griffin, and Karen M. Mudar, eds., *A Gendered*

Past: A Critical Bibliography of Gender in Archaeology, University of Michigan Museum of Anthropology, Technical Report 25 (Ann Arbor: University of Michigan Press, 1993). In the preface to the first of these, a guide to conference presentations between 1964 and 1992, Claassen notes that only 24 of the 284 entries she lists date to before January 1988, and only two of these have appeared in print.

It is important to note that my discussion focuses on developments in North America. In 1988, when the working conference organized by Gero and Conkey took place, a special issue of *Archaeological Review from Cambridge* was produced, "Women and Archaeology," based in part on the Cambridge Feminist Archaeology Workshops of 1987–88 and on presentations made at the annual Theoretical Archaeology Group Conference held in the U.K. in 1987 (with antecedents in 1982 and 1985): Karen Arnold, Roberta Gilchrist, Pam Graves, and Sarah Taylor, *Women in Archaeology*, Special Issue, *Archaeological Review from Cambridge* 7.1 (1988): 1. A year later a summary text was published describing what might be gleaned from existing publications about the status and roles of women in European prehistory: Margaret Ehrenberg, *Women in Prehistory* (Norman: University of Oklahoma Press, 1989). In addition, a group of Norwegian women in archaeology organized a conference on sex roles in prehistory in 1979. The proceedings were published eight years later: Reidar Bertelsen, Arnvid Lillehammer, and Jenny-Rita Naess, eds., *Were They All Men? An Examination of Sex Roles in Prehistoric Society*, AmS-Varia 17 (Stavanger: Arkeologisk Museum i Stavanger, 1987). Norwegian women archaeologists have been meeting regularly since this time and have published a journal since 1985, *Norwegian Women in Archaeology* (the Norwegian acronym is KAN), an initiative that is now being broadened to include other Scandinavian archaeologists.

52. An increasingly large number of sympathetic colleagues have since taken up these questions and, in some cases, organized conferences and symposia on gender research in archaeology. One such conference, the Chacmool conference (a thematic conference held annually at the University of Calgary) for 1989, drew over 100 submissions on the archaeology of gender in response to an open call for papers, a very substantial increase in submissions over previous years. Those who participated came from all over the United States, New Zealand, Norway and Sweden, the U.K., and various Western European countries; only 17 submissions were from Canadian scholars. The proceedings appeared two years later: Dale Walde and Noreen Willows, eds., *The Archaeology of Gender*, Proceedings of the 1989 Chacmool Conference (Calgary: Archaeological Association of the University of Calgary, 1991). In the same year *Historical Archaeology* published a special issue on gender in historical

archaeology: Donna Seifert, ed., *Gender in Historical Archaeology*, Special Issue, *Historical Archaeology* 25.4 (1991). In the meantime, two conferences on these topics were organized in North Carolina, one of which has resulted in published proceedings: Cheryl Claassen, ed., *Exploring Gender through Archaeology: Selected Papers from the 1991 Boone Conference*, Monographs in World Archaeology, no. 11 (Madison: Prehistory Press, 1992). A group of Australian archaeologists also organized a conference on gender research and equity issues in archaeology the proceedings of which are in press: Hilary duCros and Laurajane Smith, eds., *Women in Archaeology: A Feminist Critique*, Occasional Papers, Department of Prehistory and Anthropology (Canberra: Australian National University, 1994). So despite the paucity of published literature, an enthusiasm about the prospects for archaeological work on gender seems to have taken hold across the whole length and breadth of the field. Further discussion of why this interest in questions about women and gender has arisen so quickly and so late in archaeology is to be found in Alison Wylie, "Gender Theory and the Archaeological Record," 31–54; and in "Feminist Critiques and Archaeological Challenges," in Walde and Willows, eds., *The Archaeology of Gender*, 17–23. A discussion of the significance of the 1989 Chacmool conference and a content analysis of submissions are to be found in Marsha Hanen and Jane Kelley, "Gender and Archaeological Knowledge," in Embree, ed., *Metaarchaeology*, 195–227.

53. Patti Lather, "Deconstructing/Deconstructive Inquiry: The Politics of Knowing and Being Known," *Educational Theory* 41 (1991): 154.

54. Sandra Harding, *The Science Question in Feminism* (Ithaca: Cornell University Press, 1986), 195.

55. Lather, "Deconstructing/Deconstructive Inquiry," 154.

56. Frances E. Mascia-Lees, Patricia Sharpe, and Colleen Ballerino Cohen, "The Postmodernist Turn in Anthropology: Cautions from a Feminist Perspective," *Signs* 15 (1989): 15.

57. Ibid. In fact, the most uncompromising critics of the relativizing tendencies inherent in radically deconstructive approaches insist that "the feminist postmodernists' plea for tolerance of multiple perspectives is altogether at odds with feminists' desire to develop a successor science that can refute once and for all the distortions of androcentrism": Mary E. Hawkesworth, "Knowers, Knowing, Known: Feminist Theory and Claims of Truth," *Signs* 14 (1989): 538.

58. For a discussion of these issues, see Helen Longino and Ruth Doell, "Body, Bias, and Behaviour: A Comparative Analysis of Reasoning in Two Areas of Biological Science," *Signs* 9 (1983): 207–8; Helen Longino, *Science as Social Knowledge: Values and Objectivity in Scientific Inquiry* (Princeton: Princeton University Press, 1990), chap. 1, "Introduc-

tion: Good Science, Bad Science," 3–15; Harding, *The Science Question in Feminism*; Wylie, "Gender Theory and the Archaeological Record," 38–44.

59. Sally Slocum, "Woman the Gatherer: Male Bias in Anthropology," in Rayna Reiter, ed., *Toward an Anthropology of Women* (New York: Monthly Review Press, 1975), 36–50.

60. Joan Kelly-Gadol, "Did Women Have a Renaissance?" in R. Bridenthal and C. Koonz, eds., *Becoming Visible: Women in European History* (Boston: Houghton Mifflin, 1977), 137–64.

61. Carol Gilligan, *In a Different Voice: Psychological Theories and Women's Development* (Cambridge, Mass.: Harvard University Press, 1982).

62. Anne Fausto-Sterling, *Myths of Gender: Biological Theories about Women and Men* (New York: Basic Books, 1985).

63. Ibid., 9.

64. Longino and Doell, "Body, Bias, and Behaviour," 227.

65. Ibid., 208.

66. Longino, *Science as Social Knowledge*, chap. 10, "Conclusion: Social Knowledge," 215–32.

67. Ibid., 188.

68. Alison Wylie, "Feminist Theories of Social Power: Some Implications for a Processual Archaeology," *Norwegian Archaeological Review* 25 (1992): 51–68.

69. See, for example, a discussion of the general issues that arise in this connection due to Nancy Fraser and Linda Nicholson, "Social Criticism without Philosophy: An Encounter between Feminism and Postmodernism," in Andrew Ross, ed., *Universal Abandon? The Politics of Postmodernism* (Minneapolis: University of Minnesota Press, 1988), 83.

70. Christopher Norris, *What's Wrong with Postmodernism: Critical Theory and the Ends of Philosophy* (Baltimore: The Johns Hopkins University Press, 1990).

71. Alison Wylie, "On 'Heavily Decomposing Red Herrings,'" 269–88.

72. Hodder, *Reading The Past*, 16.

73. Shanks and Tilley, *Re-constructing Archaeology*, 104.

74. Ibid., 245.

75. Ibid., 104.

76. Wylie, "On 'Heavily Decomposing Red Herrings,'" 269–88.

77. One example is the constructivist position originally defended by Pickering in *Constructing Quarks*: "In principle the decisions which produce the world [evidentially as well as theoretically] are free and unconstrained. They could be made at random, each scientist choosing by the toss of a coin at each decision point what stance to adopt" (Andrew

Pickering, *Constructing Quarks* [Edinburgh: Edinburgh University Press, 1984], 406). Pickering has since moved between an uncompromising rejection of "constraint talk" in which he reaffirms a strongly relativist constructivism, and what seem an amendment of his earlier position in which, for example, he distances himself from the stance taken by H. M. Collins in *Changing Order: Replication and Induction in Scientific Practice* (London: Sage, 1985). For the former position, see Andrew Pickering, "Living in the Material World: On Realism and Experimental Practice," in David Gooding, Trevor Pinch, and Simon Schaffer, eds., *The Uses of Experiment: Studies in the Natural Sciences* (Cambridge: Cambridge University Press, 1989), 275–97; and Andrew Pickering, "Beyond Constraint: The Temporality of Practice and the Historicity of Knowledge," paper presented at the conference "Philosophical and Historiographical Problems about Small-Scale Experiments," University of Toronto, March 1990. And for the latter, see Andrew Pickering, "Essay Review: Forms of Life Science, Contingency and Harry Collins," *British Journal for the History of Science* 20 (1987): 213–21.

78. Longino and Doell, "Body, Bias, and Behaviour," 208–10. See also Longino, *Science as Social Knowledge*, chap. 6, "Research on Sex Differences," 103–32.

79. Dudley Shapere, "The Concept of Observation in Science and Philosophy," *Philosophy of Science* 49 (1982): 505.

80. Dudley Shapere, "Observation and the Scientific Enterprise," in P. Achinstein and O. Hannaway, eds., *Observation, Experiment, and Hypothesis in Modern Physical Science* (Cambridge, Mass.: MIT Press, 1985), 22, 36.

81. Longino, *Science as Social Knowledge*, chap. 4; "Values and Objectivity," 62–82. Shapere argues, in this connection, that "though it is true that the background information employed in science is *not certain* . . . it is not for that reason uncertain. . . . The mere possibility of doubt arising is not itself a reason for [global] doubt": "The Concept of Observation," 514–15. In a subsequent discussion he summarizes these points: "What is referred to as 'observational' is far from being the pure, uninterpreted 'given,' distinguishable from all theoretical infusions, that much of classical and also much of modern empiricism claimed it must be," but what counts as observational "nevertheless plays the very same epistemic roles assigned to observation by the empiricist tradition . . . namely of being the basis of testing beliefs and of acquiring new knowledge about nature": "Observation and the Scientific Enterprise," 25–26. This last statement is perhaps misleading inasmuch as Shapere does intend to distance his account of observation from those views, including positivist and empiricist theories of science, that have insisted that "a good reason [must] be an

absolute guarantee or at least be based ultimately on absolute guarantees" (ibid., 22). By his own account, observation (in the sense he explicates with reference to physicists' claims to observe events at the center of the sun) clearly plays a rather different epistemic role than that ascribed to it by many empiricist theories of science. Presumably he intends to reconstruct a defensible, moderate (fallibilist) empiricism that recognizes both the constructed and the constraining nature of observation.

82. See, for example, Peter Galison, *How Experiments End* (Chicago: University of Chicago Press, 1987), 7–9. Also relevant are Peter Galison, "Multiple Constraints, Simultaneous Solutions," in A. Fine and J. Leplin, eds., *PSA 1988* (East Lansing: Philosophy of Science Association, 1988), vol. 2, pp. 157–63; Peter Galison, "Philosophy in the Laboratory," *Journal of Philosophy* 85 (1988): 525–27; Ian Hacking, "On the Stability of the Laboratory Sciences," *Journal of Philosophy* 85 (1988): 507–14; Ian Hacking, "Philosophers of Experiment," *PSA 1988*, vol. 2, pp. 147–56; and Ian Hacking, "The Participant Irrealist at Large in the Laboratory," *British Journal for the Philosophy of Science* 39 (1988): 277–94.

83. Shapere, "Observation and the Scientific Enterprise," 29.

84. Longino and Doell, "Body, Bias, and Behaviour," 209–10.

85. Kosso in amplification of Shapere: Peter Kosso, "Dimensions of Observability," *British Journal for the Philosophy of Science* 39 (1988): 449–67.

86. Ibid., 455.

87. Michael B. Schiffer, "Toward the Identification of Formation Processes," *American Antiquity* 48 (1983): 675–706.

88. Longino and Doell, "Body, Bias, and Behaviour," 209–10. Longino and Doell also extend this metaphor of "distance" to some quite distinct considerations of independence.

89. Binford, *Working at Archaeology*, 135; see also Binford and Sabloff, "Paradigms, Systematics, and Archaeology."

90. Ian Hacking, *Representing and Intervening: Introductory Topics in the Philosophy of Natural Science* (Cambridge: Cambridge University Press, 1983), 183–85.

91. Kosso, "Dimensions of Observability," 456.

92. Longino and Doell, "Body, Bias, and Behaviour," 210.

93. Hacking, *Representing and Intervening*, 185.

94. It is striking that, while recognizing the importance of independence in this sense for archaeological testing, Binford does not acknowledge the possibility that independence may come in degrees and that there is considerable scope for a limited "degree of nepotism" that does not collapse into "straightforward self-accounting," a point made most compellingly by Kosso, "Dimensions of Observability," 457. I discuss the importance of this insight for archaeological practice below.

95. Peter Kosso, "Science and Objectivity," *Journal of Philosophy* 86 (1989): 253–56.

96. Hacking, *Representing and Intervening*, chap. 11, "Microscopes," 186–209.

97. The "miracle" argument in question here is, simply, that it would be highly implausible that independent means of detection should produce convergent results if the body or structure under "observation" did not exist: Hacking, *Representing and Intervening*, 202. As Kosso puts this point, "the chances of these independent theories all independently manufacturing the same fictitious result is small enough to be rationally discounted": "Science and Objectivity," 247. Although Kosso is mainly concerned with arguments that exploit the independence between the background knowledge used to constitute observational evidence and the claims this evidence is used to support or refute—he develops a formal measure of independence of this sort—he also considers the role played in stabilizing evidential claims (and securing their objectivity) by the use of multiple lines of evidence that bear on a single subject. In this connection he considers in passing the way in which evidence is used to establish claims about "ancient history." Although this is presented as an extension of Hacking's discussion of multiple methods for detecting the same entity or determining the value of a constant, I would suggest that there is an important difference between cases in which the variety of evidence is mutually reinforcing because it bears on the *same* entity or aspect of the subject domain, and cases where it bears on different aspects of a subject domain presumed to be interdependent in some specific way. In the former, Hacking's and Kosso's cases, considerations of *disunity in the sources* of interpretive inference are crucial (i.e., in the background knowledge used to constitute data as evidence), while in the latter, considerations of *unity in the subject* (as opposed to its self-identity) play a central role as well.

98. Gero and Conkey, eds., *Engendering Archaeology*.

99. The summary of archaeological cases that is presented is elaborated in Wylie, "The Interplay of Evidential Constraints and Political Interests," 15–34.

100. Patty Jo Watson and Mary C. Kennedy, "The Development of Horticulture in the Eastern Woodlands of North America: Women's Role," in Gero and Conkey, eds., *Engendering Archaeology* 255–75.

101. Ibid., 262.   102. Ibid., 266.

103. Ibid., 262–64.   104. Ibid., 264.

105. Christine A. Hastorf, "Gender, Space, and Food in Prehistory," in Gero and Conkey, eds., *Engendering Archaeology*, 132–59.

106. Ibid., 139.

107. Ibid., 152.

108. Elizabeth Brumfiel, "Weaving and Cooking: Women's Production in Aztec Mexico," in Gero and Conkey, eds., *Engendering Archaeology*, 224–53.

109. Ibid., 241.

110. Russell G. Handsman, "Whose Art Was Found at Lepenski Vir? Gender Relations and Power in Prehistory," in Gero and Conkey, eds., *Engendering Archaeology*, 360.

111. Ibid., 338–39.

112. Ibid., 340.

113. Ibid., 343.

114. Margaret W. Conkey with Sarah H. Williams, "Original Narratives: The Political Economy of Gender in Archaeology," in Micaela di Leonardo, ed., *Gender, Culture and Political Economy: Feminist Anthropology in the Post-modern Era* (Berkeley and Los Angeles: University of California Press, 1991), 121. This paper was not Conkey's contribution to the 1988 conference, but several of its main points were discussed in this context.

115. Kosso, "Dimensions of Observability."

116. Watson and Kennedy, "The Development of Horticulture," 266.

117. Ibid., 262.

118. Clark Glymour, *Theory and Evidence* (Princeton: Princeton University Press, 1980). For an application to archaeological cases, see Alison Wylie, "Bootstrapping in Un-natural Sciences: An Archaeological Case," in A. Fine and P. Machamer, eds., *PSA 1986* (East Lansing: Philosophy of Science Association, 1986), vol. 1, 314–22.

119. Longino and Doell, "Body, Bias, and Behaviour," 206–27.

120. Alison Wylie, "The Reaction against Analogy," in Michael B. Schiffer, ed., *Advances in Archaeological Method and Theory* (New York: Academic Press, 1985), vol. 8, 63–111; Alison Wylie, "'Simple' Analogy and the Role of Relevance Relations: Implications of Archaeological Practice," *International Studies in the Philosophy of Science* 2 (1988): 134–50.

121. Ehrenberg, *Women in Prehistory*, chap. 2, "The Earliest Communities," 38–76; chap. 3, "The First Farmers," 77–107.

122. This is terminology introduced by Julian S. Weitzenfeld, "Valid Reasoning by Analogy," *Philosophy of Science* 51 (1984): 137–49. It is applied to archaeological cases in Wylie, "'Simple' Analogy," 134–50.

123. For example, as Gero argues in her own contribution to *Engendering Archaeology*, archaeologists should be prepared to question a range of assumptions that associate specific types of stone-tool production and use primarily with male activities and labor. Ethnohistoric and experimental evidence concerning the physical requirements of such production and the

typical, even predominant, patterns of use documented among foragers undermines many standard assumptions about the gender associations of these tools and activities: Joan M. Gero, "Genderlithics: Women's Roles in Stone Tool Production," in Gero and Conkey, eds., *Engendering Archaeology*, 164–76.

124. Binford, *Working at Archaeology*, 135; Lewis R. Binford, "In Pursuit of the Future," in D. J. Meltzer, D. D. Fowler, and J. A. Sabloff, eds., *American Archaeology Past and Future* (Washington, D.C.: Smithsonian Institution, 1986), 472.

125. Kosso, "Dimensions of Observability," 456; Hacking, *Representing and Intervening*, 183–85.

126. Conkey in Conkey and Williams, "Original Narratives"; Handsman, "Whose Art Was Found at Lepenski Vir?"

127. In fact, this seems to be the conclusion that Longino and Doell draw in analysis of the inferential distance involved in making any ascription of social, functional significance to tools associated with early hominid sites when they argue that "any speculation regarding the behaviour and social organization of early humans remains just that": "Body, Bias, and Behaviour," 217. It is perhaps significant that they cite, in this connection, three prominent early New Archaeologists, including Binford, who were among the most outspoken critics of any use of analogical inference, given their deductivist commitments. I have argued that the position taken by the New Archaeologists (especially Binford) on this issue represents a largely rhetorical reaction against the practices they associate with "traditional" archaeology. It is inaccurate as an account of analogical inference and has proven largely unsustainable in archaeological practice: Wylie, "The Reaction against Analogy," 63–111; Wylie, " 'Simple' Analogy," 134–50. Thus, while I would wholeheartedly endorse Longino and Doell's insistence that the limits of reconstructive inference be clearly recognized, I question their generalized pessimism about the cases they consider.

128. Longino and Doell, "Body, Bias, and Behaviour," 209–10.

Cat, Cartwright, and Chang, Otto Neurath

In citing Neurath's work, we use the following abbreviations:

ES   *Empiricism and Sociology*, ed. Marie Neurath and Robert S. Cohen (Dordrecht: Reidel, 1973).

PP   *Philosophical Papers, 1913–1946*, ed. and trans. Robert S. Cohen and Marie Neurath (Dordrecht: Reidel, 1983).

GpmS   *Gesammelte philosophische und methodologische Schriften*, ed. R. Haller and H. Rutte (Vienna: Holder, Pichler, and Tempsky, 1981).

Citations of these works give the title of the article followed by the year of original publication (in parentheses), source, and page number(s).

1. See, for example, "Pseudorationalism of Falsification" (1935), *PP*, 121–31; and "Individual Sciences, Unified Science, Pseudorationalism" (1936), *PP*, 132–38.

2. Neurath, "Sociology in the Framework of Physicalism" (1931), *PP*, 75 (emphasis original); Neurath, "Empirical Sociology" (1931), *ES*, 362; *GpmS*, 540.

3. Neurath, "Sociology in the Framework of Physicalism" (1931), *PP*, 66; emphases original.

4. Neurath, "Radical Physicalism and the 'Real World'" (1934), *PP*, 101.

5. Neurath, "Protocol Statements" (1932/33), *PP*, 92. This metaphor of the boat occurs in two other places in Neurath's writings: "Probleme der Kriegwirtschaftlehre," *Zeitschrift für die Gesamte Staatwissenschaft* 69 (1913): 457; "Anti-Spengler" (1921), *ES*, 199.

6. Neurath, "Empirical Sociology" (1931), *ES*, 404.

7. Ibid.

8. Neurath, "Sociology in the Framework of Physicalism" (1931), *PP*, 61. It is rather puzzling why he used the expression "in the sense of contemporary physics," but the examples make his meaning clear.

9. Neurath, "Empirical Sociology" (1931), *ES*, 405.

10. Neurath, "Universal Jargon and Terminology" (1941), *PP*, 218.

11. Neurath, "Sociology in the Framework of Physicalism" (1931), *PP*, 59; emphases original.

12. Neurath, "Empirical Sociology" (1931), *ES*, 329.

13. Neurath, "Sociology in the Framework of Physicalism" (1931), *PP*, 59; emphases original.

14. See ibid., 65–66; also "Physicalism" (1931), *PP*, 54–55.

15. Neurath, "The Orchestration of the Sciences by the Encyclopedism of Logical Empiricism" (1946), *PP*, 233.

16. Neurath, "Sociology in the Framework of Physicalism" (1931), *PP*, 62.

17. Neurath, "Unified Science and Its Encyclopedia" (1937), *PP*, 176–77; emphases original.

18. Neurath, "The New Encyclopedia of Scientific Empiricism" (1937), *PP*, 194.

19. Neurath, "Unified Science and Its Encyclopedia" (1937), *PP*, 176.

20. For a detailed account of the Bavarian Revolution, see Allan Mitchell, *Revolution in Bavaria, 1918–1919* (Princeton: Princeton University Press, 1965). See also Richard Grunberger, *Red Rising in Bavaria*

(London: Barker, 1973); and Heinrich Ströbel, *The German Revolution and After,* tr. H. J. Stenning (London: Jarrolds, 1923).

21. Quoted in Mitchell, *Revolution,* 118.

22. *Die Neue Zeit* 1:24–25; quoted in Mitchell, *Revolution,* 120.

23. See Neurath, *Wesen und Weg der Sozialisierung: Gesselschaftliches Gutachten, vorgetragen dem Münchener Arbeiterrat am 25. Januar 1919* (Munich: Callway, 1919); reprinted in Neurath, *Durch die Kriegswirtschaft zur Naturalwirtschaft* (Munich: Callway, 1919), 209–20.

24. See Mitchell, *Revolution,* 257; see also Lujo Brentano, *Mein Leben* (Jena: Diederich, 1931).

25. "Bayrische Sozialisierungserfahrungen," *Neue Erde* 2 (1919): 43. This is translated in part into English; see *ES,* 18–19.

26. Brentano, *Mein Leben,* 364.

27. The text of this plan can be found in *Die Sozialisierung Sachens. Drei Vorträge* (Chemnitz: Arbeiter- und Soldatenratsverlag, 1919).

28. Neurath, "Die Sozialisierung und die wirtschaftlichen Räte," *Neue Erde* 1 (1919): 270.

29. This is described in "Bayerischer Landtag Sozialisierungs Außschuß," *Münchner Neueste Nachrichten,* 25 Mar. 1919. Here Neurath was probably referring to the work of Josef Popper-Lynkeus.

30. "Memories of Otto Neurath," *ES,* 17.

31. See *München-Augsburger Abendzeitung,* 1919, nos. 139, 140 (26 Mar.).

32. See Dr. Linter's statement at Neurath's trial, Akt. Nr. I 2139 vom Staatsarchiv München, proceedings against Neurath before the Standgericht München, 1919; also *München-Augsburger Abendzeitung,* 1919, no. 159 (7 Apr.).

33. For details of Neurath's activities in this period, see *München-Augsburger Abendzeitung,* 1919, nos. 139–68 (26 Mar.–13 Apr.).

34. See Neurath, "Die wirtschaftlichen Räte im Programm der bayerischer Vollsozialisierung," *Der Kampf* 13 (1920): 136–41. See also Neurath's statements in his trial proceedings.

35. Neurath, "Experiences of Socialization in Bavaria" (1920), *ES,* 27.

36. Ibid., 25.

37. For a detailed description of the fate of the leaders of the Bavarian Soviet Republic, see Grunberger, *Red Rising,* 142ff.

38. Neurath, "Zur Theorie der Sozialwissenschaften" (1910), *GpmS,* 24.

39. Ibid., 45.

40. Ibid., 46.

41. Neurath, "An International Encyclopedia of Unified Science" (1936), *PP*, 141.

42. Neurath, "Experiences of Socialization in Bavaria" (1920), *ES*, 19.

43. Neurath, "Sociology in the Framework of Physicalism" (1931), *PP*, 59.

44. Ibid., 68; *GpmS*, 534.

45. Neurath, "Empirical Sociology" (1931), *ES*, 407.

46. Neurath, "Individual Sciences, Unified Science, Pseudorationalism" (1936), *PP*, 133.

47. Neurath, "Wirtschaftsplan, Planwirtschaft, Landesverfassung und Völkerordnung," *Der Kampf* 13 (1920): 226.

48. Neurath, "Vollsozialisierung und gemeinwirtschaftliche Anstalten," *Der Kampf* 15 (1922): 55.

49. Neurath, "Sociology in the Framework of Physicalism" (1931), *PP*, 62; *GpmS*, 537.

50. Neurath et al., "Wissenschaftliche Weltauffassung: Der Wiener Kreis" (1929), *ES*, 305. This text was the product of a joint effort between Neurath, Hahn, and Carnap, but Neurath did most of the writing.

51. Quoted in Elisabeth Nemeth, *Otto Neurath und der Wiener Kreis* (Frankfurt: Campus Verlag, 1981), 51.

52. Neurath, "Through War Economy to Economy in Kind" (1919), *ES*, 155.

53. A. Rabinbach, *The Crisis of Austrian Socialism: From Red Vienna to the Civil War (1927–1934)* (Chicago: University of Chicago Press, 1983), 28.

54. Hans Ziesel, "The Austromarxists in Red Vienna: Reflections and Recollections," in A. Rabinbach, ed., *The Austrian Socialist Experiment: Social Democracy and Austromarxism, 1918–1934* (Boulder: Westview Press, 1985), 123.

55. See Otto Bauer, *The Austrian Revolution* (New York: Franklin, 1925), 98–129, 162–203.

56. See Charles Gulick, *Austria from Habsburg to Hitler* (Berkeley and Los Angeles: University of California Press, 1948), 100–101.

57. "Verhendhugen der Parteiteger der österreichschen Sozialdemokratie in Hainfeld" (1889), quoted in M. Blum, *The Austro-Marxists (1890–1918): A Psychological Study* (Lexington: University of Kentucky Press, 1985), 108.

58. See Tom Bottomore, "Introduction," and Otto Bauer, "What Is Austro-Marxism?" in Tom Bottomore and Patrick Goode, eds., *Austro-Marxism* (Oxford: Clarendon Press, 1988), 1–44, 45–48.

59. The meaning of this German word is roughly the same as that of *Bildungsgemeinschaft*.

60. "Die Nationalitätenfrage und die Sozialdemokratie" (1907); quoted in Rabinbach, *Crisis*, 16.

61. Rabinbach, *Crisis*, 16.

62. "Bücherschaue," *Der Kampf* 1 (Mar. 1907): 94; "Die Arbeiterbibliotek," *Der Kampf* 1 (Oct. 1907): 48; and "Eine Parteischule für Deutschösterreich," *Der Kampf* 10 (Jan. 1910): 173–75; among others.

63. Max Adler, *Lehrbuch der materialischen Geschichtsauffassung* (Berlin: Laub, 1930), 73.

64. Max Adler, *Der Sozialismus und die Intellektuellen* (Vienna: Brand, 1910), 52.

65. Max Adler, "Zur Frage der Organization des Proletariat der Intelligenz," *Die Neue Zeit* 13, no. 21 (Feb. 1895): 690; quoted in Blum, *Austro-Marxists*, 101.

66. Neurath, "Museums of the Future" (1933), *ES*, 220.

67. Quoted in Peter Galison, "Aufbau/Bauhaus," *Critical Inquiry* 16 (1990): 714–15.

68. This is Max Adler's term for people who will have gone through the proper socialist education, the members of the new socialized humanity.

69. Neurath, "Personal Life and Class Struggle" (1928), *ES*, 258–59.

70. Ibid., 251.

71. Neurath, "Museums of the Future" (1933), *ES*, 221.

72. Ibid., 222.

73. Neurath, "Personal Life and Class Struggle" (1928), *ES*, 295.

74. Ibid., 297.

75. Neurath, "Visual Education: Humanization versus Popularization" (1945), *ES*, 229.

76. Neurath, "Unified Science and Psychology," in B. McGuiness, ed., *Unified Science* (Dordrecht: Reidel, 1987), 23.

77. Neurath et al., "Wissenschaftliche Weltauffassung: Der Wiener Kreis" (1929), *ES*, 317–18.

Lenoir and Ross, The Naturalized History Museum

1. See Bruno Latour and Steve Woolgar, *Laboratory Life: The Construction of Scientific Facts*, rev. ed. (Princeton: Princeton University Press, 1986); Martin J. S. Rudwick, *The Great Devonian Controversy: The Shaping of Scientific Knowledge among Gentlemanly Specialists* (Chicago: University of Chicago Press, 1985); Steven Shapin and Simon Schaffer, *Leviathan and the Air-Pump: Hobbes, Boyle, and the Experimental Life* (Princeton: Princeton University Press, 1985).

2. In addition to the works cited above, see Timothy Lenoir, "Models

and Instruments in the Development of Electrophysiology, 1845–1912," *Historical Studies in the Physical and Biological Sciences* 17 (1986): 1–54; M. Norton Wise, "Mediating Machines," *Science in Context* 2 (1988): 77–114; Simon Schaffer, "Glass Works: Newton's Prisms and the Uses of Experiment," in David Gooding, Trevor Pinch, and Simon Schaffer, eds., *The Uses of Experiment* (Cambridge: Cambridge University Press, 1989), 67–104; Crosbie Smith and M. Norton Wise, *Energy and Empire: A Biographical Study of Lord Kelvin* (Cambridge: Cambridge University Press, 1989); David Gooding, *The Making of Meaning* (Dordrecht: Nijhoff, 1989).

3. Trevor Pinch, "Towards an Analysis of Scientific Observation: The Externality and Evidential Significance of Observation Reports in Physics," *Social Studies of Science* 15 (1985): 3–35; Harry Collins, *Changing Order* (London: Sage, 1985).

4. Michael Lynch, *Art and Artifact in Laboratory Science: A Study of Shop Work and Shop Talk in a Research Laboratory* (London: Routledge and Kegan Paul, 1985); Andy Pickering, "Living in the Material World," in *The Uses of Experiment*, 275–98; Timothy Lenoir, "Practical Reason and the Construction of Knowledge: The Lifeworld of Haber-Bosch," in Ernan McMullin, ed., *The Social Dimension of Science* (Notre Dame: University of Notre Dame Press, 1992), 158–97; Andrew Warwick, "Cambridge Mathematics and Cavendish Physics: Cunningham, Campbell and Einstein's Relativity, 1905–1911," parts 1 and 2, *Studies in History and Philosophy of Science* 23 (1992): 625–56, 24 (1993): 1–25.

5. Simon Schaffer, "Late Victorian Metrology and Its Instrumentation: A Manufactory of Ohms," in Robert Bud and Susan E. Cozzens, eds., *Invisible Connections: Instruments, Institutions, and Science* (Bellingham: SPIE Optical Engineering Press, 1992), 23–56. In a similar vein, Joseph Rouse has explored standardization as a means for multiplying contexts: *Knowledge and Power: Toward a Political Philosophy of Science* (Ithaca: Cornell University Press, 1987), 111–26.

6. See Madeleine Akrich and Bruno Latour, "A Summary of a Convenient Vocabulary for the Semiotics of Human and Nonhuman Assemblies," in Wiebe E. Bijker and John Law, eds., *Shaping Technology/Building Society: Studies in Sociotechnical Change* (Cambridge, Mass.: MIT Press, 1992), 259–64. Also see Madeleine Akrich, "The De-Scription of Technical Objects," ibid., 205–24; Bruno Latour, *We Have Never Been Modern* (Cambridge, Mass.: Harvard University Press, 1993); Donna Haraway, "The Promises of Monsters: A Regenerative Politics for Inappropriate/d Others," in Lawrence Grosberg, Cary Nelson, and Paul Treichler, eds., *Cultural Studies* (New York: Routledge, 1992), 295–337; N. Katherine Hayles, "Constrained Constructivism: Locating Scientific Inquiry in the Theater of Representation," in George Levine, ed., *Realism and Representa-*

tion: Essays on the Problem of Realism in Relation to Science, Literature, and Culture (Madison: University of Wisconsin Press, 1993), 27–43.

7. See Timothy Lenoir, "Was the Last Turn the Right Turn? The Semiotic Turn and A. J. Greimas," *Configurations* 2 (1994): 119–36. The semiotics of Roland Barthes is the starting point for such studies. Examples in science studies include the work of Haraway. See sources cited in n. 8 below.

8. See Gillian Beer, *Darwin's Plots: Evolutionary Narrative in Darwin, George Eliot, and Nineteenth-Century Fiction* (London: Routledge and Kegan Paul, 1983); Donna Haraway, *Primate Visions: Gender, Race, and Nature in the World of Modern Science* (New York: Routledge, Chapman and Hall, 1989); idem, *Simians, Cyborgs and Women: The Reinvention of Nature* (New York: Routledge, Chapman and Hall, 1991). A particularly valuable collection of essays on this topic is Levine, ed., *Realism and Representation*. See there esp. Gillian Beer, "Wave Theory and the Rise of Literary Modernism," 193–213; Simon Schaffer, "Augustan Realities: Nature's Representatives and Their Cultural Resources in the Early Eighteenth Century," 279–320; and Ludmilla Jordanova, "Museums: Representing the Real?" 255–74.

9. Lothar P. Witteborg, "Design Standards in Museum Exhibits," *Curator* 1 (1958): 29.

10. Most work in museum studies tends toward either silence or naivete on these questions. Recent and welcome exceptions are Ludmilla Jordanova, "Objects of Knowledge: A Historical Perspective on Museums," and Stephen Bann, "On Living in a New Country," in Peter Vergo, ed., *The New Museology* (London: Reaktion Books, 1989), 22–40, 99–118.

11. Joel Porte, ed., *Emerson in His Journals* (Cambridge, Mass.: Harvard University Press, 1982), 110–11.

12. Patricia Parker has suggestively discussed "rhetorics of property" in seventeenth-century writing, which she argues were intimately connected with the European approach to the New World. Emerson's journal entry provides a later, and interestingly inverted, example of the trope Parker describes as viewing the New World prior to conquering or colonizing it: Emerson comes from the New World, one site of appropriation for the creators of natural-history exhibits, to the European centers of natural history in order to learn how to "see" nature. See Patricia Parker, "Rhetorics of Property: Exploration, Inventory, Blazon," in *Literary Fat Ladies: Rhetoric, Gender, Property* (London: Methuen, 1987), 126–54.

13. Jonathan Culler, *Framing the Sign: Criticism and Its Institutions* (Norman: University of Oklahoma Press, 1988), 153–67.

14. Ibid., 164.

15. Roland Barthes, *Mythologies*, tr. Annette Lavers (New York: Hill and Wang, 1957), 116.

16. Ibid., 129.

17. Recent efforts in art history, paralleling ours here, to use semiotic models to capture the workings of political and social power often omitted from traditional discussions of museum images, are included in Norman Bryson, ed., *Calligram: Essays in New Art History from France* (Cambridge: Cambridge University Press, 1988).

18. See Adrian J. Desmond, "Designing the Dinosaur: Richard Owen's Response to Robert Edmond Grant," *Isis* 70 (1979): 224–34; "Artisan Resistance and Evolution in Britain, 1819–1848," *Osiris*, ser. 2, 3 (1987): 77–110; and *The Politics of Evolution: Morphology, Medicine, and Reform in Radical London* (Chicago: University of Chicago Press, 1989).

19. See J. F. C. Harrison, *Quest for the New Moral World: Robert Owen and the Owenites in Britain and America* (New York: Scribner's, 1969).

20. See, for example, Jack Morrell and Arnold Thackray, *Gentlemen of Science: Early Years of the British Association for the Advancement of Science* (Oxford: Clarendon Press, 1981).

21. Richard Owen, "Report on British Fossil Reptiles, Part II," *Report of the British Association for the Advancement of Science, 1841* (1842), 200.

22. Ibid., 109.

23. Ibid., 203–4.

24. Adrian J. Desmond, "Richard Owen's Reaction to Transmutation in the 1830s," *British Journal for the History of Science* 18 (1985): 25–50, esp. 45.

25. Adrian J. Desmond, "Central Park's Fragile Dinosaurs," *Natural History* 83 (October 1974): 65.

26. On 31 December 1853, Hawkins held such an event to inaugurate the exhibit; invitations read, "Mr. B. Waterhouse Hawkins solicits the honour of Professor ———'s company at dinner, *in the iguanodon*": see ibid., 66.

27. Richard Owen, *Geology and Inhabitants of the Ancient World* (London: Bradbury and Evans, 1854), 5.

28. Ibid., 31.

29. Beer, *Darwin's Plots*, 30–36.

30. Richard Owen, *On the Extent and Aims of a National Museum of Natural History* (London: Saunders, Otley, 1862), 2–3.

31. Ibid.

32. Ibid.

33. Ibid., 125.

34. Ibid., 68.

35. Ibid.

36. Ibid., 69.

37. Ibid., 10–11.

38. Ibid., 113.

39. Ibid.

40. Ibid., 85.

41. Ibid., 117.

42. Ibid., 126.

43. Ibid., 114.

44. Mark Girouard, from whose history of the architecture of the museum this discussion is derived, notes the similarity of this design to Bramante's unexecuted but famous plan for the dome of St. Peter's Cathedral in Rome: *Alfred Waterhouse and the Natural History Museum* (New Haven: Yale University Press, 1981), 26–27 *et passim*.

45. Haraway, *Primate Visions*, 26–58; Ronald Rainger, *An Agenda for Antiquity: Henry Fairfield Osborn and Vertebrate Paleontology at the American Museum of Natural History* (Tuscaloosa: University of Alabama Press, 1991).

46. Haraway, *Primate Visions*, 56.

47. Hawkins narrowly missed having his dinosaur reconstructions exported to this site: see Desmond, "Central Park's Fragile Dinosaurs," for details of the controversy.

48. Henry F. Osborn, *The Hall of the Age of Man* (New York: American Museum Press, 1921), 21.

49. Ibid., 31.

50. Henry F. Osborn, *The American Museum and Education* (New York: American Museum Press, 1925), 4–5.

51. Ibid., 5. The American Museum extended such educational efforts beyond its walls: the same issue of *Natural History* that heralds the opening of the Hall of the Age of Man offers an article by Mrs. John I. Northrop, president of New York City's School Nature League, entitled "Nature and the City Child." Mrs. Northrop praises the American Museum's Department of Public Education for supplying traveling cases of specimens to schools. Implying that such specimens are the indispensable signifieds of what are otherwise mere words learned in natural-history classes, she insists, "If we cannot take the children of New York City to the country, we must bring all the country that is transportable to the children": see *Natural History* 10 (May–June 1920): 265–76.

52. Henry F. Osborn, *Men of the Old Stone Age* (New York: Scribner's, 1921), 317.

53. Support for this account of titanothere extinction is provided in Osborn's massive two-volume monograph, *The Titanotheres of Ancient Wyoming, Dakota and Nebraska*, Department of the Interior and U.S. Geological Survey Monograph 55, vol. 2 (Washington, D.C.: USGPO, 1929), 883–88.

54. Osborn, *Men of the Old Stone Age*, 502.

55. Henry F. Osborn, preface to Madison Grant, *The Passing of the Great Race; or, The Racial Basis of European History* (New York: Scribner's, 1918), vii–xiii.

56. Ibid., 261–62.

57. Ibid., 110.

58. Osborn, *Men of the Old Stone Age*, 272.

59. Henry F. Osborn, "Address of Welcome," in *Eugenics, Genetics, and the Family: Scientific Papers of the Second International Congress of Eugenics* (Baltimore: Williams and Wilkins, 1923), 2. An account of the history of modern eugenics movements is available in Daniel J. Kevles, *In the Name of Eugenics: Genetics and the Uses of Human Heredity* (Berkeley and Los Angeles: University of California Press, 1985).

60. Ralph Waldo Emerson, *Natural History of the Intellect and Other Papers* (Boston: Houghton Mifflin, 1921), 23.

Rouse, Beyond Epistemic Sovereignty

This essay is a substantially abridged and revised version of my earlier "Foucault and the Natural Sciences," in John Caputo and Mark Yount, eds., *Foucault and the Critique of Institutions* (State College: Pennsylvania State University Press, 1993), 137–62.

1. Steven Shapin and Simon Shaffer, *Leviathan and the Air-Pump* (Princeton: Princeton University Press, 1985), 332.

2. Karl Popper, *Conjectures and Refutations* (New York: Basic Books, 1962).

3. Paul Feyerabend, *Against Method* (London: New Left Books, 1975), 20.

4. Roy Bhaskar, *Scientific Realism and Human Emancipation* (London: Verso, 1986); Richard Boyd, "How to Be a Moral Realist," in G. Sayre-McCord, ed., *Essays on Moral Realism* (Ithaca: Cornell University Press, 1988), 181–228.

5. Thomas Kuhn, *The Structure of Scientific Revolutions*, 2d ed. (Chicago: University of Chicago Press, 1970).

6. Peter Galison, "Aufbau/Bauhaus: Logical Positivism and Architectural Modernism," *Critical Inquiry* 16 (1990): 709–52.

7. Jordi Cat, Nancy Cartwright, and Hasok Chang, "Otto Neurath: Politics and the Unity of Science," this volume, 347–69.

8. Michel Foucault, *Discipline and Punish*, tr. Alan Sheridan (New York: Random House, 1977); Michel Foucault, *The History of Sexuality, Volume 1*, tr. Robert Hurley (New York: Random House, 1978); Michel Foucault, *Power/Knowledge* (New York: Random House, 1980).

9. I believe that such a dynamic conception of scientific knowledge also requires that we reject the choice between scientific realism and empiricist or constructivist antirealisms, but I shall not argue that here. The avoidance of both realism and antirealism is discussed in Joseph Rouse, *Knowledge and Power* (Ithaca: Cornell University Press, 1987), chap. 5, "Against Realism and Anti-Realism," pp. 127–65.

10. Foucault, *History of Sexuality*, 88–89.

11. Ibid., 86–87.   12. Ibid., 88.

13. Ibid., 138.   14. Ibid., 88.

15. The typical philosophical response to claims that method and adjudicating evidence are theory-laden has been not to abandon the standpoint of sovereignty, but to reconstitute it at a different location. Hence, one finds accounts of rationality at the scale of research programs or scientific domains (Lakatos, Laudan, Shapere), or at the level of a metatheoretical explanation of theory-dependent instrumental success (scientific realism).

16. Foucault, *History of Sexuality*, 93.

17. Ibid.   18. Ibid., 92.

19. Ibid., 95.   20. Ibid., 94.

21. Ibid., 105.

22. For an account of Mendel that emphasizes the *continuity* of his researches with those of contemporary plant breeders, in terms both of his research practices and skills, and of his goals, see Robert Olby, "Mendel No Mendelian?" *History of Science* 17 (1979): 53–72.

23. Rouse, *Knowledge and Power*, chap. 4, "Local Knowledge," 69–126.

24. I argue in more detail for the temporal situatedness and dispersion of all articulations of scientific knowledge in *Knowledge and Power*, chap. 4 (esp. pp. 120–25), and in "The Narrative Reconstruction of Science," *Inquiry* 33 (1990): 179–96.

25. Rouse, "The Narrative Reconstruction of Science."

26. Ibid.

27. I am using the term "resistance" in a broader sense than might be expected. Obviously, there is resistance to a knowledge claim when there are people who refuse to accept it, and attempt to counter it by producing counterevidence or other arguments, or even by trying to bypass it or studiously ignore it. But there is also resistance to a claim when the purported objects of that claim do not behave in accordance with it. It is important not to separate these two aspects of resistance to knowledge too sharply, because of the ways they reinforce, and even constitute, one another. What *counts* as successful accord between knowledge and its objects is often itself contested, and the outcome of such conflicts is typically the result of successful negotiations (both negotiating with those who oppose the claim about the standards of success, and renegotiating the practices and procedures through which the objects of knowledge make themselves manifest). The concept of "resistance" that one needs in order to understand the dynamics of scientific knowledge is one that does not respect any sharp distinction between actions by people and behavior

by things. Detailed arguments for such a conception of "resistance" are provided by Bruno Latour, *Science in Action* (Cambridge, Mass.: Harvard University Press, 1987); and Andy Pickering, "Living in the Material World," in D. Gooding, T. Pinch, and S. Schaffer, eds., *The Uses of Experiment* (Cambridge: Cambridge University Press, 1989), 275–97.

28. Anyone who doubts that a useful approach to understanding science could not plausibly be organized under the heading of sex should look closely and imagine extrapolating the often-ironic discussion of the sexuality of particle accelerators and detectors in Sharon Traweek, *Beamtimes and Lifetimes* (Cambridge, Mass.: Harvard University Press, 1988).

29. Steve Fuller (*Philosophy of Science and Its Discontents* [Boulder: Westview Press, 1989]) has made a similar point, in objecting to what he calls the "textbook fallacy": the family, the economy, cognition (or "science"), education, etc., may be naively introduced to students as if they were discrete domains, when in fact they are overlapping categorizations. Thus, he points out that even in the context of textbook naivete, "it is unlikely that a discussion of the family will be restricted to the formation and maintenance of *gemeinschaftlich* bonds. In addition, the reader is likely to find an analysis of the family as an economic unit, as a vehicle for transmitting political ideology, and the like" (p. 16). One should add that in the end, such analyses of the family cannot be confined to discrete subsections either. And of course Fuller shares my principal point, that it would be an error to think that knowledge can be isolated as a sovereign realm any more than can the family.

30. Such criticisms have been articulated most influentially by Charles Taylor, "Foucault on Freedom and Truth," in *Foucault: A Critical Reader*, ed. David Hoy (Oxford: Blackwell, 1986), 69–102; Jurgen Habermas, "Taking Aim at the Heart of the Present," in Hoy, ed., *Foucault*, 103–8; and Nancy Fraser, *Unruly Practices* (Minneapolis: University of Minnesota Press, 1989); they are sufficiently widespread almost to go without saying in many contexts (e.g., Hilary Putnam, *Reason, Truth and History* [Cambridge: Cambridge University Press, 1981]).

31. Foucault, *History of Sexuality*, 88.

32. There is a more technical sense in which I would talk about questions of legitimation as on the same "level" of analysis as the inquiries they examine, which is worked out in some versions of "disquotational" accounts of truth, for which "p is true" is materially equivalent to "p." (See Paul Horwich, *Truth* [Oxford: Blackwell, 1990]; Arthur Fine, *The Shaky Game: Einstein, Realism and the Quantum Theory* [Chicago: University of Chicago Press, 1986].)

33. Steve Fuller (*Social Epistemology* [Bloomington: Indiana University Press, 1988], chap. 4, "Bearing the Burden of Proof: On the Frontier

of Science and History," pp. 99–116) offers an illuminating discussion of the role of the burden of proof in scientific and philosophical argument, and the ways in which the burden of proof can be shifted.

34. I take this notion of an "alignment" to be a commentary upon Foucault's discussion of networks or chains of power/knowledge and their strategic and tactical interaction, but the term itself is taken from Thomas Wartenburg (*The Forms of Power* [Philadelphia: Temple University Press, 1990]), whose account of the dynamic and heterogeneous character of social power has interesting parallels with the epistemological dynamics I have been developing here. Wartenburg's own view, while developed in original and insightful ways to which I am indebted, is closer to Foucault's position than his own interpretation of Foucault would suggest.

35. Wartenburg (*The Forms of Power*) offers a very informative account of such feminist strategies as part of a discussion of new social movements, to exemplify the insights offered by his dynamic account of power.

Keller, Scientific Subjectivity

1. The term "postvital culture" is borrowed from Richard Doyle, "On Living: Vital and Post-vital Rhetorics in Molecular Biology," Ph.D. dissertation, Dept. of Rhetoric, University of California, Berkeley, 1993.

2. See, e.g., Freeman Dyson, *Disturbing the Universe* (New York: Harper and Row, 1979).

3. See, e.g., "From Secrets of Life to Secrets of Death," in Mary Jacobus, E. F. Keller, and Sally Shuttleworth, eds., *Body/Politics: Women and the Discourses of Science* (New York: Routledge, 1990), 177–91; "Secrets of God, Nature, and Life," *History of the Human Sciences* 3 (1990): 229–41.

4. Brian Rotman, *Signifying Nothing: The Semiotics of Zero* (New York: Macmillan, 1987).

5. Lorraine Daston, "Baconian Facts, Academic Civility, and the Prehistory of Objectivity," *Annals of Scholarship* 9.1 (1991): 337–64; also "The Objectivity of Interchangeable Observers, 1830–1930," paper presented at the History of Science Society Meeting, Seattle, 28 Oct. 1990; Lorraine Daston and Peter Galison, "The Image of Objectivity," *Representations* 40 (1992): 81–128; Peter Dear, "From Truth to Disinterestedness in the Seventeenth Century," *Social Studies of Science* 22 (1992); Theodore Porter, "Quantification and the Accounting Ideal in Science," paper presented at the History of Science Society Meeting, Seattle, 28 Oct. 1990.

6. The remainder of this essay is adapted from my article "The Paradox of Scientific Subjectivity," *Annals of Scholarship* 9.2 (1992): 135–54.

7. Porter, "Quantification and the Accounting Ideal in Science."

8. Rotman, *Signifying Nothing*, 28.

9. Steven Shapin, "Pump and Circumstance: Robert Boyle's Literary Technology," *Social Studies of Science* 14 (1984): 481–520, on 510.

10. René Descartes, "Traité de l'homme," in *Oeuvres* (Paris: Pléiade, 1953).

11. Freeman Dyson, "The World, the Flesh, and the Devil," Third Bernal Lecture, Birkbeck College, London, 16 May 1972 (unpublished).

12. J. D. Bernal, *The World, the Flesh, and the Devil* (Bloomington: Indiana University Press, 1929), 48.

13. Ibid., 57.   14. Ibid., 67.

15. Ibid., 60.   16. Ibid., 73.

17. Ibid., 74.   18. Ibid., 79–80.

19. Dyson, "The World, the Flesh, and the Devil."

20. Francis Crick, *Of Molecules and Men* (Seattle: University of Washington Press, 1966).

21. Ibid., 12.

22. Francis Crick, "Eugenics and Genetics," in G. Wolstenholme, ed., *Man and His Future* (Boston: Little, Brown, 1963), 27.

23. It should be noted, at least parenthetically, that the easy slide between the first two of these is surely as significant as the distance marked between "I/we" and "they": indeed, the use of the first-person-singular pronoun is highly unusual, employed here merely in the service of a small hedge—employed simultaneously to stand in for the singular, to evade personal responsibility, and to invoke the anonymous authority of a seamless community not, of course, to be confused with "people."

24. Crick, "Eugenics and Genetics," 284.

25. Ibid., 295.

26. This quote and that following are from Daniel Hillis, "Intelligence as Emergent Behavior," in *Artificial Intelligence*, Special Issue, *Daedalus* (winter 1988): 189.

27. In a sense, of course, neither does Hillis. His response to my argument is worth quoting, if only to underscore the coexistence in his own vision of modern (even Cartesian) and postmodern conceptions of the subject (pers. comm., 29 Nov. 1990):

I was touched that you understood exactly what I was saying, even if you did not seem to empathize with it entirely.

I sympathize with your argument that we have historically tried to cast away the observer, piece by piece. But I interpret this very differently than you. To me, with each step we are not denying the essential "I," but we are rather trying to find it. We say, "what is important about me is not my position in space nor my position in society, nor even is it the substance of my body." This is not an attempt to whittle our self down to nothing, but rather to expose its essential core. Each time we shed some unessential symbol of our own self-importance we gain new freedom.

The dualism still lurking in Hillis's vision perhaps finds its clearest expression in the proximity and interdependence of the two principal by-products of AI currently under development—namely the twin technologies of virtual reality and robotics.

28. The explicit connections between the ambitions to escape the body (where "human thought will live free of bones and flesh") and the scientific and technological drive to decode the mysteries of biological reproduction (the ultimate "secret of life") have yet to be fully drawn. But a beginning (focusing especially on the symbolic role of "woman" in scientific discourse) can be found in Keller, "From Secrets of Life to Secrets of Death" and "Secrets of God, Nature, and Life." And for another perspective on many of the same issues, see the work of Donna Haraway, especially *Simians, Cyborgs, and Women: The Reinvention of Nature* (London: Routledge, Chapman and Hall, 1991).

29. Hans Moravec, *Mind Children* (Cambridge, Mass.: Harvard University Press, 1988), 1.

30. Jean Baudrillard, *The Ecstasy of Communication*, tr. Bernard and Caroline Schutze (New York: Semiotext(e), 1988).

31. Donna Haraway, "The Promises of Monsters," in L. Grossberg, C. Nelson, and P. Treichler, eds., *Cultural Studies* (New York: Routledge, 1992), 295–337, on 297.

32. Rotman, *Signifying Nothing*, 28.

Haraway, Feminist Diffractions

For support in preparing this essay, I am grateful for Academic Senate Faculty Research Grants from the University of California, Santa Cruz, and to the University of California Humanities Research Institute, sponsor of the Conference on Located Knowledges at the University of California, Los Angeles, 8–10 April 1993, where I first delivered this paper.

The epigraph sources for this essay are as follows: the grandmother and community defense–system software designer Malkah, speaking in Marge Piercy, *He, She, and It* (New York: Ballantine, 1993), 25; Steven Shapin and Simon Schaffer, *Leviathan and the Air-Pump: Hobbes, Boyle, and the Experimental Life* (Princeton: Princeton University Press, 1985), 65.

1. Commerce is a variant of conversation, communication, intercourse, passage. As any good economist will tell you, commerce is a procreative act.

2. Traweek was studying the legitimate sons of Robert Boyle; her

physicists' detector devices are the mechanical issue of his air pump, as well. Humans and nonhumans have progeny in the odd all-masculine reproductive practices of technoscience. Sharon Traweek, *Beamtimes and Lifetimes* (Cambridge, Mass.: Harvard University Press, 1988), 162: "I have presented an account of how high energy physicists construct their world and represent it to themselves as free of their own agency, a description, as thick as I could make it, of an extreme culture of objectivity: a culture of no culture, which longs passionately for a world without loose ends, without temperament, gender, nationalism, or other sources of disorder—for a world outside human space and time."

3. Shapin and Schaffer, *Leviathan and the Air-Pump*, 25.

4. Grasping the irreducible specificity of this point is in the bud stage in recent science-studies scholarship; the implications have hardly begun to be assimilated. When they are assimilated, the consequences for the relentlessly Eurocentric and Euro-Americancentric accounts of both science and science studies—in their contents, methods, languages, audiences, questions, and more—should be immense.

5. Shapin and Schaffer, *Leviathan and the Air-Pump*, 77.

6. Ibid.

7. Ibid., 79.

8. Ibid., 336.

9. See the discussion below on strong objectivity and Sandra Harding's antiracist, feminist standards for critical rationality in the experimental way of life.

10. Shapin and Schaffer, *Leviathan and the Air-Pump*, 79.

11. Elizabeth Potter, "Making Gender/Making Science: Gender Ideology and Boyle's Experimental Philosophy," in Bonnie Spanier, ed., *Making a Difference* (Bloomington: Indiana University Press, forthcoming). All citations are from the unpaginated manuscript.

12. See the series of essays and counteressays that begins with H. M. Collins and Steven Yearley, "Epistemological Chicken," in Andrew Pickering, ed., *Science as Practice and Culture* (Chicago: University of Chicago Press, 1992), 301–26. Bruno Latour, Steve Woolgar, and Michel Callon were the other combatants. Some of the interlocutors seemed better-humored than others. The stakes were what got to count as the really real.

13. Margo Hendricks, "Obscured by Dreams: Race, Empire and Shakespeare's *A Midsummer Night's Dream*," prepublication manuscript, 1993, University of California, Santa Cruz, Literature Board.

14. Potter, "Making Gender/Making Science," prepublication manuscript. A possible resource for Boyle's positive valuation of masculine modesty existed in English literature in the King Arthur narratives. See Bonnie Wheeler, "The Masculinity of King Arthur: From Gildas to the

Nuclear Age," *Quondam et Futurus: A Journal of Arthurian Interpretations* 2, no. 4 (winter 1992): 1–26. Wheeler argues that the first reference to the Arthur figure in the sixth century referred to him as a *vir modestus*, and the qualifier followed Arthur through his many literary incarnations. This tradition may have been culturally available to Boyle and his peers looking for effective new models of masculine reason. Shapin and Schaffer do not ask, because gender seems self-evident and ahistorical in their book. *Modestus* and *modestia* referred to measure, moderation, solicitude, studied equilibrium, and reticence in command. This constellation moves counter to the dominant strand of Western heroism, which emphasizes self-glorification by the warrior hero. The *vir modestus* was a man characterized by high status and disciplined ethical restraint. *Modestia* linked high class, effective power, and masculine gender. Wheeler sees in the King Arthur figure "one alternative norm of empowered masculinity for post-heroic culture" (p. 1).

15. Note how the status of being visible (opaque, bodily, polluted), rather than transparent, glides into being perceived as "subjective"; i.e., reporting only on the self, biased, not objective. Colored and sexed persons simply have to do a lot of work to become sufficiently transparent to count as objective witnesses to the world, rather than to their "bias." To be the object of vision, rather than the self-invisible source of vision, is to be evacuated of agency. Recall the trope of the eye of God, in Linnaeus's vision of the second Adam as the authorized manner of the new plants and animals revealed by eighteenth-century explorations. Nature can be seen and warranted; it is not itself the witness to itself. This narrative epistemological point is part of the apparatus for repeatedly placing "white" women and people of "color" in nature. Only as objects can they enter science; their only subjectivity in science is called bias and special interest, unless they become honorary honorable men. This is an ethnospecific story of representation, requiring surrogacy and ventriloquism as part of its technology. The self-acting agent who is the modest witness is also "agent" in another sense—as the delegate for the thing represented, as its spokesperson and representative. Agency, optics, and recording technologies are old bedfellows.

16. "From this perspective the proper subject of gender and science thus becomes the analysis of the web of forces that supports the historic conjunction of science and masculinity, and the equally historic disjunction between science and femininity. It is, in a word, the conjoint making of 'men,' 'women,' and 'science,' or more precisely how the making of 'men' and 'women' has affected the making of 'science'": Evelyn Fox Keller, "Gender and Science: 1990," in *The Great Ideas Today* (Chicago: Encyclopedia Britannica, 1990), 74. If "gender" here means "kind," and

thus includes *constitutively* the complex lineages of racial, sexual, class, and national formations in the production of differentiated men, women and science, I could not agree more.

17. The classic texts are Carolyn Merchant, *The Death of Nature: Women, Ecology, and the Scientific Revolution* (New York: Harper and Row, 1980); and Evelyn Fox Keller, *Reflections on Gender and Science* (New Haven: Yale University Press, 1985).

18. The veil is the chief epistemological element in orientalist systems of representation, including much of technoscience. The point of the veil is to promise that something is behind it. The veil guarantees the worth of the quest more than what is found. The metaphoric system of discovery that is so crucial to the discourse about science depends on there being things hidden to be discovered. How can one have breakthroughs if there is no resistance, no trial of the hero's resolve and virtue? The explorer is a hero, another aspect of epistemological manly valor in technoscience narratives. See Meyda Yeganogolou, "Veiled Fantasies: Towards a Feminist Reading of Orientalism," Ph.D. dissertation, Sociology Board, University of California, Santa Cruz, 1993. Feminist narratologists have spent a lot of time on these issues. Science-studies scholars should spend a little more time with feminist, multicultural, and postcolonial narratology and film theory.

19. Steve Woolgar, *Science: The Very Idea!* (London: Tavistock, 1988); and Steve Woolgar, ed., *Knowledge and Reflexivity* (London: Sage, 1988).

20. Remember that the author is a fiction, a position, and an ascribed function. There are other Latours, in and out of print, who offer a much richer tropic tool kit than that in *Science in Action*.

21. Bruno Latour, *Science in Action: How to Follow Scientists and Engineers through Society* (Cambridge, Mass.: Harvard University Press, 1987), 92.

22. Ibid., 15.

23. Steven Shapin, "Following Scientists Around," *Social Studies of Science* 18 (1988): 533–50.

24. See the fall 1990 newsletter of the Society for the Social Study of Science (*Technoscience* 3, no. 3, pp. 20, 22) for language about "going back to nature." A session of the 4S October meetings was titled "Back to Nature." Malcolm Ashmore's abstract, "With a Reflexive Sociology of Actants, There Is No Going Back," offered "fully comprehensive insurance against going back," instead of other competitors' less good "ways of not going back to Nature (or Society or Self)." All this occurred in the context of a crisis of confidence among many 4S scholars that their very fruitful research programs of the last ten years were running into dead ends. They were. I cannot refrain from commenting on the generic

misogyny in the conventional terror of "going back" to a fantastic nature (figured by science critics as "objective" nature; literary academicians figure the same terrible dangers slightly differently; for both groups such a nature is definitively presocial, monstrously not-human, and a threat to their careers). In these adolescent boys' narratives (a hard-to-avoid, conventionalized genre, which can be and has been written by men or women, including myself, and not a set of personal authorial intentions or lapses by Ashmore), Mother Nature always waits to smother the newly individuated hero. The author forgets this weird mother is a generic creation; the forgetting, or the inversion, is basic to ideologies of scientific objectivity and of nature as "Eden under glass." (The phrase is from Susanna Hecht and Alexander Cockburn, *The Fate of the Forest: Developers, Destroyers, and Defenders of the Amazon* [New York: Verso, 1989].) Such forgetting plays a significant role in some of the best (most reflexive) science studies, structurally muting their intended critical potential.

25. Martin Bernal, *Black Athena* (London: Free Association Books, 1987); Luce Irigaray, *The Speculum of the Other Woman*, tr. Gillian C. Gill (Ithaca: Cornell University Press, 1985).

26. Sandra Harding, *Whose Science? Whose Knowledge? Thinking from Women's Lives* (Ithaca: Cornell University Press, 1991).

27. "A stronger, more adequate notion of objectivity would require methods for systematically examining all of the social values shaping a particular research process, not just those that happen to differ between members of a scientific community. Social communities, not either individuals, or 'no one at all,' should be conceptualized as the 'knowers' of scientific knowledge claims. Culturewide beliefs that are not critically examined within scientific processes end up functioning as evidence for or against hypotheses": Sandra Harding, *The "Racial" Economy of Science: Toward a Democratic Future* (Bloomington: Indiana University Press, 1993), 18. Harding maintains that democracy-enhancing projects and questions are most likely to meet the strongest criteria for reliable scientific knowledge production, with built-in critical reflexivity.

28. Katie King, "Feminism and Writing Technologies," *Configurations* 2 (1994): 89–106.

29. On situated knowledges and diffraction, see Donna J. Haraway, "Situated Knowledges: The Science Question in Feminism as a Site of Discourse on the Privilege of Partial Perspective," *Feminist Studies* 14 (1988): 575–99; and "The Promises of Monsters: Reproductive Politics for Inappropriate/d Others," in Larry Grossberg, Cary Nelson, and Paul Treichler, eds., *Cultural Studies* (New York: Routledge, 1992), 295–337.

30. Susan Leigh Star, "Power, Technology, and the Phenomenology of Conventions: On Being Allergic to Onions," in *Power, Technology and*

the Modern World, ed. John Law, Special Issue, Sociological Review Monograph, no. 38 (Oxford: Blackwell, 1991), 26–56.

31. Ibid., 39.
32. Ibid., 43.
33. Ibid., 52.

Stump, Afterword

1. F. Suppe, *Structure of Scientific Theories* (Urbana: University of Illinois Press, 1977), 618.

2. For example, R. Giere, "History and Philosophy of Science: Intimate Relationship or Marriage of Convenience?" *British Journal for the Philosophy of Science* 24 (1973): 282–97; E. McMullin, "History and Philosophy of Science: A Marriage of Convenience?" in R. S. Cohen et al., eds., *PSA 1974* (East Lansing: Philosophy of Science Association, 1976), 585–602; R. Burian, "More than a Marriage of Convenience: On the Inextricability of History and Philosophy of Science," *Philosophy of Science* 44 (1977): 1–42.

3. D. Shapere, *Reason and the Search for Knowledge* (Dordrecht: Reidel, 1984); I. Hacking, *Representing and Intervening* (Cambridge: Cambridge University Press, 1983), and in this volume; P. Galison, *How Experiments End* (Chicago: University of Chicago Press, 1987); "History, Philosophy, and the Central Metaphor," *Science in Context* 2 (1988): 197–212; and "Multiple Constraints, Simultaneous Solutions," in A. Fine and J. Leplin, eds., *PSA 1988*, vol. 2 (East Lansing: Philosophy of Science Association, 1989), 157–63; A. Fine, *The Shaky Game: Einstein, Realism, and the Quantum Theory* (Chicago: University of Chicago Press, 1986), 112–50; J. Dupré, "The Disunity of Science," *Mind* 92 (1983): 321–46; *The Disorder of Things: Metaphysical Foundations of the Disunity of Science* (Cambridge, Mass.: Harvard University Press, 1993); and in this volume; T. Nickles, "From Natural Philosophy to Metaphilosophy of Science," in P. Achinstein and R. Kargon, eds., *Kelvin's Baltimore Lectures and Modern Theoretical Physics: Historical and Philosophical Perspectives* (Cambridge, Mass.: MIT Press, 1987), 507–41.

4. S. Bromberger, "Why-Questions," in R. G. Colodny, ed., *Mind and Cosmos* (Pittsburgh: University of Pittsburgh Press, 1966), 86–108; A. Garfinkel, *Forms of Explanation: Rethinking the Questions in Social Theory* (New Haven: Yale University Press, 1981); B. van Fraassen, *The Scientific Image* (Oxford: Clarendon Press, 1980).

5. F. Kafka, "A Little Fable," in *The Great Wall of China: Stories and Reflections* (New York: Schocken, 1946), 151.

Select Bibliography

Source citations for the essays in this volume may be found in the Notes, pp. 453–526. The list below is intended as an introduction to the issues discussed in the book overall. It consists of titles of general interest cited by the contributors to this volume, as well as additional items drawn from a variety of bibliographic sources.

Alcoff, Linda, and Elizabeth Potter, eds. *Feminist Epistemologies*. New York: Routledge, 1993.

Anderson, Alun. "Neuroscientists Struggle to Achieve a Critical Mass." *Science* 256 (1992): 468.

Aronowitz, Stanley. *Science as Power: Discourse and Ideology in Modern Society*. Minneapolis: University of Minnesota Press, 1988.

Barnes, Barry. *On the Nature of Power*. Cambridge: Polity, 1988.

Barnes, Barry, and David Edge, eds. *Science in Context: Readings in the Sociology of Science*. Cambridge, Mass.: MIT Press, 1982.

Barnes, Barry, and Steven Shapin, eds. *Natural Order*. London: Sage, 1979.

Bass, Robert E. *Some Features of Organization in Nature: A Contribution to Unified Science*. Toledo: Adamson, 1991.

Beatty, John. "Evolutionary Anti-reductionism: Historical Reflections." *Biology and Philosophy* 5 (1990): 197–210.

Berger, Peter L., and Thomas Luckmann. *The Social Construction of Reality: A Treatise in the Sociology of Knowledge*. Garden City: Doubleday, 1966.

Biagioli, Mario. *Galileo Courtier: The Practice of Science in the Culture of Absolutism*. Chicago: University of Chicago Press, 1993.

Bijker, Wiebe E., Thomas P. Hughes, and Trevor J. Pinch, eds. *The Social Construction of Technological Systems: New Directions in the Sociology and History of Technology*. Cambridge, Mass.: MIT Press, 1989.

Bleecken, Stefan. "Die Einheit der Wissenschaft—Abschied von einer Illusion." *Merkur—Deutsche Zeitschrift für Europäisches Denken* 46 (1992): 1096–1108.

Boucher, D. *Texts in Context: Revisionist Methods for Studying the History of Ideas.* Dordrecht: Nijhoff, 1985.

Brozek, Josef. "Disunity versus Diversity." *American Psychologist* 45 (1990): 983.

Buchwald, J., ed. *Scientific Practice: Theories and Stories of Doing Physics.* Chicago: University of Chicago Press, 1995.

Carnap, Rudolf. "Logical Foundations of the Unity of Science." In O. Neurath, R. Carnap, and C. W. Morris, eds., *International Encyclopedia of Unified Science*, vol. 1, no. 1, pp. 42–62. Chicago: University of Chicago Press, 1938.

Cartwright, Nancy, J. Cat, K. Fleck, and T. Uebel, eds. *Between Science and Politics: The Philosophy of Otto Neurath.* Cambridge: Cambridge University Press, forthcoming.

Cat, Jordi, H. Chang, and N. Cartwright. "Otto Neurath: Unification as the Way to Socialism." In J. Mittelstrass, ed., *Einheit der Wissenschaften*, 91–110. Berlin: de Gruyter, 1991.

Causey, Robert L. *The Unity of Science.* Dordrecht: Reidel, 1977.

Churchland, Patricia Smith. *Neurophilosophy: Toward a Unified Science of the Mind-Brain.* Cambridge, Mass.: MIT Press, 1986.

Churchland, Paul M. "A Deeper Unity: Some Feyerabendian Themes in Neurocomputational Form." In G. Munevar, ed., *Beyond Reason: Essays on Paul Feyerabend* 1–23. Dordrecht: Kluwer, 1990.

Coulter, Jeffrey. "Decontextualized Meanings: Current Approaches to 'Verstehende' Investigations." *Sociological Review* 19 (1971): 301–33.

Cousins, Mark, and Athar Hussain. "The Question of Ideology: Althusser, Pecheux and Foucault." In John Law, ed., *Power, Action and Belief: A New Sociology of Knowledge?* 158–79. Boston: Routledge and Kegan Paul, 1986.

Dewey, John. "Unity of Science as a Social Problem." In O. Neurath, R. Carnap, and C. W. Morris, eds., *International Encyclopedia of Unified Science*, vol. 1, no. 1, pp. 29–38. Chicago: University of Chicago Press, 1938.

Dupré, John. *The Disorder of Things: Metaphysical Foundations of the Disunity of Science.* Cambridge, Mass.: Harvard University Press, 1993.

———. "The Disunity of Science." *Mind* 92 (1983): 321–46.

Eldredge, Niles. *Unfinished Synthesis: Biological Hierarchies and Modern Evolutionary Thought.* Oxford: Oxford University Press, 1985.

Ereshefsky, Marc, ed. *The Units of Evolution: Essays on the Nature of Species.* Cambridge, Mass.: MIT Press, 1992.

European Commission, Joint Research Centre. "A New Vision of Unity in Standards and Training." *Nature* 338 (1989): 732–33.

Fabel, A. "The Phenomenon of a Discovery—The Unity of a New Science and the Perennial Wisdom." *Journal of Social and Biological Structures* 14 (1991): 1–13.

Feyerabend, Paul. *Against Method*. London: New Left Books, 1975.

———. *Science in a Free Society*. London: New Left Books, 1978.

Fine, Arthur. *The Shaky Game: Einstein, Realism, and the Quantum Theory*. Chicago: University of Chicago Press, 1986.

Fisher, Donald. "Boundary Work and Science: The Relation between Power and Knowledge." In Susan E. Cozzens and Thomas F. Gieryn, eds., *Theories of Science in Society*, 98–119. Bloomington: Indiana University Press, 1990.

Foucault, Michel. *Power/Knowledge: Selected Interviews and Other Writings, 1972–1977*. Tr. Colin Gordon, Leo Marshall, John Mepham, and Kate Soper. New York: Random House, 1980.

Friedman, Robert Marc. "Text, Context, and Quicksand: Method and Understanding in Studying the Nobel Science Prizes." *Historical Studies in the Physical Sciences* 20 (1989): 63–77.

Fuller, Steve. *Rhetoric and the End of Knowledge: The Coming of Science and Technology Studies*. Madison: University of Wisconsin Press, 1993.

———. *Social Epistemology*. Indiana University Press, 1988.

Fyfe, Gordon, and John Law, eds. *Picturing Power: Visual Depiction and Social Relations*. London: Routledge and Kegan Paul, 1988.

Galison, Peter. "History, Philosophy, and the Central Metaphor." *Science in Context* 2 (1988): 197–212.

———. *How Experiments End*. Chicago: University of Chicago Press, 1987.

———. *Image and Logic: The Material Culture of Twentieth-Century Physics*. Forthcoming.

Gemes, Ken. "Exclamation, Unification, and Content," *Noûs* 28 (1994): 225–40.

Giere, Ronald. *Explaining Science: A Cognitive Approach*. Chicago: University of Chicago Press, 1988.

———. *Understanding Scientific Reasoning*. New York: Holt, Rinehart and Winston, 1984.

Gooding, David, Trevor Pinch, and Simon Schaffer, eds. *The Uses of Experiment: Studies in the Natural Sciences*. Cambridge: Cambridge University Press, 1989.

Goosens, W. K. "Reduction by Molecular Genetics." *Philosophy of Science* 45 (1978): 73–95.

Green, Christopher D. "Is Unified Positivism the Answer to Psychology's Disunity?" *American Psychologist* 47 (1992): 1057–58.

Hacking, Ian. "Disunified Science." In Richard Q. Elvee, ed., *The End of Science? Attack and Defense*, 33–52. Lanham: University Press of America, 1992.

——. *Representing and Intervening: Introductory Topics in the Philosophy of Natural Science*. Cambridge: Cambridge University Press, 1983.

Hagstrom, Warren O., ed. *The Scientific Community*. New York: Basic Books, 1965.

Haraway, Donna. *Primate Visions: Gender, Race, and Nature in the World of Modern Science*. London: Routledge, Chapman and Hall, 1989.

——. *Simians, Cyborgs, and Women: The Reinvention of Nature*. London: Routledge, Chapman and Hall, 1991.

Harding, Sandra. *The Science Question in Feminism*. Ithaca: Cornell University Press, 1986.

——. *Whose Science? Whose Knowledge? Thinking from Women's Lives*. Ithaca: Cornell University Press, 1991.

Harsanyi, John C. "Models for the Analysis of Balance of Power in Society." In Ernest Nagel, Patrick Suppes, and Alfred Tarski, eds., *Logic, Methodology and Philosophy of Science*, 442–62. Stanford: Stanford University Press, 1960.

Harwood, Jonathan. *Styles of Scientific Thought: The German Genetics Community, 1900–1933*. Chicago: University of Chicago Press, 1993.

Holton, Gerald. *The Scientific Imagination: Case Studies*. Cambridge: Cambridge University Press, 1978.

——. *Thematic Origins of Scientific Thought: Kepler to Einstein*. Cambridge, Mass.: Harvard University Press, 1973.

Hoyningen-Huene, Paul. "The Interrelations between the Philosophy, History and Sociology of Science in Thomas Kuhn's Theory of Scientific Development." *British Journal for the Philosophy of Science* 43 (1992): 487–501.

Hull, David. *The Philosophy of Biological Science*. Engelwood Cliffs: Prentice-Hall, 1974.

——. *Science as a Process: An Evolutionary Account of the Social and Conceptual Development of Science*. Chicago: University of Chicago Press, 1988.

Hunter, Michael, and Simon Schaffer, eds. *Robert Hooke: New Studies*. Woodbridge: Boydell, 1989.

Jasanoff, Sheila. "Contested Boundaries in Policy-relevant Science." *Social Studies of Science* 17 (1987): 195–230.

——. *Risk Management and Political Culture*. New York: Russell Sage Foundation, 1986.

Johnston, Ron. "Contextual Knowledge: A Model for the Overthrow of the Internal/External Dichotomy in Science." *Australia and New Zealand Journal of Sociology* 12 (1976): 193–203.

Kaiser, M. "Philosophers Adrift—Comments on the Alleged Disunity of Method." *Philosophy of Science* 60 (1993): 500–512.

Keller, Evelyn Fox. *Reflections on Gender and Science.* New Haven: Yale University Press, 1985.

Keller, Evelyn Fox, and E. A. Lloyd, eds. *Keywords in Evolutionary Biology.* Cambridge, Mass.: Harvard University Press, 1992.

Kevles, Daniel J. *The Physicists: The History of a Scientific Community in Modern America.* New York: Knopf, 1978.

Kincaid, Harold. "Molecular Biology and the Unity of Science." *Philosophy of Science* 57 (1990): 575–93.

Kitcher, Philip. *The Advancement of Science: Science without Legend, Objectivity without Illusions.* Oxford: Oxford University Press, 1993.

——. "Explanatory Unification." *Philosophy of Science* 48 (1981): 507–31.

——. "1953 and All That: A Tale of Two Sciences." *Philosophical Review* 93 (1984): 335–73.

Knight, David M. *Humphry Davy: Science and Power.* Oxford: Blackwell, 1992.

Knorr Cetina, Karin D. *The Manufacture of Knowledge: An Essay on the Constructivist and Contextual Nature of Science.* Oxford: Pergamon, 1981.

Knorr Cetina, Karin D., and Aaron Cicourel, eds. *Advances in Social Theory and Methodology: Toward an Integration of Micro- and Macrosociologies.* Boston: Routledge and Kegan Paul, 1981.

Knorr Cetina, Karin D., Roger Krohn, and Richard Whitley, eds. *The Social Process of Scientific Investigation.* Sociology of the Sciences Yearbook. Dordrecht: Reidel, 1980.

Krumhansl, J. A. "Unity in the Science of Physics." *Physics Today* 44 (1991): 33–38.

Kumar, Deepak. *Science and Empire: Essays in Indian Context, 1700–1947.* Delhi: Anamika Prakashan, 1991.

Latour, Bruno. *The Pasteurization of France.* Tr. A. Sheridan and J. Law. Cambridge, Mass.: Harvard University Press, 1988.

——. *Science in Action: How to Follow Scientists and Engineers through Society.* Cambridge, Mass.: Harvard University Press, 1987.

Laudan, Rachel, and Larry Laudan. "Dominance and the Disunity of Method: Solving the Problems of Innovation and Consensus." *Philosophy of Science* 56 (1989): 221–38.

Law, John. "On Power and Its Tactics: A View from the Sociology of Science." *Sociological Review* 34 (1986): 1–37.

Law, John, ed. *Power, Action and Belief: A New Sociology of Knowledge?* London: Routledge and Kegan Paul, 1986.

Lenoir, Timothy. *The Strategy of Life: Teleology and Mechanics in Nineteenth-Century German Biology.* Dordrecht: Reidel, 1982.

Lenzen, V. "Procedures of Empirical Science." In O. Neurath, R. Carnap, and C. W. Morris, eds. *International Encyclopedia of Unified Science*, vol. 1, no. 6, pp. 279–340. Chicago: University of Chicago Press, 1938.

Lindberg, David C. *The Beginnings of Western Science: The European Scientific Tradition in Philosophical, Religious, and Institutional Context, 600 B.C. to A.D. 1450*. Chicago: University of Chicago Press, 1992.

Lindberg, David C., and Robert S. Westman, eds. *Reappraisals of the Scientific Revolution*. Cambridge: Cambridge University Press, 1990.

Livesey, Steven J., ed. *Theology and Science in the Fourteenth Century: Three Questions on the Unity and Subalternation of the Sciences from John of Reading's "Commentary on the Sentences."* Leiden: Brill, 1989.

Lloyd, Elisabeth. *The Structure and Confirmation of Evolutionary Theory*. Westport: Greenwood, 1988.

Longino, Helen E. "Multiplying Subjects and the Diffusion of Power." *Journal of Philosophy* 88 (1991): 666–74.

Lynch, Michael. *Art and Artifact in Laboratory Science: A Study of Shop Work and Shop Talk in a Research Laboratory*. London: Routledge and Kegan Paul, 1985.

——. *Scientific Practice and Ordinary Action: Ethnomethodology and Social Studies of Science*. Cambridge: Cambridge University Press, 1992.

MacDonald, Graham. "The Possibility of the Disunity of Science." In G. MacDonald and C. Wright, eds., *Fact, Science and Morality: Essays on A. J. Ayer's "Language, Truth and Logic,"* 219–47. Oxford: Blackwell, 1987.

MacIntyre, Alasdair. "Relativism, Power and Philosophy." *Proceedings and Addresses of the American Philosophical Society* 59 (1985): 5–22.

MacKinnon, Edward. "Niels Bohr on the Unity of Science." In Peter D. Asquith and Ronald N. Giere, eds., *PSA 1980*, vol. 2, pp. 224–44. East Lansing: Philosophy of Science Association, 1981.

Maddox, John. "Choice and the Scientific Community." *Minerva* 2 (1954): 141–59.

Maienschein, Jane. *Transforming Traditions in American Biology, 1880–1915*. Baltimore: The Johns Hopkins University Press, 1991.

Margolis, Joseph. *Culture and Cultural Entities: Toward a New Unity of Science*. Dordrecht: Reidel, 1984.

——. *Science without Unity—Reconciling the Human and the Natural Sciences*. Oxford: Blackwell, 1987.

Maull, Nancy L. "Unifying Science without Reduction." *Studies in History and Philosophy of Science* 8 (1977): 143–62.

Mayr, Ernst. *Animal Species and Evolution*. Cambridge, Mass.: Harvard University Press, 1963.

——. *Populations, Species, and Evolution.* Cambridge, Mass.: Harvard University Press, 1970.

Mayr, Ernst, and William B. Provine, eds. *The Evolutionary Synthesis.* Cambridge, Mass.: Harvard University Press, 1980.

McGuiness, B., ed. *Unified Science.* Dordrecht: Reidel, 1987.

Mellor, D. H. "Natural Kinds." *British Journal for the Philosophy of Science* 28 (1977): 299–312.

Merchant, Carolyn. *The Death of Nature: Women, Ecology, and the Scientific Revolution.* New York: Harper and Row, 1980.

Mills, C. Wright. *Power, Politics, and People.* New York: Ballantine, 1963.

Morrison, Margaret. "A Study in Theory Unification: The Case of Maxwell's Electromagnetic Theory." *Studies in History and Philosophy of Science* 23 (1992): 103–45.

Mukerji, Chandra. *A Fragile Power.* Princeton: Princeton University Press, 1990.

Nagel, Ernest. *The Structure of Science.* Indianapolis: Hackett, 1961.

Nelkin, Dorothy, and Lawrence Tancredi. *Dangerous Diagnostics: The Social Power of Biological Information.* New York: Basic Books, 1989.

Neurath, Otto. "Individual Sciences, Unified Science, Pseudorationalism." In Robert S. Cohen and Marie Neurath, eds., *Philosophical Papers, 1913–1946,* 132–38. Dordrecht: Reidel, 1983.

——. "Unified Science as Encyclopedic Integration." In O. Neurath, R. Carnap, and C. W. Morris, eds., *International Encyclopedia of Unified Science,* vol. 1, no. 1, pp. 1–27. Chicago: University of Chicago Press, 1938.

Neurath, Otto, Rudolf Carnap, and Charles W. Morris, eds. *International Encyclopedia of Unified Science.* Chicago: University of Chicago Press, 1938.

Nickles, Thomas. "Theory Generalization, Problem Reduction, and the Unity of Science." In R. S. Cohen, C. A. Hooker, A. C. Michalos, and J. W. Evra, eds., *PSA 1974,* 33–78. East Lansing: Philosophy of Science Association, 1976.

——. "Two Concepts of Intertheoretic Reduction." *Journal of Philosophy* 70 (1973): 181–220.

Oleson, Alexandra, and John Voss, eds. *The Organization of Knowledge in Modern America, 1860–1920.* Baltimore: The Johns Hopkins University Press, 1979.

Oppenheim, Paul, and Hilary Putnam. "Unity of Science as a Working Hypothesis." In H. Feigl, M. Scriven, and G. Maxwell, eds., *Concepts, Theories, and the Mind-Body Problem,* Minnesota Studies in the Philosophy of Science 2, 3–36. Minneapolis: University of Minnesota Press, 1958.

Papanicolaou, Andrew C., and Pete A. Y. Gunter, eds. *Bergson and Modern Thought: Towards a Unified Science*. New York: Harwood, 1987.

Paul, Harry W. *From Knowledge to Power: The Rise of the Science Empire in France, 1860–1939*. Cambridge: Cambridge University Press, 1985.

Pellegrin, Pierre. *Aristotle's Classification of Animals: Biology and the Conceptual Unity of the Aristotelian Corpus*. Berkeley and Los Angeles: University of California Press, 1986.

Perkins, David N. "Are Cognitive Skills Context-bound?" *Educational Researcher* 18 (1989): 16–25.

Pickering, Andrew, ed. *Science as Practice and Culture*. Chicago: University of Chicago Press, 1992.

Proctor, Robert. *Value-free Science? Purity and Power in Modern Knowledge*. Cambridge, Mass.: Harvard University Press, 1991.

Putnam, Hilary. *Reason, Truth and History*. Cambridge: Cambridge University Press, 1981.

Ravetz, Jerome R. *The Merger of Knowledge with Power: Essays in Critical Science*. London: Mansell, 1990.

Ringer, Fritz K. *The Decline of the German Mandarins: The German Academic Community, 1890–1933*. Cambridge, Mass.: Harvard University Press, 1990.

Rorty, Richard. "Is Science a Natural Kind?" In Ernan McMullin, ed., *Construction and Constraint: The Shaping of Scientific Rationality*, 49–74. Notre Dame: University of Notre Dame Press, 1988.

Rosenberg, Alexander. *Instrumental Biology; or, The Disunity of Science*. Chicago: University of Chicago Press, 1994.

———. *The Structure of Biological Science*. Cambridge: Cambridge University Press, 1985.

Rouse, Joseph. "The Dynamics of Power and Knowledge in Science." *Journal of Philosophy* 88 (1991): 658–65.

———. *Knowledge and Power: Toward a Political Philosophy of Science*. Ithaca: Cornell University Press, 1987.

Ruse, Michael. *The Philosophy of Biology*. London: Hutchinson, 1973.

———. "Reduction in Genetics." In R. S. Cohen, C. A. Hooker, A. C. Michalos, and J. W. Evra, eds., *PSA 1974*, 633–52. East Lansing: Philosophy of Science Association, 1976.

Schaffer, Simon. "Glass Works: Newton's Prisms and the Uses of Experiment." In David Gooding, Trevor Pinch, and Simon Schaffer, eds., *The Uses of Experiment: Studies in the Natural Sciences*, 67–104. Cambridge: Cambridge University Press, 1989.

Schaffner, Kenneth F. "Approaches to Reduction." *Philosophy of Science* 34 (1967): 137–57.

——. "Reductionism in Biology: Prospects and Problems." In R. S. Cohen, C. A. Hooker, A. C. Michalos, and J. W. Evra, eds., *PSA 1974*, 613–32. East Lansing: Philosophy of Science Association, 1976.

Scharf, David C. "Quantum Measurement and the Program for the Unity of Science." *Philosophy of Science* 56 (1989): 601–23.

Shapere, Dudley. *Reason and the Search for Knowledge*. Dordrecht: Reidel, 1984.

Shapin, Steven. "A Course in the Social History of Science." *Social Studies of Science* 10 (1980): 231–58.

Shapin, Steven, and Simon Shaffer. *Leviathan and the Air-Pump: Hobbes, Boyle and the Experimental Life*. Princeton: Princeton University Press, 1985.

Shapin, Steven, and Arnold Thackray. "Prosopography as a Research Tool in History of Science: The British Scientific Community, 1700–1900." *History of Science* 12 (1974): 1–28.

Smocovitis, V. B. "Disciplining Evolutionary Biology: Ernst Mayr and the Founding of the Society for the Study of Evolution and *Evolution* (1939–1950)." *Evolution* 48 (1994): 1–8.

——. "Unifying Biology: The Evolutionary Synthesis and Evolutionary Biology." *Journal of the History of Biology* 25 (1992): 1–65.

Sober, Elliott, ed. *Conceptual Issues in Evolutionary Biology*. Cambridge, Mass.: MIT Press, 1994.

Spiegel-Rosing, I., and D. J. de Solla Price, eds. *Science, Technology and Society*. London: Sage, 1977.

Staats, Arthur. "Unified Positivism and Unification Psychology." *American Psychologist* 46 (1991): 889–912.

Suppes, Patrick. "The Plurality of Science." In P. D. Asquith and I. Hacking, eds., *PSA 1978*, vol. 2, pp. 3–16. East Lansing: Philosophy of Science Association, 1978.

Swenson, Lloyd S., Jr. *Genesis of Relativity: Einstein in Context*. New York: Franklin, 1979.

Tolstoy, Ivan. *The Knowledge and the Power: Reflections on the History of Science*. Edinburgh: Canongate, 1990.

Toren, Nina. *Science and Cultural Context: Soviet Scientists in Comparative Perspective*. New York: Lang, 1988.

Tuanna, Nancy, ed. *Feminism and Science*. Bloomington: Indiana University Press, 1989.

Uebel, Thomas E. *Overcoming Logical Positivism from Within: The Emergence of Neurath's Naturalism in the Vienna Circle's Protocol Sentence Debate*. Amsterdam: Rodopi, 1992.

van Fraassen, Bas. *The Scientific Image*. Oxford: Clarendon Press, 1980.

Viney, Wayne. "The Cyclops and the Twelve-Eyed Toad: William James and the Unity-Disunity Problem in Psychology." *American Psychologist* 44 (1989): 1261–65.

Walter, Maila. *Science and Cultural Crisis: An Intellectual Biography of Percy Williams Bridgman (1882–1961)*. Stanford: Stanford University Press, 1990.

Waters, C. Kenneth. "Why the Antireductionist Consensus Won't Survive the Case of Classical Mendelian Genetics." In A. Fine, M. Forbes, and L. Wessels, eds., *PSA 1990*, vol. 1, pp. 125–39. East Lansing: Philosophy of Science Association, 1990.

Weinberg, Steven. *Dreams of a Final Theory*. New York: Pantheon, 1992.

Whitley, Richard. "The Context of Scientific Investigation." In Karin D. Knorr Cetina, Roger Krohn, and Richard Whitley, eds., *The Social Process of Scientific Investigation*, Sociology of the Sciences Yearbook, 297–322. Dordrecht: Reidel, 1980.

Wimsatt, William. "Reductive Explanation: A Functional Account." In R. S. Cohen, C. A. Hooker, A. C. Michalos, and J. W. Evra, eds., *PSA 1974*, 671–711. Dordrecht: Reidel, 1976.

Young, Robert M. "The Historiographic and Ideological Context in Nineteenth-Century Debate on Man's Place in Nature." In M. Teich and R. M. Young, eds., *Changing Perspectives in the History of Science*, 344–438. London: Heinemann, 1973.

Zissler, Dieter. "In der Mannigfaltigkeit die Einheit zu erkennen: Über Natur und Naturwissenschaft im Werk Ernst Jüngers." *Text und Kritik* 105 (1990): 125–40.

Index

In this index an "f" after a number indicates a separate reference on the next page, and an "ff" indicates separate references on the next two pages. A continuous discussion over two or more pages is indicated by a span of page numbers, e.g., "57–58." *Passim* is used for a cluster of references in close but not continuous sequence.

Library of Congress Cataloging-in-Publication Data

The disunity of science : boundaries, contexts, and power / edited by
    Peter Galison and David J. Stump.
        p.   cm. — (Writing science)
    Includes bibliographical references and index.
    ISBN 0-8047-2436-9 (alk. paper). — ISBN 0-8047-2562-4 (pbk. :
alk. paper)
        1. Science—Philosophy.   2. Science—Social aspects.   I. Galison,
Peter Louis.   II. Stump, David J.   III. Series.
Q175.D6636   1996
501—dc20   95-9064
                                                                CIP

Original printing 1996
Last figure below indicates year of this printing:
05   04   03   02   01   00   99   98   97   96

⊗ This book is printed on acid-free, recycled paper.

6121

DATE DUE